T0297845

CAMBRIDGE LIBRARY COLLECTION

Books of enduring scholarly value

Life Sciences

Until the nineteenth century, the various subjects now known as the life sciences were regarded either as arcane studies which had little impact on ordinary daily life, or as a genteel hobby for the leisured classes. The increasing academic rigour and systematisation brought to the study of botany, zoology and other disciplines, and their adoption in university curricula, are reflected in the books reissued in this series.

The Landscape Gardening and Landscape Architecture of the Late Humphry Repton, Esq.

By the beginning of the nineteenth century, landscape gardening had divided into at least two branches. The geometric style promoted strictly ordered gardens, while the natural style, for which the period is known, preserved characteristics of untamed vistas. Edited by a former professional rival, John Claudius Loudon (1783–1843), this one-volume collection of the works of Humphry Repton (1752–1818) first appeared in 1840. Featuring more than 250 engravings, it illuminates the principal styles and contemporary debates of landscape design. Including perspective tricks to disguise differing water levels, and instructions on the use of cattle as a natural measure of scale, Repton's writings reflect the attention to detail that was involved in planning and executing major projects. The collection is prefaced with a biographical notice believed to have been written by the architect John Adey Repton (1775–1860), who collaborated with his father on many schemes.

Cambridge University Press has long been a pioneer in the reissuing of out-of-print titles from its own backlist, producing digital reprints of books that are still sought after by scholars and students but could not be reprinted economically using traditional technology. The Cambridge Library Collection extends this activity to a wider range of books which are still of importance to researchers and professionals, either for the source material they contain, or as landmarks in the history of their academic discipline.

Drawing from the world-renowned collections in the Cambridge University Library and other partner libraries, and guided by the advice of experts in each subject area, Cambridge University Press is using state-of-the-art scanning machines in its own Printing House to capture the content of each book selected for inclusion. The files are processed to give a consistently clear, crisp image, and the books finished to the high quality standard for which the Press is recognised around the world. The latest print-on-demand technology ensures that the books will remain available indefinitely, and that orders for single or multiple copies can quickly be supplied.

The Cambridge Library Collection brings back to life books of enduring scholarly value (including out-of-copyright works originally issued by other publishers) across a wide range of disciplines in the humanities and social sciences and in science and technology.

The Landscape Gardening
and
Landscape Architecture
of the Late
Humphry Repton, Esq.

Being His Entire Works on These Subjects

EDITED BY J.C. LOUDON

CAMBRIDGE
UNIVERSITY PRESS

CAMBRIDGE
UNIVERSITY PRESS

University Printing House, Cambridge, CB2 8BS, United Kingdom

Published in the United States of America by Cambridge University Press, New York

Cambridge University Press is part of the University of Cambridge.
It furthers the University's mission by disseminating knowledge in the pursuit of
education, learning and research at the highest international levels of excellence.

www.cambridge.org
Information on this title: www.cambridge.org/9781108066174

© in this compilation Cambridge University Press 2013

This edition first published 1840
This digitally printed version 2013

ISBN 978-1-108-06617-4 Paperback

Many of the illustrations in this book are printed as overlays.
The 'before' and 'after' views can be seen online at www.cambridge.org/9781108066174

Selected books of related interest, also reissued in the
CAMBRIDGE LIBRARY COLLECTION

Amherst, Alicia: *A History of Gardening in England* (1895) [ISBN 9781108062084]

Anonymous: *The Book of Garden Management* (1871) [ISBN 9781108049399]

Blaikie, Thomas: *Diary of a Scotch Gardener at the French Court at the End of the Eighteenth Century* (1931) [ISBN 9781108055611]

Candolle, Alphonse de: *The Origin of Cultivated Plants* (1886) [ISBN 9781108038904]

Drewitt, Frederic Dawtrey: *The Romance of the Apothecaries' Garden at Chelsea* (1928) [ISBN 9781108015875]

Evelyn, John: *Sylva, Or, a Discourse of Forest Trees* (2 vols., fourth edition, 1908) [ISBN 9781108055284]

Farrer, Reginald John: *In a Yorkshire Garden* (1909) [ISBN 9781108037228]

Field, Henry: *Memoirs of the Botanic Garden at Chelsea* (1878) [ISBN 9781108037488]

Forsyth, William: *A Treatise on the Culture and Management of Fruit-Trees* (1802) [ISBN 9781108037471]

Haggard, H. Rider: *A Gardener's Year* (1905) [ISBN 9781108044455]

Hibberd, Shirley: *Rustic Adornments for Homes of Taste* (1856) [ISBN 9781108037174]

Hibberd, Shirley: *The Amateur's Flower Garden* (1871) [ISBN 9781108055345]

Hibberd, Shirley: *The Fern Garden* (1869) [ISBN 9781108037181]

Hibberd, Shirley: *The Rose Book* (1864) [ISBN 9781108045384]

Hogg, Robert: *The British Pomology* (1851) [ISBN 9781108039444]

Hogg, Robert: *The Fruit Manual* (1860) [ISBN 9781108039451]

Hooker, Joseph Dalton: *Kew Gardens* (1858) [ISBN 9781108065450]

Jackson, Benjamin Daydon: *Catalogue of Plants Cultivated in the Garden of John Gerard, in the Years 1596–1599* (1876) [ISBN 9781108037150]

Jekyll, Gertrude: *Home and Garden* (1900) [ISBN 9781108037204]

Jekyll, Gertrude: *Wood and Garden* (1899) [ISBN 9781108037198]

Johnson, George William: *A History of English Gardening, Chronological, Biographical, Literary, and Critical* (1829) [ISBN 9781108037136]

Knight, Thomas Andrew: *A Selection from the Physiological and Horticultural Papers Published in the Transactions of the Royal and Horticultural Societies* (1841) [ISBN 9781108037297]

Lindley, John: *The Theory of Horticulture* (1840) [ISBN 9781108037242]

Loudon, Jane: *Instructions in Gardening for Ladies* (1840) [ISBN 9781108055659]

Mollison, John: *The New Practical Window Gardener* (1877) [ISBN 9781108061704]

Paris, John Ayrton: *A Biographical Sketch of the Late William George Maton M.D.* (1838) [ISBN 9781108038157]

Paxton, Joseph, and Lindley, John: *Paxton's Flower Garden* (3 vols., 1850–3) [ISBN 9781108037280]

Repton, Humphry and Loudon, John Claudius: *The Landscape Gardening and Landscape Architecture of the Late Humphry Repton, Esq.* (1840) [ISBN 9781108066174]

Robinson, William: *The English Flower Garden* (1883) [ISBN 9781108037129]

Robinson, William: *The Subtropical Garden* (1871) [ISBN 9781108037112]

Robinson, William: *The Wild Garden* (1870) [ISBN 9781108037105]

Sedding, John D.: *Garden-Craft Old and New* (1891) [ISBN 9781108037143]

Veitch, James Herbert: *Hortus Veitchii* (1906) [ISBN 9781108037365]

Ward, Nathaniel: *On the Growth of Plants in Closely Glazed Cases* (1842) [ISBN 9781108061131]

For a complete list of titles in the Cambridge Library Collection please visit:
http://www.cambridge.org/features/CambridgeLibraryCollection/books.htm

S. Shelley del.ᵗ   H.B Hall. sc.ᵗ

HUMPHRY REPTON, ESQ.

London. Published 1839, by Longman & Cº Paternoster Row.

THE

LANDSCAPE GARDENING

AND

LANDSCAPE ARCHITECTURE

OF THE LATE HUMPHRY REPTON, ESQ.

BEING HIS ENTIRE WORKS ON THESE SUBJECTS.

A NEW EDITION:

WITH AN HISTORICAL AND SCIENTIFIC INTRODUCTION, A SYSTEMATIC
ANALYSIS, A BIOGRAPHICAL NOTICE, NOTES, AND A COPIOUS
ALPHABETICAL INDEX.

By J. C. LOUDON, F.L.S., &c.

ORIGINALLY PUBLISHED IN ONE FOLIO AND THREE QUARTO VOLUMES, AND
NOW COMPRISED IN ONE VOLUME OCTAVO.

ILLUSTRATED BY
UPWARDS OF TWO HUNDRED AND FIFTY ENGRAVINGS.

LONDON:

PRINTED FOR THE EDITOR, AND SOLD BY LONGMAN & CO.
AND A. & C. BLACK. EDINBURGH.

MDCCCXL.

INTRODUCTION.

GARDENING, as an Art of Culture, since the commencement of the present century, has made rapid progress ; but,
as an Art of Taste, it has been comparatively stationary.
One of the principal causes of this state of things, is the
abundance of good and cheap books on subjects belonging
to the former department, and the scarcity and high price
of those treating of the latter. To remedy this evil, it is
contemplated to publish a series of four or five volumes,
which shall include a reprint of all the best works on
Landscape Gardening which have hitherto appeared ; illus-
trating these works, where it appears desirable, with notes,
commentaries, and engravings.

The art of laying out the grounds which immediately
surround a country residence, may be displayed in two very
distinct styles : the first of which is called the Ancient,
Roman, Geometric, Regular, or Architectural Style ; and
the second, the Modern, English, Irregular, Natural, or
Landscape Style. Both these styles are, in different stages
of society, equally congenial to the human mind. The
Geometric Style was most striking and pleasing, and most
obviously displayed wealth and taste, in an early state of
society, and in countries where the general scenery was
wild, irregular, and natural, and man, comparatively, uncultivated and unrefined ; while, on the other hand, in modern
times, and in countries subjected to cultivation, and covered

with enclosures, rows of trees, and roads, all in regular lines, or forms, and where society is in a higher state of cultivation, the natural, or irregular style, from its rarity in such a country, and from the sacrifice of profitable lands requisite to make room for it, becomes equally a sign of wealth and taste. Of each of these styles, circumstances, either geographical or national, have given rise to two or more modifications; and these, in the language of art, may be called Schools. Thus, the Geometric Style, in Italy, owing to the hilliness of the country, and the national taste of the inhabitants for architecture, is characterized by flights of steps in the open air, terrace-walls, vases, and statues. The same style in France, where estates are much more extensive, the surface of the country more even, and the inhabitants less fond of architecture, is characterized by long avenues :

> " Woods and long rows of trees my pen invite :
> Groves ever please ; but most when placed aright.
>   * * * * * *
> Thus Normandy extends her guard of trees
> Against the wind which blows from British seas.
> High sylvan avenues the coast surround,
> Divide large farms, and ample lordships bound :"
>
> RAPIN on Gardens.

while in Holland, a perfectly flat country, it is distinguished by long, straight canals, and grassy terraces. Thus we have the Italian, the French, and the Dutch Schools, of the Geometric Style. These schools are exemplified in various French and Italian works ; the best of which, however, may be compressed into an octavo volume, which will form one of the series which we contemplate.

The Modern, or Landscape Style, when it first displayed itself in English country residences, was distinctly marked by the absence of everything having the appearance of a terrace, or of architectural forms, or lines, immediately adjoining the house. The house, in short, rose abruptly from the lawn; and the general surface of the ground was characterized by smoothness and bareness. This constituted the first school of the Landscape Style; and, as it appears to have been introduced by Kent, it may not improperly bear his name, and be called Kent's School. The publications which illustrate this school are chiefly those of Shenstone, G. Mason, Whately, and Mason the poet, all of which may be included in one closely-printed octavo volume, forming another of the proposed series.

The smooth, bare, and almost bald appearance, which characterized Kent's School, soon gave rise to one distinguished by roughness and intricacy, which may be called the Picturesque School; and the principles of which will be found in the writings of the Rev. W. Gilpin and Mr. Uvedale Price. These writings are full of the most valuable instruction for the gardener, relative to the general composition of landscape scenery, and landscape architecture; and may, very properly, form another volume of the series.

The rage for destroying avenues and terraces having subsided, and the propriety of uniting a country house with the surrounding scenery, by architectural appendages, having been pointed out, in a masterly manner, by Uvedale Price, Kent's School gave way; not, however, as may be supposed, to the Picturesque School (which, though adopted in many instances, in some parts of an estate, yet, in very few cases was exclusively employed), but to what may be called Repton's School, and which may be considered as combining all that was excellent in the former schools, and, in fact, as consisting of the union of an artistical knowledge of the subject with good taste and good sense. The principles of this school, and their application in practice, are exhibited in the five volumes published by Mr. Repton (see p. 22), which, with copies of all the numerous plates by which they were illustrated, are included in the present volume, forming one of the series.

As all arts are necessarily progressive, and, as the spread of Kent's and Repton's Schools was materially accelerated by the taste for landscape painting, pictures, and poetry, which prevailed, more or less, during the last century among the higher classes of society in this country, so the present prevailing taste for botany and horticulture, and the introduction, from other countries, of many new plants which thrive in the open air in our climate, have called for such a change in the manner of laying out and planting grounds as shall display these new plants to a greater advantage than hitherto. This change has given rise to a school which we call the Gardenesque; the characteristic feature of which, is the display of the beauty of trees, and other plants, individually. According to the practice of Kent and Repton, and, more especially, to that of all the followers of the Picturesque School, trees, shrubs, and flowers were indiscriminately mixed, and crowded together, in shrubberies or other plantations; and they were generally left to grow up and destroy one another, as they would have done in a natural forest; the weaker becoming stunted, or distorted, in such a manner as to give no idea of their natural forms and dimensions; though forming picturesque groups and masses highly pleasing to the admirers of natural landscape. According to the Gardenesque School, on the contrary, all the trees and shrubs planted are arranged in regard to their kinds and dimensions; and they are

planted at first at, or, as they grow, thinned out to, such
distances apart as may best display the natural form and
habit of each : while, at the same time, in a general point
of view, unity of expression and character are aimed at,
and attained, as effectually as they were under any other
school. In short, the aim of the Gardenesque is to add, to
the acknowledged charms of the Repton School, all those
which the sciences of gardening and botany, in their present
advanced state, are capable of producing.

The Gardenesque School of Landscape has been more
or less adopted in various country residences, from the
anxious wish of gardeners and botanical amateurs to display
their trees and plants to the greatest advantage. Perhaps it
may be said to have always existed in botanic gardens ; and
to have been first applied, in the case of a country residence,
by the present Duke of Marlborough, when Marquis of
Blandford, at White Knights. It may now be seen in its
most decided character, as far as respects trees and shrubs,
wherever Arboretums have been properly planted: as, for
example, at Chatsworth ; and, in the case of flowers,
wherever there is a flower-garden in an airy situation, and
the flowers are grown in beds, unmixed with trees and
shrubs. The Gardenesque School of Landscape is particularly adapted for laying out the grounds of small villas;
and it is nowhere better exemplified than in the villa of
W. Harrison, Esq., at Cheshunt, described in detail, with
numerous engravings, in the fifteenth volume of the *Gardener's Magazine.* An entire volume is not required to
describe this school; but one of our proposed series will
be devoted to giving a systematic view of the whole art of
Landscape Gardening, including all the styles and schools :
and, among the latter, the Gardenesque. In this volume

will be shewn how all the materials of the art, such as ground, wood, water, rocks, buildings, &c., may be managed according to the Geometric and Landscape Styles, and to all the different schools which we have enumerated of these styles. This volume will also contain a history of the art, and of its literature; and a descriptive catalogue of all the remarkable country residences that have ever been formed, so far as we can obtain any account of them.

As the number of engravings in the other volumes we contemplate, will be much less numerous than in that now published, it is expected that the price of all the five will not exceed £5; and, as each volume will be complete in itself, the possessor of only one of them never need consider himself as having an incomplete work. The original works, which will thus be included in five octavo volumes, would have cost, at their published prices, above £50.

Our first idea was to amalgamate the contents of the proposed five volumes into one general treatise; but a little reflection convinced us that the different schools would not be so distinctly marked by this mode of proceeding, and that the result would have a tendency to present only one system of laying out grounds to the young gardener, instead of several. There is a beautiful unity of system and manner of thinking in most of the works which we intend to reprint, which would have been, in a great measure, destroyed, by breaking them up into fragments, and scattering these under the different heads, which must necessarily have been done in forming a general treatise. The great advantage of treating of the different schools separately, and so as strongly to impress each on the mind of the young gardener, is, that he will thereby acquire a knowledge how to effect the same object according to different systems; and hence, in practice, he will be able either to adopt the style or school best calculated for the situation, climate, and circumstances in which he is placed, or to adopt and combine such parts of different styles and schools as may best attain the object in view in the given locality. This we consider to be the most effectual mode of preventing mannerism, or the adoption of one style, school, or system, as better than all the others, and employing it indiscriminately in every situation, though under widely different circumstances. This last mode was always adopted in the time of Kent and Brown; and hence that sameness which characterizes the artificial features of all the places laid out by those artists. The only safeguard against the continuance of this system, especially among gardeners, is the dissemination of a knowledge of different styles and schools; by which the idea that any one of them is better than another will be neutralized, and the true art of laying out grounds shewn to consist in the choice and application of a school, or of parts of different schools, adapted to the particular case under consideration. Art and Nature would thus be more harmoniously combined, and country residences produced of a more distinctive and interesting character.

Such is the plan and the intention of our series of five volumes; which, if carried into execution, with such improvements as may from time to time suggest themselves, will form, we think, as complete an Encyclopædia of Landscape Gardening as the present state of our knowledge, in that art, will admit.

After this slight outline of our general design, it remains only to say a few words respecting the arrangement of that part of it which constitutes the present volume. The reader is first presented with a systematic

Table of Contents, by which all that is said on any particular subject, in different parts of the work, may be immediately referred to. Next follows a List of the Engravings, a glance over which may assist an inquirer in discovering those designs in the book which may best suit his purpose; and then a general Table of Contents, in the order in which the different subjects are printed. The Biographical Notice was furnished by a member of Mr. Repton's family; and, as portraying the life of a man not less eminent for his artistical genius and taste, than for his goodness of heart, and amiability of character, it will, we are certain, be read with very great interest. The different works are printed in the order in which they were published, between 1795 and 1816. At the end of the volume is a copious Alphabetical Index. The Notes, by the Editor, are generally in brackets, thus—[ ]—with his initials added. In the copies with the engravings coloured, the colouring is a faithful imitation of that in the plates in Mr. Repton's volumes, as originally published.

J. C. L.

BAYSWATER,
November, 1839.

THE CONTENTS,

SYSTEMATICALLY CLASSED ACCORDING TO THE SUBJECTS.

HISTORICAL, BIOGRAPHICAL, AND CRITICAL REMARKS.

GENERAL PRINCIPLES.

The Study of the Subject, and the Communication of Ideas as to proposed Improvements, by the Landscape Gardener to his Employer, or to the Public.

Roads, Approaches, Walks, and Drives.

Fences and Gates.

Animated Nature.

Times of the Day, Weather, &c.

Seasons of the Year.

II.—APPLICATION OF GENERAL PRINCIPLES TO THE FORMATION OF THE COMPONENT PARTS OF COUNTRY RESIDENCES.

Dwelling-houses, Mansions, &c.

Terraces and other Architectural Appendages to a House in the Country.

The Surrounding Country.

III.—THE APPLICATION OF GENERAL PRINCIPLES IN THE UNION OF THE COMPONENT PARTS OF COUNTRY RESIDENCES, IN THE FORMATION OF DIFFERENT CHARACTERS OF HOUSES AND GROUNDS, FROM THE COTTAGE TO THE PALACE, AND IN LAYING-OUT PUBLIC GARDENS.

Situations and Characters.

The Cottage Ornée.

The Ferme Ornée.

The Villa.

The Mansion Residence.

The Baronial Residence.

The Palace.

Public Gardens.

c

LIST OF ENGRAVINGS,

CLASSED ACCORDING TO THE SUBJECTS.

DIAGRAMS.

ARCHITECTURAL ELEVATIONS.

THE CONTENTS, AS PRINTED.

SKETCHES AND HINTS

ON

LANDSCAPE GARDENING.

d

OBSERVATIONS ON THE THEORY AND PRACTICE

OF

LANDSCAPE GARDENING.

AN INQUIRY INTO THE CHANGES OF TASTE
IN
LANDSCAPE GARDENING.

DESIGNS FOR THE PAVILLON
AT
BRIGHTON.

FRAGMENTS ON THE THEORY AND PRACTICE

OF

LANDSCAPE GARDENING.

BIOGRAPHICAL NOTICE

OF

THE LATE HUMPHRY REPTON, ESQ.

(Written expressly for Loudon's Edition of
Repton's Landscape Gardening.)

THE following records of some of the principal events in the life of the late Humphry Repton are gleaned from a collection of MSS. in his own handwriting, containing interesting details of his public and private life; and it is somewhat difficult to select, from materials so varied, such as may best convey to the reader a general idea of the history and character of the author of the works contained in this volume.

The papers alluded to were left as a valued memorial for his children; it may be imagined, therefore, that they contain details of a private nature, which would be found devoid of interest to the world. Mr. Repton, indeed, possessed a mind as keenly alive to the ludicrous, as it was open to all that was excellent, in the variety of characters with whom his extensive professional connexions brought him acquainted; and he did not fail to observe and note down many curious circumstances, and traits of character, in themselves highly amusing, but, for obvious reasons, unfit subjects for publication. We may remark, however, that not one taint of satire or ill-nature ever sullied the wit which flowed spontaneously from a mind sportive sometimes even to exuberance. After all such curtailments have been made, however, there still remains so much that might be considered curious or interesting to the general reader, that it is difficult to limit

B

this Sketch within the bounds necessary for the present publication.

To those who have had an opportunity of personal acquaintance with the subject of these Memoirs, it cannot but be interesting to trace the early history and gradual progress of one whose genius and varied accomplishments eventually enabled him to rise to the highest eminence in his profession; or, perhaps, it would be speaking more correctly to say, that his talents enabled him to exalt into an honourable profession, that pursuit which, before his time, had been looked upon but as an occupation for the gardener or nurseryman.

The term " Landscape Gardener" was first adopted by Mr. Repton: and Mr. Uvedale Price, in his well-known " *Essay on the Picturesque,*" accuses him of " assuming a title of no small pretensions." His reply to this shews that he was fully aware of the difficulties that were to be conquered, when he first " determined to make that pursuit his profession which had hitherto been only his amusement." He remarks: —" It is the misfortune of every liberal art, to find amongst its professors some men of uncouth manners; and, since my profession has more frequently been practised by mere day-labourers, and persons of no education, it is the more difficult to give it that rank in the polite arts which I conceive it ought to hold. But I am more particularly called upon to support its respectability, since you attack the very existence of that profession, at the head of which both you and Mr. Knight have the goodness to say I am deservedly placed."

That Mr. Repton succeeded in this, his laudable ambition, can be denied by none who remember him in the days of his celebrity, when he was looked up to, by the highest ranks of society, as the acknowledged arbiter of taste, from whose fiat there was no appeal. The name of " Landscape Gardener" may *now*, indeed, be considered as " one of no small pretensions," since it is expected that he who so calls himself, should be a person not only gifted by Nature with the love of all that is beautiful, but that culture and education should have refined his taste, and improved his powers of judgment; while a knowledge of the habits of polished life, to be acquired only by an admittance into the best society, must have taught him how to combine those thousand little nameless circumstances which render the mansions and surrounding pleasure-grounds of our " Homes of England" the admiration of all who can appreciate that truly English word " comfortable."

Unfortunately, the monumental works of the landscape gardener are not like those of the architect, which live to future ages, and become a lasting record of the taste and genius of their contriver. Time makes unrelenting havoc with designs which, during the first ten or twenty years, may have afforded unmixed satisfaction. Young trees will out-grow their situations, while old ones will be uprooted by age or accident; flower-gardens which owed their charm to the light but fragile trellis ornament, or the constant culture of their elegant parterres, will fall into decay, or be neglected by their owners; while the facility with which any alterations may be made, aiding the love of change which is natural to most minds, in the course of years leaves no trace of that master-hand which had first laid the foundation of future improvement.* It is, therefore, by Mr. Repton's printed works alone that his well-earned fame can be properly appreciated; and, in the republication of those works, their present Editor is conferring a benefit on the public, by bring-ing them forward in a form which will render them more accessible to the general reader.

Humphry Repton was born at Bury St. Edmund's, May 2nd, 1752. His father was Mr. John Repton, who, for many years, held the honourable and lucrative situation of Collector of Excise. From many trifling anecdotes recorded by his son (who never mentions him but in a tone of deep rever-ence, blended with warm affection), we gather that he was a man of high religious principles, and great benevolence of heart; with manners peculiarly dignified, and possessing a firmness of mind, which frequently led him to suppress any outward demonstration of the best feelings of his nature. During his residence in Bury, he married Martha, daughter of Mr. John Fitch, of Moor Hall, in the county of Suffolk, a descendant of Sir Thomas Fitch, who was created a Knight-

---

* On visiting places which were known to be more particularly formed by Mr. Repton's taste, the writer of this Sketch has questioned the presid-ing gardener, but has generally been answered—" It was *all my master's laying out.*"

Banneret on the field of Agincourt. She was a woman of singular personal beauty, and sweetness of disposition, and was in possession of a small landed property at Stoke, near Clare. This happy union was blessed with three children; the eldest, a girl, named Dorothy (who was married to Mr. John Adey, a much-esteemed solicitor, at Aylsham, in Norfolk); the second, Humphry, the subject of these Memoirs; and the third, John, who rented and passed the greater part of his life at Oxnead Hall, in the same county.

At a very early age, Humphry was sent to the Grammar-school at Bury: he says, " I was too young to recollect much of those *happy days*, as they are always deemed by men, but of which children think differently, since the fear of the rod and the ferula, with the labour of the lesson and the task, are not less evils, while they last, than the fears of riper age, or the anxieties of manhood; perhaps the true difference between the life of a child and that of a man should be estimated by his power of enjoying pleasures, rather than in his experience of evils. The character of the future man may be traced in the boy; he may become a great man, or a rich man; but whether a happy man or not, will much depend upon the degree of natural cheerfulness with which Heaven has originally endowed him; since it is hardly in the power of Fate to confer much happiness on the man of gloomy disposition, nor lasting misery on one of cheerful temperament. The former will sigh upon a throne, while the latter may smile in a dungeon. Through the darkening medium of care, we see imaginary ills in the future, and even the brightness of the past is clouded: but care seldom clouds the views of childhood; which, forgetful of the past, and regardless of the future, enjoys the present moment; and much of the secret of happiness through life lies in the modification of this short sentence." From this school he was removed to the Grammar-school at Norwich, in which city his parents then resided;— and thus seven years passed in laying the foundation of classical knowledge; and he was rapidly rising to a high station amongst his schoolfellows, when, as he expresses it, " My father thought proper to put the stopper in my vial of classic literature; having determined to make me a rich, rather than a learned man. Perhaps wisely considering, that

if Solomon himself had not been the richest, the world would
scarcely give him credit for having been the wisest man."
Large fortunes were at that time made by the exportation of
the Norwich manufactures; and his father imagined, that, by
sending him abroad, and directing his education in a new
channel, he might, in time, rival those fortunate men who die
possessed of £100,000. In the summer of 1764, therefore, his
father and sister accompanied him from Harwich to Helvoet-
sluys, that the foundation of his future greatness might be laid
by his learning Dutch in a school in Holland. This formed
a new epoch in his early life, and he dwells upon its details
with a minuteness and feeling which must be interesting to
all who can enter into the situation of a boy about to be left
in a foreign country, without the power of expressing himself
in its language ; and who was, for the first time, to be entirely
separated from the dearest objects of his love. The customs
of a Dutch school, so unlike all he had ever met with in
England, are also naturally portrayed; but it would extend
our present Sketch too much were we to make extracts. The
description of the scenery bordering the canals, as viewed
from a trekschuit, or canal boat, we give, as having reference
to the subject of the present volume.

   " At that time it was the pride of every possessor of a few
acres, or even square yards, of ground, to display his riches and
his taste to the view of passengers who have scarcely any other
mode of travelling except in a trekschuit. This display was
different in different places; sometimes it consisted of a parterre hanging to the water, in which the design traced on the
ground was like a pattern for working muslin on embroidery.
The outline might, perhaps, be marked with an edging of box,
and, in some few instances, small grass-plots were introduced;
but, generally, the effect of these gardens (as they were called)
was produced without any vegetation; yet, by a contrast of
colours, and a variety of forms, the eye of the stranger was
amused, while that of taste might smile at the absurdity.
Instead of filling these beds with earth, or mould, in which
plants might grow, one part was filled with red brick-dust,
another with charcoal, a third with yellow sand, a fourth with
chalk, a fifth with broken china, others with green glass,
others with spars and ores, and, in short, with materials of

every colour or kind that might imitate the gardens of precious stones, described in fairy tales, or the Arabian Nights. Such fanciful ground-plans were surrounded by clipped hedges, intermixed with statues and vases of lead, painted in gaudy colours, and often richly gilt;—but sometimes they were flat boards, on which were painted men and women, to imitate the action and colours of nature. In other gardens, a taste less extravagant prevailed. The lofty trees, though always planted in rows, and always cut to preserve the exact limit of their shade, were accompanied by ornaments of sculpture in marble; and the vases were enriched with real flowers, instead of gilded pine-apples. In many places only partial views of the gardens were opened to the canal;—but these were always studied in their effect to the passengers, by a long perspective, not like that of an avenue, which is the same from one end to the other; but frequently, by arches, or other contrivances, the eye was led across many different compartments of an extensive garden, and the view was generally terminated by a scene painted at the back of some seat, which gave imaginary extent beyond the real boundary. The whole interior of these gardens was as formal and fantastic as these occasional vistas. Nature was never consulted, they were works of art; and the lofty clipped hedges, and close overarching trees, were as carefully kept by the shears, as the walks were by the scythes and rollers. All was neatness; the effect of incessant labour. A Dutch merchant's accounts and his garden were kept with the same degree of accuracy and attention. I have been more particular with this description, because it so strongly confirms the opinion I have conceived, that no degree of care or forethought in the father can avail the son in the future pursuits of his life. How few have reached to any great eminence in the profession to which, by the fond parent, they were originally designed! Could it be expected that the future landscape gardener of England should have studied in the parterres and clipped vistas of a Dutchman?"

The school in which he was placed by his father, was situated in the small village of Workum, and here he passed what would have been called a miserable twelvemonth, by any one of less buoyant spirits than his own; but he was one of

those enviable beings who are so well described by Wordsworth,
as peculiarly—

> " Blest with a kindly faculty to blunt
> The edge of adverse circumstance, and turn
> Into their contraries, the petty plagues
> And hind'rances with which they stand beset."

And his cheerful endurance was rewarded by a fortunate
occurrence about this time, which made an entire change in
his situation during the remainder of his stay in Holland.

With Mr. Zachary Hope, of Rotterdam, had been placed
a sum sufficient to defray his school expenses; and a half-
yearly payment had regularly been remitted by him to
Workum, with some general inquiry as to the health and progress of the little Englishman. For this civility it was deemed
necessary that the young gentleman should call and express his
thanks. To most boys of thirteen this would have been an
awful undertaking, but he possessed a naturally frank and open
disposition, which, combined with the advantage of a strikingly
handsome person, seldom failed to prepossess strangers in his
favour. Perhaps these advantages were aided by the interesting situation of a boy thus thrown upon the kindness of strangers. From whatever cause it arose, however, this call of
civility ended in an invitation to remain two days: and, during
that short time, he became so great a favourite, that it was
declared " impossible to part with young Repton:" and thus,
for five months, he was domesticated in Mr. Hope's family, a
sharer in all the advantages of education with his only son,
enjoying every pleasure and luxury which wealth could pro-
cure, and honoured by the friendship of other branches of
that numerous and respectable name, which, both at Amster-
dam and Rotterdam, had established a kind of rank which
vied with the proudest families of other countries. With
these kind friends he visited the celebrated watering-place of
Spa; and here he again enjoyed the benefit to be derived
from an introduction into the best society both of Englishmen
and foreigners. By the latter, " le petit Anglais" was now
especially noticed with kindly interest. Whether all these
circumstances tended to forward his father's views in sending
him to study Dutch, in order to make him a rich man, may be
somewhat doubted; but it may be imagined that they im-

proved the natural quickness of his intellect, gave a polish to his manners (not likely to have been acquired under the tuition of his good old schoolmaster, Mynheer Algidius Zimmerman), and expanded his mind by so early an intercourse with the world. These were advantages of inestimable value in his future career, which have always been acknowledged by himself with gratitude, when he recalled the happy days passed with his early friend, Mr. Zachary Hope.

But our limits will not permit us to dwell longer on this early period of Mr. Repton's life. Two years were spent in a school at Rotterdam (as he was removed from that at Workum, at Mr. Hope's suggestion), from whence he was enabled to pass much of his time with the same invaluable friends. On his return to Norwich, now a youth nearly sixteen years of age, a considerable premium was advanced, that he might, during the next seven years, become initiated in all the mysteries of trade; and when we consider how utterly useless was all such knowledge to him in after life, it is somewhat amusing to find him learnedly descanting upon the nature of calimancoes, Mecklenburgs, worsted satins, and other articles which fashion has now discarded from the list of modern dress and furniture. The records of this part of his life, however, lead us to infer, that the exercise of his talents for poetry, music, and drawing, occupied more of his time than was quite consistent with the views of his affectionate, though, in this instance, not very discriminating, parent. A taste for the latter of these accomplishments seems to have been more especially a part of his nature; and he enumerates it amongst his many sources of gratitude to Heaven, that he was blessed " with a poet's feelings and a painter's eye;" for he says, " it was to my early facility, and love of the art of drawing, that I am indebted, not only for success in my profession, but for more than half the enjoyments of my life. When I look back to the many hundred evenings passed in the circle of my own family—drawing and representing to others what I saw in my own imagination, I may reckon this art among the most delightful of my joys."

It is not surprising that a young man, possessing such attractions of mind and person, should become a distinguished favourite with the fair belles of a provincial city. He was an undoubted acquisition at a ball; and, in a private concert, his fine voice and sweet-toned flute were not to be dispensed with. He honestly confesses—" In these days of my puppy-age, every article of my dress was most assiduously studied; and while I can now smile with contempt on the singular hat, or odd-shaped pantaloons of some dandy of the present day, I recall to mind the white coat, lined with blue satin, and trimmed with silver fringe, in which I was supposed to captivate all hearts on one memorable occasion."

These feelings of youthful vanity, however, became sobered down by an ardent and lasting attachment to a young lady of the name of Clarke; and, after a union of forty years, he was able, with truth, to say, " I fixed my hopes where *I have never been disappointed.*" As his father very prudently objected to his marrying before he was twenty-one, this engagement of three years was not fulfilled till the 3rd of May (just three days after his coming of age). In consequence of this event, his father supplied him with a capital sufficient to commence business as a general merchant; and he turned his whole attention to the occupation thus marked out for him. For the first few years he seems to have been tolerably successful; but the casualties of ships lost at sea, and failures in speculations, the details of which it would be tedious to relate, together with the death of both his beloved parents, within a year of each other, made him still more disgusted with a pursuit so little in accordance with his natural taste and inclination, and by which he foresaw his income was likely to be diminished, rather than increased. He therefore determined to retire into the country, and went to reside at Sustead, a most sequestered spot, not many miles distant from Aylsham (where his sister resided in the house which had been left to them by their father). Here passed five years of uninterrupted domestic happiness. The improvement of his garden, as may be expected, was his favourite occupation; the beauties of Nature were his delight; and the investigation of her wonders his amusement. Not an insect or flower passed unnoticed by his inquiring mind; and in these pursuits he was encouraged by the frequent visits of his friend and schoolfellow, Mr. (afterwards Sir) James

Edward Smith, who, when in after-life he became President
of the Linnæan Society, thus writes to him:—" Not all the
black devils, nor even *blue* ones, which last, thank God, are
known to me *nomine tantum*, can restrain me from thanking
you for the last paragraph in your letter, which recalls youth
and Sustead, and my first botanizing days, when I hoarded
up a hazel-twig gathered in your grounds. There I first
began to emerge from the still pool of life, where fate had
dropped me. I hope to remind you of some things you may
have forgotten, if we meet in town this spring."

An extract from one of the numerous letters Mr. Repton
wrote at this time, will best convey an idea of his pursuits,
and the perfect happiness of his mind:—

" To EDWARD CHAMBERLAYNE, Esq.,
" Treasury.

" Dear Sir,—While I thank you for the present of lamb,
let me beg you not to send costly presents in return for mine,
which cost nothing but the ' sweat of my brow,'—amply
repaid by health, spirits, and exercise. You must remember
that I have no other way of shewing you I am alive and
happy, than by sending you some part of that abundance
with which Providence blesses my farm-yard, or my gun. I
don't wonder you should be at a loss to find Sustead in your
map! It is so small a parish, that I am obliged to enact the
various parts of churchwarden, overseer, surveyor of the
highways, and esquire of the parish. Let me add, landlord
of the inn, by receiving you in the only one there is in the
place;—for there is not even an alehouse to disturb my
peace. I am impatient to shew you the alterations in my
house and lands. The wet hazy meadows, which were deemed
incorrigible, have been drained, and transformed to flowery
meads. Your gun must, for a time, have rest; but, at this
delightful season, you, like me, can want no other inducement
to enjoy the air, than what Nature so bountifully provides for
her admirers in every hedge and field. Come and see how
happy we are! If there be any cause for discontent, it arises
from my considering that I have not yet deserved to be
so happy. I am taking that at the beginning of life, which
all look forward to at its close, as the reward of industry. I
am, however, not idle. I read much, having the use of my

neighbour Wyndham's library. Sometimes he is so good as to assist me in my studies; and when I was forbidden to read, from a late disorder of my eyes, he would often sit and read to me; for he says he likes my snug study at Sustead better than the old rambling library at Felbrigg. He has introduced me to Mr. Joseph Banks and other learned men, and his library has introduced me to Buffon, De Reaumur, &c.; and they have brought me acquainted with all the insects in my neighbourhood."

No pursuit seems to have afforded him so much pleasure as that of making drawings of the seats of every nobleman and gentleman within his reach. Many of these he presented to their respective owners, and others he gave to illustrate the "*History of Norfolk;*" a work then publishing in ten volumes, and for which he also supplied the letter-press describing the Hundreds of North and South Erpingham. From this he derived no other benefit than that of possessing a copy of the work: but while many a long winter's evening passed happily in gratifying his natural taste, he little dreamed that he was gaining a facility in expressing his ideas with his pencil, that would be of such important service to him in after-life.

The daily increasing intimacy with a person of mind so superior as that of Mr. Wyndham, could not fail to excite a spirit of emulation, and to call into exercise powers which, but for this, might have lain dormant in a spot so secluded from the world.

In the spring of 1783, the appointment which made his friend Secretary to the Lord Lieutenant of Ireland, excited hopes of more solid advantage; and an increasing family made him feel that he ought not to lose the opportunity of exchanging a life of idleness for one of more active exertion. He therefore wrote a letter to Mr. Wyndham, to which he received the following answer:—

"Dear Sir,—You may think it, perhaps, a sufficient attention to your letter, that I answer it by the return of the post; but I have done more for your wishes, by answering them in my own mind before you made them known to me. It happens, very whimsically, that your proposal is just an echo to a wish I was about to express to you (if you will

allow me an image, when talking of Irish affairs, that makes
the echo come first). From the moment this business was
determined (with the determination of which I will not profess myself over happy), having got myself into a scrape, my
first thought was, how I might bring my friends in with me;
and in that light I had very early designs upon you. Nothing
delayed the discovery of my wishes, but some difficulties, not
quite removed, respecting the situation I might have to offer
you; and some uncertainty of your willingness to accept any
offer I might have to make. As the latter of these is now at
an end, and no impediment exists in your own likings, other
difficulties, I trust, may be got over; and I think I may positively say, that some situation shall be found, which shall
afford me the advantage and satisfaction of your company
and assistance, with a fair prospect of benefit to yourself. If
you, as soon as it is convenient, will come to town, you
may be of great immediate use to me; and we can then more
conveniently talk of other matters.

                    " Yours, with best compliments to Mrs. Repton,
    " May 5."   " W. W."

The result of this letter was, that he accompanied Mr.
Wyndham to Ireland, as his confidential Secretary; and the
succeeding two months were passed in all the excitement
which scenes so new were likely to create. Public and private
dinners at which he was present, anecdotes of Mrs. Siddons
(with whom he lived in the same house), descriptions of Irish
scenery and manners, and animated sketches illustrative of the
character and opinions of his patron—all furnish materials for
observation, which, however interesting, must be here passed
over.

Mr. Wyndham, from some unexplained cause of dissatisfaction, threw up his situation at the end of a month; and,
with only a few hours' notice, set off for England, intrusting
entirely to his friend's management and discretion, not only
the settlement of his private affairs, but likewise the fulfilment
of his official duties, until another Secretary should be sent
from England: and, under these circumstances, Mr. Repton
passed another six weeks in Ireland, as Mr. Wyndham's representative, invited to all the dinners at Phœnix Park, and
other parties in the first circles; and assisting his Excellency

in many of his official duties, until the arrival of Mr. Pelham (Wyndham's successor); when, having settled everything to the entire satisfaction of his patron, he prepared for his return to the quiet of his own beloved home. The following characteristic conclusion to a letter written at that time to his wife, will best shew the manner in which he bore the disappointment of those bright hopes of distinction and emolument, which, at the commencement of this expedition, had promised so much of both.

"And now, my dearest Mary, what have I been doing? I have learned to love my own home; I have gained some knowledge of the world; some of public business, and some of hopeless expectancies; I have made some valuable acquaintances; I have formed some connexions with the great; I have seen a fine country, in passing through Wales, and have made some sketches; I have lost very little money; I shall have got the brogue; and you will have got a tabinet gown. So ends my Irish expedition."

But the motives which had made him seek for active employment were by no means lessened after his return to his peaceful and happy home; on the contrary, they recurred with daily increasing strength; and the utter failure of all his schemes to make his country pursuits profitable, in a pecuniary point of view, made him determine upon a vigorous effort towards retrenchment, only to be accomplished by the painful sacrifice of removing his family from their "little earthly paradise;" and the cottage at Harestreet, Essex, was fixed upon as their temporary residence. This humble dwelling subsequently became so endeared to him, as the scene of some trials and many blessings, that he never afterwards sought any other home. In the last chapter of the "*Fragments*," a view from its windows illustrates the principles of " quantity, appropriation, and foreground;" and we refer the reader to his own observations on that home in which he resided for forty years, as any farther remark would but weaken the simple pathos of a passage so descriptive of his feelings.*

* This cottage, with its surrounding very small garden, became, under Mr. Repton's superintending taste, a specimen of what may be accomplished by ingenuity of contrivance. The passing traveller has often admired, with

At this time (1784), an introduction to Mr. Palmer, who was then beginning to form his project for substituting mail-coaches for the tardy system of carrying letters then in use, gave rise to a new source of hope and exertion. Mr. Repton's active energy of character and quickness of perception, enabled him at once to see all the advantages likely to arise from the project of his new friend; and while he lent his personal exertions, and directed his best energies towards the forwarding of those projects, his sanguine disposition led him to embark the greater part of his small remaining capital, in a scheme which, by its results, has shewn that he did not miscalculate the immense benefit it was capable of bestowing upon the Government, as well as on the public. After many years, Mr. Palmer received a considerable, though scarcely adequate, pecuniary reward, in acknowledgment of *his* claims upon that public and that Government; but Mr. Repton, without whose assistance those plans could scarcely ever have been brought to perfection, received no recompense for his services, and the pecuniary losses which he sustained.* Thus baffled in an enterprise which had promised as much of personal advantage as of public utility, it is not surprising that his spirits, for a short time, were depressed by disappointment. But a firm, though unostentatious trust in the constant superintendence of Providence, was, at all times, a peculiar characteristic of his mind; adding a heartfelt cheerfulness to his lightest moments of mirth, and affording a bright ray of consolation in the gloomiest moments of sorrow. And in after-life he loved to dwell, with a deep sense of gratitude, on those circumstances of his life which, at the time of their occurrence, had been most painful; but which had invariably terminated in some unforeseen advantage. Had he been

a lingering eye, its pretty exterior, and those who were admitted to its happy fireside, could not but acknowledge, that comfort, with a certain degree of elegance, may be contained within a very limited space. But it is here necessary to remark, that the whole of this place has been altered so much, that the modernized remains of trellis-work, and two lime-trees deprived of their gracefully spreading arms, are all that now are left of " Repton's Cottage," at Harestreet.

* The late Sir Francis Freeling was so fully sensible of the active part which Mr. Repton had taken in this project, that, a few years since, he was anxious to obtain from his family any copy of his original plans for the conveyance of letters; considering them most valuable, as laying the foundation for the future success of that now astonishing source of revenue.

successful in his mercantile transactions, or in his farming experiments, or in his mail-coach enterprise, the peculiar tendency of his genius must have remained undeveloped. The possibility of turning to advantage that natural taste for improving the beauties of scenery, which had formed one of the dearest pleasures of his rural life, suggested itself to his mind one night when anxiety had driven sleep from his pillow. This scheme, which at first seems to have entered his imagination with almost the vague uncertainty of a dream, assumed a more substantial form, when, with the return of day, he meditated upon its practicability. With his usual quickness of decision, he arose the next morning; and, with fresh energy of purpose, spent the whole of that day in writing letters to his various acquaintances, in all parts of the kingdom, explaining his intention of becoming a " Landscape Gardener;" and he lost not a moment in bending his whole mind to the acquisition of such technical knowledge as was necessary for the practical purposes of such a profession.

That success attended these exertions, it is unnecessary to add. The works contained in the volume to which this brief notice of their author is prefixed, sufficiently testify the result; but, in tracing the history of such persons as have risen to any eminence, in whatever may have been their pursuit, we generally find that much as our interest may have been excited by descriptions of the early indications of genius, or the various apparently trifling circumstances which have tended to aid its development, or the obstacles which so frequently beset its path, much of that interest fades away when the subject of it has realized our expectations, by overcoming those obstacles; whether, in tracing the fortunes of an imaginary being, or a character of real life, the years of prosperity are slightly passed over, and the Biographer feels at a loss to say more, than that his hero was successful, was eminent, was rich, or was happy. Thus, in the present instance, the interest which is excited by the remainder of Mr. Repton's memoranda, arises less from any account of himself, than from the observations he has made on others; and, in attempting to trace his professional career, the task of selection and curtailment becomes more difficult.

From the Prince, who did him the honour of consulting

him in the improvements of his palace, to the humble citizen,
who, in his villa near town, asked for his assistance to arrange
his rows of poplars, or to exclude the dusty road by his fir
plantation,—all afford traits of character, both curious and
amusing;—but to insert such anecdotes would be inconsistent
with our present purpose, of giving a Sketch of Mr. Repton's
own character and life alone. It is somewhat singular, that,
amongst all his professional visits to almost every part of
England, in the county of Norfolk he had fewer engagements than in any other; although he has recorded that his
first professional visit was to Catton, the seat of Jeremiah
Ives, Esquire, and his second to Holkham, the seat of
J. W. Coke, Esquire (since Earl of Leicester): " two of my
earliest friends," as he remarks, " who have both outlived the
eighteenth century, and have seen many changes, though
none in the regard I have ever felt for each." Another of
his earliest patrons was the late Duke of Portland, on whose
amiable character he expatiates with the warmth of one who
not only had an intimate acquaintance with his domestic
virtues, but who felt that much of his success in life had been
owing to the early recommendation and steady kindness of
that great and good man. We cannot forbear transcribing
part of the account of his first visit to his Grace:—

" When I first saw the Duke of Portland, in 1789, he
explained to me his wishes concerning Welbeck, in a manner
so clear and decided, that all diffidence of my own skill was
removed. I had then had little practical experience, and felt
a painful degree of anxiety at every new concern ; afraid of
committing myself, and doubting my own powers to suggest
new ideas. The Duke's gracious manners, and his remarks,
which evinced his taste and judgment in my own pursuits,
alarmed my sensibility at first, but soon convinced me that I
should have no difficulty in conveying or explaining my
opinion to him. At the close of a full hour's conversation, I
begged his Grace would inform me to whom I was indebted
for the honour of this introduction: after a short pause he
said, ' I have been endeavouring to recollect, but cannot name
any one person in particular. Whenever I consult any professional gentleman, in whom I wish to place implicit confidence, both in his skill and integrity, it of course leads me

to make a strict inquiry of all who are likely to know him;
and, as I have uniformly received the same answer, it is to
yourself alone that you are indebted.' "

There was an evident sincerity, and an elegance in the
expression of this flattering compliment, that was calculated
to make a deep impression upon a sensitive mind; and it
naturally laid the foundation for that affectionate regard,
amounting almost to veneration, which, during an intercourse
of twenty years, every circumstance contributed to strengthen.
In describing the character of the Duke of Portland, Mr.
Repton says, "few men so well understood that genuine po-
liteness which springs from the wish to see others comfortable,
extending itself to the *act* of making them so; and in all his
words, actions, conversations, and correspondence, there was
a marked attention and civility which seemed the effect of
nature rather than of art—in fact, it was both; it was an habitual benevolence, which is as visible in trifles as in the most
important matters, and as evident in a short note as in a
volume. Of all the persons from whom I have ever received
letters, there was none who wrote with more marked attention to the value of my time than the Duke of Portland; and it is always flattering to a professional man to
suppose his time fully occupied. * * * * The last time I
saw his Grace, was at Burlington House, 27th February,
1806; in his usual kind manner, he requested me to spend
a day or two at Bulstrode, before he met me in Easter
week; but begged I would suit my own convenience in the
time, and, even in meeting him at Easter, not to interfere with any other engagement. I said that I should
always consider his early notice and patronage as the source
of all the fame I had acquired in my profession; and, therefore, I should never let any engagement whatever prevent
my attending first to his Grace's wishes, whenever he would
have the goodness to let me know them. He very cordially
and kindly replied, ' Mr. Repton, that very circumstance
makes me sometimes unwilling to put you to inconvenience.' "

That Mr. Repton's professional life was one of unmixed
enjoyment and uninterrupted success, might, perhaps, be

D

imagined by any one who only perused the letters of flattering commendation which he was daily in the habit of receiving ; of these many have been carefully preserved, as gratifying testimonials, from persons whose names alone are sufficient to render their words of praise invaluable.

But that there were also vexations and disappointments attendant on his profession, as there must be on every other pursuit in life, is evident, from such remarks as are interspersed in his notes. We extract the following, in explanation of our meaning :—" In short, of many hundred plans, digested with care, thought, and attention, few were ever so carried into execution, that I could be pleased with my own works." And again : "Like every pursuit of an active mind, mine has been of more use to others than to myself. At the end of a very few years I had reached the top of my ladder, which I had, in a great measure, reared up myself; and being conspicuously placed on its summit, it is natural to expect that I should become the mark for envy and rivalship. I saw myself attacked in the public papers for blunders at places I had never visited, or for absurdities introduced before I visited them; and I heard opinions quoted as mine, which I had never advanced, and was blamed for errors which I had never advised." * * * * " The established professor in every art is seldom contradicted, while present to defend the propriety of his plans ; but taste, as it is called, is so universal, that every one sets up for a connoisseur, and each is so jealous of his own opinion, that the greater the number who canvassed my plans, the greater was the departure from them, since every one boasted of a little taste of his own." * * * * " Whatever have been the causes, I found all my plans counteracted at Harewood ; and years have elapsed since I have seen that place, but I have heard that my design for the magnificent arch at the entrance—a design on which I had been complimented and flattered by the 'immortal Pitt'—has been adopted only in the general outline; the columns, instead of being detached, are sunk, half buried in a wall, and mounted on pedestals ; while the whole building is placed very differently from what I intended it. Instead of being at the end of the village of Harewood, it is removed to an unmeaning distance, isolated and detached, without any relation to the house or village. This vindication I am called on to assert, in consequence of many of my plans having been misunderstood and misrepresented."

Of the disappointment he experienced in the early part of his professional career, from a misunderstanding with one in whom he placed the greatest confidence, it is here unnecessary to speak. He never himself alluded to this subject without feelings of deep regret, yet untinged with anger; but towards the close of his professional life, when his ambition was about to be gratified, by the patronage of the highest personage in the kingdom, it was painful to find himself superseded by that very friend, who, in earlier life, had participated in his bright visions of future fame.

Mr. Repton's plans for the Pavilion at Brighton are now before the public, who must decide whether those that have been substituted in their stead, are improvements or otherwise. But the original designs had been received by the Prince, with these unqualified words of approbation :—" Mr. Repton, I consider the whole of this work as perfect, and will have every part of it carried into immediate execution ; not a tittle shall be altered,—even you yourself shall not attempt any improvement." A difficulty arose, however, as to the means of supplying the funds for the commencement of the works, which caused a delay that terminated as we have before shewn.

But we should do injustice to Mr. Repton's character did we leave it to be imagined, that these painful circumstances are dwelt upon by him in the spirit of complaint; on the contrary, it was one of his favourite maxims, that as there are more beautiful flowers and useful herbs in the world, than there are noxious or unsightly weeds, so the proportion of good in every person's life greatly outweighs that of evil, could we but persuade men to measure each with equal justice. Of this maxim his own life certainly affords us an example. Nature had bestowed on him one of her rarest gifts; a heart totally devoid of selfishness. This displayed itself in every trifling circumstance, as well as in the more important concerns of daily life. To give pleasure to another, was but adding to his own snare of happiness; and, with an even flow of spirits that shed light and cheerfulness on all around him, he was peculiarly blessed in his own family circle. For more than thirty years of his life, success, beyond his hopes, attended him in the profession he had marked out for himself; and in the exercise of which, he not only felt pleasure himself, but frequently had the power of promoting it in others. And to these blessings was added that of *health*, which had never known a day's interruption, till the unfortunate night of January the 29th, 1811; when, returning with his daughters from a ball given by Sir Thomas Lennard, his carriage was overturned, owing to an accumulation of snow in the road;—he received an injury in the spine, from which he never entirely recovered. For many weeks this accident confined him to his bed, deprived of all power of motion. In a situation so trying to one of his active disposition, his mind still retained its energy; and his patient endurance of suffering, and cheerfulness of spirits, never deserted him for a moment. It was many months ere he was able to resume his usual pursuits; and there is little doubt that the loss of his accustomed exercise laid the foundation of that complaint which, for the remaining years of his life, occasioned him, at times, great agony; and which his physician pronounced to be *Angina Pectoris*. It was well known to himself (and he did not conceal it from those most dear to him), that the termination of this disease would be as sudden as it must be fatal;—but the stroke was so long delayed, that hope had almost raised a doubt in the minds of his friends as to the truth of that awful fate which he himself never forgot was hanging over him. On the morning of the 24th of March, 1818, he came down to breakfast, not more unwell than usual (the act of dressing had, for some time, been attended with a few moments of spasm in the chest), but he no sooner reached the breakfast-room, than he fell into the arms of his servant, and expired without a groan. So instantaneous was his death, that before his son could hasten from the adjoining room, his spirit had fled for ever.

Perhaps there is no stronger proof of Mr. Repton's love for the beauties of nature, than the wish he had latterly expressed, that his remains might be deposited in a " garden of roses." To gratify this innocent fancy, he himself selected the small enclosure on the south side of the picturesque church of Aylsham, in Norfolk: a simple Gothic monument records his name and age, followed by some lines written by himself:—

"The tomb of Humphry Repton, who died March 24th, 1818.

'Not like the Egyptian tyrants—consecrate,
Unmixt with others shall my dust remain;
But mouldering, blended, melting into earth,
Mine shall give form and colour to the rose;
And while its vivid blossoms cheer mankind,
Its perfum'd odour shall ascend to heaven.'"

In the same grave reposes that gentle being, who, for five-and-forty years, had been the beloved participator in all his joys and griefs: she did not long survive him; and it was her last request that they should not be separated in death. Out of a family of sixteen children, but seven reached the age of manhood;—and at their parents' death, only four sons and a daughter remained of this numerous family. The latter still resides with her eldest brother, John Adey, who is, perhaps, less known as a follower of his father in the art of landscape gardening, than as one deeply versed in antiquarian lore; from which peculiar taste he has gathered an accurate knowledge of ancient Gothic architecture, many specimens of which are to be found in the engravings that accompany his father's works. The second son, Edward, who was originally intended for his father's profession, having evinced a preference for the Church, was sent to Magdalen College, Oxford, and is now a Prebendary of Westminster. The third son, William, is in the law, and possesses the small paternal estate at Aylsham; and the fourth son, George Stanley, who has for many years practised as an architect, married the eldest daughter of the late Lord Chancellor Eldon, only a few months previous to the death of his father. One of the most ardent wishes of that father's heart was gratified, by living to see his children united and happy: and we cannot better close this Notice, than by transcribing his own concluding words, so expressive of the ruling feeling of his mind:—" My ship of life is sinking, and it is time to quit it; these pages will serve to shew how actively I have performed the voyage—how I have glided through calms, and struggled through tempests. I have touched at every port, and where have we met with happiness unalloyed? or, where found a

man not disappointed? Nowhere! Yet still I must repeat,
that there is more of good than of evil; and for this redun-
dancy, all our gratitude must, at last, resolve itself into that
reiterated aspiration from my heart—*Laus Deo.*" A. B.

The following is a List of Mr. Repton's published works, of which,
those only that treat on landscape gardening are reprinted in the present
volume:—

The Hundred of North Erpingham, in the History of Norfolk, &c.
1781. 8vo.

The Bee; or, a Critique on the Exhibition of Paintings at Somerset
House. 1788. 8vo.

Variety; a Collection of Essays. 1788. 12mo.

The Bee; a Critique on the Shakspeare Gallery. 1789. 8vo.

Letter to Uvedale Price, Esq. on Landscape Gardening. 1794. 4to.

Sketches and Hints on Landscape Gardening; collected from Designs
and Observations now in the possession of the different Noblemen and Gentlemen for whose use they were originally made : the whole tending to establish fixed Principles in the Art of laying out Ground; 16 coloured Plates.
Lond. 1795. Oblong quarto. 2*l.* 12*s.* 6*d.*

Observations on the Theory and Practice of Landscape Gardening;
including some remarks on Grecian and Gothic Architecture; collected
from various MSS. in the possession of different Noblemen and Gentlemen :
the whole tending to establish fixed Principles in the respective Arts; with
many Plates. Lond. 1803. 4to. 5*l.* 5*s.*

Odd Whims; being a republication of some Papers in Variety; with a
Comedy, and other Poems. Lond. 1804. 2 vols. 12mo.

An Inquiry into the Changes of Taste in Landscape Gardening; to
which are added, some Observations on its Theory and Practice, including a
Defence of the Art. Lond. 1806. 8vo. 5*s.*

Designs for the Pavilion at Brighton; humbly inscribed to His Royal
Highness the Prince of Wales. Including an Inquiry into the Changes of
Architecture, as it relates to the Palaces and Houses in England. By
H. Repton, Esq. with the assistance of his Sons, John Adey Repton, F.A.S.,
and G. S. Repton, Architects. Lond. 1808. Folio. 6*l.*

On the supposed Effects of Ivy upon Trees. Trans. Linn. Soc. 1810.
Vol. xi. p. 27.

Fragments on the Theory and Practice of Landscape Gardening; including some Remarks on Grecian and Gothic Architecture; collected from
various Manuscripts in the possession of the different Noblemen and Gentlemen for whose use they were originally written : tending to establish
fixed Principles in the respective Arts. By H. Repton, Esq. assisted by his
Son, J. Adey Repton, F.A.S Lond. 1816. 4to. 6*l.*

SKETCHES AND HINTS

ON

LANDSCAPE GARDENING:

COLLECTED FROM

DESIGNS AND OBSERVATIONS

NOW IN THE

POSSESSION OF THE DIFFERENT NOBLEMEN AND GENTLEMEN

FOR WHOSE USE THEY WERE ORIGINALLY MADE:

THE WHOLE TENDING TO ESTABLISH FIXED PRINCIPLES IN

THE ART OF LAYING OUT GROUND.

BY H. REPTON, ESQ.

(Originally published in 1795, in one volume, oblong quarto.)

DEDICATION.*

TO THE KING.

SIRE,

Your Majesty's gracious patronage of
this Volume, while it impresses me with the deepest gratitude, excites in me a desire that the Work were more worthy
of the Royal favour. If it should appear that, instead of displaying new doctrines, or furnishing novel ideas, it serves
rather, by a new method, to elucidate old established principles, and to confirm long received opinions, I can only plead
in my excuse, that true taste, in every art, consists more in
adapting tried expedients to peculiar circumstances, than in
that inordinate thirst after novelty, the characteristic of uncultivated minds, which, from the facility of inventing wild
theories, without experience, are apt to suppose, that taste is
displayed by novelty, genius by innovation, and that every
change must necessarily tend to improvement.

That Your Majesty may long continue to be the Patron
of liberal arts, the encourager of polite literature, and the
great arbiter of true taste in this country, must ever be the
prayer of those who delight in contemplating the genius and
industry of Great Britain, fostered by our glorious constitution, under the benign protection of Your Majesty.

Permit me, SIRE, to subscribe myself, with the most
profound humility,

Your Majesty's
most dutiful Subject and Servant,

H. REPTON.

Harestreet, near
Romford, Dec. 6, 1794.

[* We have reprinted this Dedication, because it contains what, it appears, Mr. Repton considered as principles : viz., 1. that true taste, in every
art, consists more in adapting tried expedients to peculiar circumstances,
than in an inordinate thirst after novelty, &c. ; and, 2. that this inordinate
thirst after novelty, is a characteristic of uncultivated minds.—J. C. L.]

ADVERTISEMENT,

EXPLAINING THE NATURE OF THIS WORK.

————

MY opinions on the *general principles of Landscape Gardening* have been diffused in separate manuscript volumes, as opportunities occurred of elucidating them in the course of my practice ; and I have often indulged the hope of collecting and arranging these scattered opinions, at some future period of my life, when I should retire from the more active employment of my profession : but that which is long delayed, is not, therefore, better executed; and the task which is deferred to declining years, is frequently deferred for ever; or, at best, performed with languor and indifference.

This consideration, added to the possibility of being anticipated by a partial publication of my numerous manuscripts, not always in the possession of those by whom I have the honour to be consulted, induced me to print the following pages, with less methodical arrangement than I originally intended. I once thought it would be possible to form a complete system of *Landscape Gardening*, classed under certain *general rules*, to which this art is as much subject as *Architecture, Music*, or any other of the *polite Arts :* but, though daily experience convinced me that such rules do actually exist, yet I have found so much variety in their application, and so

E

much difficulty in selecting proper examples, without greatly
increasing the number of expensive plates, that I have preferred this mode of publishing a volume of HINTS and
SKETCHES; being detached fragments, collected from my different works. It never was my intention to publish the whole of
any one *Red Book*; nor to multiply my examples, by referring
to a number of different books, when a single engraving would
answer the purpose: I have, therefore, availed myself of
the honour conferred upon me by his Grace the Duke of
Portland, in permitting me to use the *Red Book of Welbeck* as
the ground-work of the present volume; though I shall, occasionally, refer to other places, in order to increase the number
of examples, without augmenting the number and expense of
plates. Thus an opportunity may sometimes occur of comparing my observations with the subjects themselves, or with
the original drawings in different libraries.

It will, perhaps, be expected that, in this advertisement, I
should give some account of the sequel of this Work, or the
number of volumes to which it may be extended; but, from
the multitude of my private engagements, I have found so much
trouble and difficulty in preparing this volume for the press,
that I dare not suggest the period, if ever it should arrive,
when I shall produce another.

H. R.

CATALOGUE OF THOSE RED BOOKS

*From whence the following Extracts are made; or which are men-
tioned as containing further Elucidations of the Subjects introduced in this First Volume.*

PLACE.	COUNTY.	A SEAT OF
Antony House	Cornwall . .	Reginald Pole Carew, Esq., M.P.
Babworth . .	Nottinghamshire	Honourable John Bridgman Simpson.
Bessacre Manor	Yorkshire . .	B. D. W. Cook, Esq.
Brandsbury .	Middlesex . .	Honourable Lady Salusbury.
Brocklesby . .	Lincolnshire .	Right Honourable Lord Yarborough.
Brookmans . .	Herts . . .	S. R. Gaussen, Esq.
Buckminster .	Leicestershire .	Sir William Manners, Bart.
Castle Hill . .	Middlesex . .	H. Beaufoy, Esq., M.P.
Catchfrench .	Cornwall . .	F. Glanville, Esq.
Claybury . .	Essex . . .	James Hatch, Esq.
Cobham Park .	Kent. . . .	Earl Darnley.
Courteen Hall .	Northamptonshire	Sir William Wake, Bart.
Crewe Hall . .	Cheshire. . .	John Crewe, Esq., M.P.
Culford . . .	Suffolk . . .	Marquis Cornwallis.
Donington . .	Leicestershire .	Earl Moira.
Ferney Hall .	Shropshire . .	Late Sam. Phipps, Esq.
Finedon . .	Northamptonshire	J. English Dolben, Esq.
Garnons . .	Herefordshire .	J. G. Cotterell, Esq. [In 1838, Sir J.G. Cotterell, Bart.]
Gayhurst . .	Bucks . . .	George Wrighte, Esq.
Glevering . .	Suffolk . . .	Chaloner Arcedeckne, Esq.
Hanslope Park	Bucks . . .	Edward Watts, Esq.
Hazells Hall .	Bedfordshire .	Francis Pym, Esq.
Herriard's House	Hampshire . .	G. Purefoy Jervoise, Esq.
Holkham . .	Norfolk . . .	T. W. Coke, Esq., M.P. [In 1838, the Earl of Leicester.]
Holme Park .	Berkshire . .	Richard Palmer, Esq.
Holwood . .	Kent. . . .	Right Honourable W. Pitt. [In 1838, John Ward, Esq.]
Lamer . . .	Herts . . .	Lieutenant-Colonel C. Drake Garrard.
Langley Park .	Kent. . . .	Sir Peter Burrell, Bart., M.P. [In 1838, E. Goodhart, Esq.]

28   REPTON'S LANDSCAPE GARDENING, &c.

PLACE.	COUNTY.	A SEAT OF
Lathom	Lancashire	Wilbraham Bootle, Esq.
Little Green	Sussex	Thomas Peckham Phipps, Esq.
Livermere Park	Suffolk	N. Lee Acton, Esq.
Milton	Cambridgeshire	Samuel Knight, Esq.
Milton Park	Northamptonshire	Earl Fitzwilliam.
Nacton	Suffolk	P. B. Broke, Esq.
Northrepps	Norfolk	Bartlet Gurney, Esq.
Ouston	Yorkshire	Bryan Cook, Esq.
Port Eliot	Cornwall	Right Honourable Lord Eliot.
Prestwood	Staffordshire	Honourable Edward Foley, M.P.
Purley	Berkshire	Anthony Morris Storer, Esq.
Riven Hall	Essex	C. C. Western, Esq., M.P.
Rudding Hall	Yorkshire	Lord Loughborough, L.H. Chancellor.
Scrielsby	Lincolnshire	Honourable the Champion Dymock.
Sheffield Place	Sussex	Right Honourable Lord Sheffield.
Stoke Pogies	Bucks	John Penn, Esq.
Stoke Park	Herefordshire	Honourable E. Foley, M.P.
Stoneaston	Somersetshire	Hippesley Coxe, Esq., M.P.
Sundridge	Kent	E. G. Linde, Esq. [In 1838, Sir Saml. Scott, Bart.]
Sunninghill	Berks	James Sibbald, Esq.
Tatton Park	Cheshire	William Egerton, Esq., M.P.
Thoresby	Nottinghamshire	Charles Pierrepont, Esq., M.P.
Trewarthenick	Cornwall	Fr. Gregor, Esq., M.P.
Tyrringham	Bucks	William Praed, Esq., M.P.
Waresley	Essex	Sir Geo. Allanson Winn, Bart., M.P.
Welbeck	Nottinghamshire	His Grace the Duke of Portland.
Wembly	Middlesex	Richard Page, Esq.
Whersted	Suffolk	Sir Robert Harland, Bart.
Widdial Hall	Herts	J. T. Ellis, Esq.

[Most, or perhaps all, of these places still exist, though, doubtless, a number of them have changed proprietors. We have endeavoured to ascertain what changes have taken place, but with very little success. This, however, is a matter of little consequence, as the real value of this work consists in the illustrative descriptions and engravings, which are altogether independent of the names of either persons or places.—J. C. L.]

INTRODUCTION.

To improve the scenery of a country, and to display its native beauties with advantage, is an ART which originated in England, and has therefore been called *English Gardening;* yet as this expression is not sufficiently appropriate, especially since Gardening, in its more confined sense of *Horticulture,* has been likewise brought to the greatest perfection in this country, * I have adopted the term *Landscape Gardening,* as most proper, because the art can only be advanced and perfected by the united powers of the *landscape painter* and the *practical gardener.* The former must conceive a plan, which the latter may be able to execute; for though a painter may represent a beautiful landscape on his canvass, and even surpass Nature by the combination of her choicest materials, yet the luxuriant imagination of the *painter* must be subjected to the *gardener's* practical knowledge in planting, digging, and moving earth; that the simplest and readiest means of accomplishing each design may be suggested; since it is not by vast labour, or great expense, that Nature is generally to be improved; on the contrary,

> " Ce noble emploi demande un artiste qui pense,
> Prodigue de génie, mais non pas de dépense."

[Which may be thus Englished:—

> " This noble employment requires an artist who thinks,
> Prodigal of genius, but not of expense."

The following paraphrase of this passage is given by Mrs. Montolieu, in her translation:—

> " Insult not Nature with absurd expense,
> Nor spoil her simple charms by vain pretence;
> Weigh well the subject, be with caution bold,
> Profuse of genius, not profuse of gold."
>
> *The Gardens.* 2nd Ed. p. 5.]

* This appears from the many valuable works on that subject; particularly the well known labours of the ingenious Mr. Speechly, gardener to the Duke of Portland; and from many other useful books produced by English kitchen gardeners.

If the knowledge of painting be insufficient without that of gardening, on the other hand, the mere gardener, without some skill in painting, will seldom be able to *form a just idea of effects before they are carried into execution.* This faculty of *foreknowing effects* constitutes the *master*, in every branch of the polite arts; and can only be the result of a correct eye, a ready conception, and a fertility of invention, to which the professor adds practical experience.

But of this art, painting and gardening are not the only foundations: the artist must possess a competent knowledge of *surveying, mechanics, hydraulics, agriculture, botany,* and the general principles of *architecture.* It can hardly be expected that a man bred, and constantly living, in the kitchen garden, should possess all these requisites; yet because the immortal BROWN was originally a kitchen gardener, it is too common to find every man, who can handle a rake or spade, pretending to give his opinion on the most difficult points of improvement. It may perhaps be asked, from whence Mr. Brown derived his knowledge?—the answer is obvious: that, being at first patronised by a few persons of rank and acknowledged good taste, he acquired, by degrees, the faculty of *prejudging effects;* partly from repeated trials, and partly from the experience of those to whose conversation and intimacy his genius had introduced him: and, although he could not design, himself, there exist many pictures of scenery, made under his instruction, which his imagination alone had painted. *

Since the art of Landscape Gardening requires the combination of certain portions of knowledge in so many different arts, it is no wonder that the professors of each should respectively suggest what is most obvious to their own experience; and thus the painter, the kitchen gardener, the engineer, the land agent, and the architect, will frequently propose expedients different from those which the landscape gardener may think proper to adopt. The difficulties which I have occasion-

---

* I must not, in this place, omit to acknowledge my obligations to Launcelot Brown, Esq., late member for Huntingtonshire, the son of my predecessor, for having presented me with the maps of the greatest works in which his late father had been consulted, both in their original and improved states.

ally experienced from these contending interests, induced me to make a complete digest of each subject proposed to my consideration, affixing the reasons on which my opinion was founded, and stating the comparative advantages to the *whole*, of adopting or rejecting certain *parts* of any plan.   To make my designs intelligible, I found that a mere *map* was insufficient; as being no more capable of conveying an idea of the *landscape*, than the *ground-plan* of a house does of its *elevation*.   To remedy this deficiency, I delivered my opinions in writing, that they might not be misconceived or misrepresented; and I invented the peculiar kind of slides to my sketches, which are here imitated by the engraver.* [Fig. 1.]

[* These slides are, doubtless, a very ingenious invention, and though they are liable to some objections, which we shall afterwards notice, yet, in the case of landscapes suited for a quarto or folio page, they are calculated to save a great deal of trouble in drawing; since one landscape, by means of one or two slides, may serve instead of two or three.   Slides of this kind are particularly useful in the case of ground plans, where the object is to shew different modes of arranging some particular part of the plan; and they may also be advantageously applied in the case of slight sketches of landscape on a large scale.   In order, therefore, that our readers may clearly understand what a slide is, we shall give an example.

Fig. 1, represents a landscape, in which it is proposed to make certain alterations.   The engraving represents the landscape as it is supposed to appear before the landscape-gardener commences his operations. On lifting up the slip of paper, or "slide," the landscape is seen as it will appear when the operations of the landscape-gardener are completed.   It is evident that two or three slides may be introduced in the same landscape, to shew the effects of different alterations, and this has been done by Mr. Repton in several

Such drawings, to shew the proposed effects, can be useful but in a very few instances; yet I have often remarked, with

cases. For example, our vignettes, Figs. 6, 7, and 8, p. 42, p. 43, and p. 44, represent a landscape, which, in Mr. Repton's original work, has two slides over it. The first, or upper slide, shews the landscape as it appears in Fig. 6; the second slide as in Fig. 7; and when both slides are lifted up, the landscape appears as in Fig. 8, p. 44.

The principal objection to these slides is, that a hard line of separation between the slide and the landscape beneath, is, in many cases, unavoidable. For example, the slide is quite unobjectionable in Mr. Repton's Plate I., which is our Fig. 1, above given (and also our Figs. 4 and 5), because the line of the top of the paling is hard in itself, as appears in our Fig. 4; but the upper slide in Mr. Repton's Plate II. (represented by our Fig. 6), produces a hard and conspicuous outline, from the difficulty of cutting off the paper close to the outline of the group of trees in the centre of the picture. At the same time, if slight sketches be employed instead of finished drawings, this objection has much less force; and in ground plans, as we have already observed, the invention is of real value.

Instead of adopting slides, we have found it more convenient, owing to the small size to which the views are reduced, to give a separate landscape for each slide. We have given these landscapes in the form of vignettes, in preference to including them in parallelograms, bounded by definite lines, for the following reasons, kindly furnished to us by a distinguished writer, who takes the signature of *Kataphusin*, in the *Architectural Magazine*; in the fifth volume of which work he has treated on the same subject, in a somewhat different manner.

It has often been a subject of astonishment to casual speculators that the multitudinous objects of an extensive landscape should be painted with accuracy so extreme, and finish so exquisite, as our every-day experience would seem to prove, upon the small space afforded by the retina of the eye. The truth is, however, that, strictly speaking, only one *point* can be clearly and distinctly seen by the fixed eye, at a given moment; and all other points included in the vision, are indistinct exactly in proportion to their distance from this central point; and when this distance has increased till the line connecting the two points subtends thirty degrees, the receding point becomes invisible. This distance of thirty degrees, therefore, may be considered as the limit of sight.

Now, if the attention be fixed exclusively on the central point, the surrounding points, being indistinct directly as their distance, will, near the limit, be barely visible; consequently the limit will not be a harsh line, but, on the contrary, will be soft and unfelt by the eye.

But this is a mode of vision very rarely employed by the eye in contemplating landscape. We prefer receiving *all* the visual rays partially, to *receiving* one perfectly; and, instead of confining the attention to the central point, distribute it,* as nearly as may be, over the whole field of vision. Partial distribution is usually and instinctively effected; perfect distribution only occasionally, when we wish to become aware of a general effect. The more general the distribution, the more severe the limit; and when the distribution is perfect, the limit is a circle, whose diameter subtends sixty degrees, whose centre is opposite the eye, and whose area is a section of the cone of rays by which the landscape is made sensible to the eye.

* This operation is partly optical, partly mental. Optical, inasmuch as a slight change takes place in the form of the eyeball; mental, because ideas which the optic nerve was not before permitted to convey to the brain, are now permitted to take their full effect.

some mortification, that it is the only part of my labours which the common observer has time or leisure to examine;

Every picture may be considered as a section of this cone, by a plane perpendicular to the horizon, and, therefore, to the central ray of the cone. Then the question is, should the intersecting plane include more than the area of the cone at the point of intersection, or exactly its area, or less than its area?

If it include more than its area, we shall not be able to see the edge of the picture, if we stand in the proper point for seeing the rest of it. All the artist's labour on the edge will, therefore, be lost on those who know where to stand; and its effect on those who do not, will be to make them stand in a wrong place.

If it include exactly the area, the edge of the picture becomes a substitute for the natural limit of sight, and everything takes its true position.

If it include less than its area, we feel that we could see, and should naturally see, more than the artist has given us; the edge of the picture becomes a cutting, interfering, distinct termination, just as the edge of a window is, when the spectator is kept twelve feet back into the room. We wish to get the edge out of our way, and to see what is behind it: and the ease, beauty, and propriety of the painting is entirely disguised or destroyed.

According to this reasoning, then, our pictures should all be circular, and of such a size that the distance of the eye from their centre should equal their diameter.

But we see that all artists, as a general principle, make their pictures parallelograms of varied proportion. This is a proof that such a form is desirable, and something very near a proof that it is proper.

We have, therefore, to investigate three questions:—I. What are the causes which render such a form desirable? II. What are the principles on which such a form is admissible? III. What are the limitations under which such a form is to be given?

I. What are the causes which render such a form desirable?

In the first place, a circle, though in itself agreeable to the eye, is the most monotonous of all figures; there is no change in it—no commencement or termination—no point upon which the eye can rest with decision; the consequence of which is that an assemblage of circles is most fatiguing and wearisome to the eye; and has, in relation to groups of other figures, very much the effect of a countenance utterly without character, and conversation altogether destitute of meaning, compared with marked features and vivid expression. Now, as it is generally very desirable to group pictures, the circle would, on this account, be a most disagreeable form: while the parallelogram admits of variety of form as well of size, according to the proportion of the sides, enters into simple and symmetrical groups, harmonizes with the right lines of walls and roof, and saves a great dea of space.

These, however, are only the upholsterer's reasons for preferring the parallelogram. The artist's are of far more weight. The first great inconvenience is that the line of sight, or horizon, must be the horizontal diameter, and this, as we shall presently see, would take away all power from the artist of indicating the elevation of the spectator; while perspective retiring lines would incline equally upwards and downwards, producing an artificial and disagreeable impression.

And, in the second place, if, as is very often—we may say generally—the case, there be no positive, continuous, horizontal line in the picture, the although it is the least part of that perfection in the art, to which these *hints* and *sketches* will, I hope, contribute.

eye, in the case of the circle, would have no criterion whatever whereby to judge of the rectitude of the verticals, it would be doubtful about its own position, and uncertain which lines it was to assume as horizontal. Nine times out of ten, therefore, the verticals would appear inclined, and the absence of the parallel terminating lines would thus be embarrassing to the artist, injurious to the drawing, and painful to the spectator.

And, lastly, the laws of composition, as far as relating to shade and colour, are very much facilitated by a rectangular form; the portions of each can be much more accurately estimated and disposed than in the circle; and the scientific forms of grouping, pyramidal, cruciform, &c., become much more evident, and, therefore, much more agreeable to the spectator. Hence it appears that the circle is practically offensive, though scientifically true; and, therefore, that if we can, by any modification of design, turn it into the parallelogram, without infringing any law of vision, it will be a most important and valuable alteration. Therefore,

II. What are the principles on which such a form is admissible?

First. It is very rarely indeed that the eye contemplates any landscape without elevating or depressing itself. In all mountain and architectural scenery it is raised; in all prospects of distant country, depressed. In this case the cone of rays enters the eye obliquely, upwards or downwards. But the plane of the picture is *always* vertical to the eye. Consequently we have the section, by a vertical plane, of a cone whose axis is inclined. This is an *ellipse* whose major axis is vertical.

Similarly: it is seldom that the eye includes the thirty degrees on *each* side of its legitimate point of sight. There is always something more attractive on one side than on the other, and it directs itself to the attractive side, * including, perhaps, forty degrees on one side; twenty degrees on the other. We have then the section of an oblique cone by a parallel plane, or an ellipse whose major axis is horizontal.

Here, then, we have a most valuable modification of the monotonous circle; we have a figure susceptible of as much variety of form as the rectangle, and whose sides, where they cut the axes, very nearly correspond to straight lines. We have the power of increasing apparent elevation of architecture, by using the vertical ellipse; or of diminishing an overwhelming mass of sky, by taking the horizontal one. All this is of infinite practical advantage.

But we may modify the form still farther, by taking the following points into consideration :—

When an artist is *composing* his picture, he supposes the distribution of sight, which may be called, for convenience, the attention of the eye, to be perfect; and considers only that indistinct and undetailed proportion of forms and colours, which is best obtained from the finished drawing by half closing, and thus throwing a dimness over the eye. But, in finishing, he works on quite a different principle. One locality is selected by him, as chiefly worthy of the eye's attention; to that locality he directs it almost exclusively, supposing only such partial distribution of sight over the rest of the drawing, as may obtain a vague idea of the tones and forms which set off and relieve the leading feature. Accordingly, as he recedes from

* We have not space to prove this more directly; but it is always acknowledged, practically, by the artist's placing his horizontal line high or low in the picture, as the eye is depressed or elevated.

I confess that the great object of my ambition is not merely to produce a *book of pictures*, but to furnish some hints

this locality, his tones become fainter, his drawing more undecided, his lights less defined, in order that the spectator may not find any point disputing for authority with the leading idea. For instance; four years ago, in the Royal Academy, there was a very noble piece of composition by Wilkie, Columbus detailing his views, respecting a western continent, to the Monks of La Rabida. The figures were seated at a table, which was between them and the spectator, their legs being seen below it. The light fell on the table, down the yellow sleeve of a secondary figure, catching, as it past, on the countenance of Columbus. This countenance and the falling light were the leading ideas; everything diminished in distinctness as it receded, and the legs below the table, were vague conceptions of legs, sketched in grey.

Occasionally, and, indeed, in most good etchings or wood-cuts, the attention is still more perfectly confined; and there, as the principal feature cannot be so perfectly finished as in a drawing, the surrounding objects are indefinite exactly in proportion, ending frequently in mere spirited shade. And this is the reason that what most people would call a *sketchy* wood-cut, is far more agreeable to a good eye than the most laboured details, because, in fact, that which is most sketchy is most natural, and has more of the properties of a finished picture.

Hence we see that the attention, in all good paintings and engravings, is distributed in a very limited degree, and chiefly concentrated upon one leading feature. Recurring, therefore, to our first principles, we find that when such concentration takes place, the limit of vision is faint, and undefined. All objects near the limit are so excessively indistinct, that a line cutting slightly upon them will not be felt. Accordingly, the artist generally cuts off an extremely small portion of the curve of his ellipse, A B, Fig. †, and including the whole of the other axis, encloses his whole figure

Fig. †

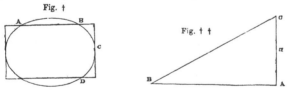

Fig. † †

between the right lines of a rectangle, whose proportion of sides, of course, indicates pretty nearly the length of the original axes, and, therefore, the whole form of the ellipse. He cuts off part of either axis, which he chooses, but very seldom curtails both. Of the included angles, B C, C D, &c., we shall speak presently.

Now we have gone through the whole of this argument merely to prove what some might be inclined to dispute,—that the edge, or frame, of the picture, though rectangular, is, *bona fide*, the representative of the natural limit of sight; it is not an arbitrary enclosure of a certain number of touches, or a certain quantity of colour, within four right lines; nor is it to be extended or diminished as the artist wishes to include more or fewer objects; it is as clearly representative of a fixed natural line as any part of the design itself, and its size and form are, therefore, regulated by laws of perspective as distinct and as inviolable.

We have, therefore, to consider, lastly, what are the limitations under which this form is to be given.

for establishing the fact, that *true* taste in *landscape garden-ing*, as well as in all the other polite arts, is not an accidental

1st. Let the height of the picture be a fixed line $= a$, in fig. † †. Draw A B, at right angles, to a. With centre c, distance 2 a, describe circle, cutting A B in B. :. $\angle$ A C B$=60$. A B is the *utmost* length of the picture which can be admitted; and A B $= \sqrt{\text{B c}^2 - a^2} = \sqrt{(2\ a)^2 - a^2} = a \sqrt{3}$. And such a length of picture as this is very rarely admissible; two-thirds of it are about the best average distance.

Hence it appears, that all such paintings as Stothard's Canterbury Pilgrimage, are panoramas, not pictures. In the Royal Academy, two years ago, there was a very sweet bit by Landseer—Highland drovers crossing a bridge; and if the picture had been confined to the breadth of the bridge itself, and a white Shetland pony looking over into the water, which was the chief light, all had been well; instead of this, we had a parallelogram of about seven feet by one, with a whole procession of figures, extending from one end to the other, the bridge in the centre, and the picture was altogether ruined.

2nd. The corners of the picture, as we have seen, are out of the ellipse, and, therefore, beyond the limit of sight. Accordingly, they might be vague and subdued in colour, and totally without objects; but as this would draw too much attention to them, the artist continues his proximate colour into them, generally, however, keeping his brush in circular sweeps, indicating the form of the ellipse. Copley Fielding's management of the angles of a breezy sea-piece is, perhaps, the best instance that can be given.

Lastly. The true distance at which the eye ought to be placed is the length of the minor axis of the ellipse; but as this minor axis is usually a little diminished, the best standard is the vertical of the equilateral $\triangle$ whose side is the major axis, or the greatest dimension of the picture. In those drawings where the composition is good, and the attention very much confined, this distance may even be exceeded.

But if, in *any* picture, it be very much exceeded, the right lines of the edge cease to be the limit of sight; they come distinctly and positively within the sphere of vision; they cut painfully upon the eye, and we feel exactly that harsh and violent impression on the eye, which, in a piece of music (for the main principles in all fine arts are essentially the same), would be caused on the ear, by the sounds suddenly and decisively ceasing in the midst of a burst of melody, instead of being guided scientifically to its close. Nothing can be more utterly destructive of all the good qualities of a picture—nothing can be more fatal to its composition, more murderous of its repose, more unjust to the artist, or more painful to the spectator—than such reduction of its just limits.

Now, in the drawing itself, there is no chance of the distance of the eye being too great; but, in engravings, diminished in a great degree from the originals, it is not unfrequently the case; and, therefore, it is most important that all engravers should be thoroughly aware of this principle, which we shall proceed to develop as shortly as possible.

When an engraving is six or eight inches in its greatest dimension, the details are generally so delicate as to compel the eye to approach within its true distance; but as a very slight alteration in position is of great consequence, and will throw the limit within the vision, it is a general rule that those pictures are best adapted for engraving which have most light on the edges, so that the termination may not be harsh. And this is *one* of the innumerable beauties of engravings from J. M. W. Turner; namely, that effect, operating on the outward senses, but an appeal to the understanding, which is able to compare, to separate, and to

the dreamy brilliancy of light which envelops them extends to their extreme limits, and their edge hardly ever cuts harshly on the paper. Martin, on the contrary, whose chief sublimity consists in lamp-black, never made a design yet which the eye could endure, if reduced to a small size.

But when, as is not unfrequently the case, it is desirable to reduce the design within still smaller limits, the eye will not be able or willing to assume a correct distance. No one ever approaches his eye within four inches of the paper; and yet, if the engraving be only four inches in diameter, this is the utmost allowable distance. Consequently, if an engraving of this size be terminated by a decided edge, this edge will cut sharply and painfully on the sight, and will make the whole drawing look as if it were pasted on the paper, or cut out of it; there will be a sense of confinement, and regularity, and parallelism, totally destructive of the good qualities of the design; and, instead of being delighted by the beauty of its studied lines, we shall be tormented by an omnipresence of right angles and straight edges. And that this is actually the case any one may convince himself by five minutes' careful observation. This evil ought to be avoided with the greatest care; it is of no slight influence, for the best and most delicate engraving would be utterly spoiled by the error. Now there is only one mode by which such a result is avoidable, and it has been long employed in obedience to the natural instinct, which is as true as any scientific principle, the introduction, namely, of the vignette, by whose indeterminate edge the eye is made to feel that it is a part of a picture, not a perfect one, which it is contemplating. All harshness is thus avoided; and we feel as if we might see more if we chose, beyond the dreamy and undecided limit, but have no desire to move the eye from its indicated place of rest. The vignette, strictly speaking, is the representation of that part of a large picture which the eye would regard with particular interest; and, as in this case, those bits of painting which are distinguished by colour, or brilliancy, or shade, would, of necessity, draw the eye more away from the central point, in one place than in another, we are at liberty to give any form we choose to the fragment, and introduce that graceful variety which enables the artist to give the ethereal spirit and the changeful character by which a good vignette is distinguished.

As examples of the power thus attained, we cannot too frequently recommend close and constant study of vignettes from J. M. W. Turner. These most exquisite morceaux are finished in water-colour, by the artist, on the scale of the engraving (so that the proportions of the light and shade are exactly the same in the copy), and are so thoroughly inimitable, that the most pure and perfectly intellectual mind may test its advancement in knowledge and taste by the new beauties which, on every such advancement, will burst out upon it in these designs.

But the point, to which we wish to direct particular attention, is this, that *no* engraving less than six inches in the greatest dimension, can, in any case, be included within defined limits; and even when they are six or seven inches across, they will hurt the eye if very dark. So that, in reducing pictures to a less size, if they fall within these limits, they must be thrown into vignettes. We should wish to see the authority of this rule more distinctly owned among engravers than it now is; for, in consequence of its violation, many exquisite engravings are utterly useless, as far as regards any pleasing effect on mind or eye. We hope, however, that if the combine, the various sources of pleasure derived from external objects, and to trace them to some pre-existing causes in the structure of the human mind. *

attention of the master engravers be once directed to it, their own sense and feeling will shew them that it is no speculative and useless limitation, but an authoritative rule, whose practice is as necessary as its principles are correct.—*Kataphusin.*   Oxford, February, 1839.   J. C. L.]

* " Where disposition, where decorum, where congruity, are concerned, —in short, wherever the best taste differs from the worst, I am convinced that the understanding operates, and nothing else."—*Burke's Preface to Sublime and Beautiful.*

CHAP. I.

CONCERNING DIFFERENT CHARACTERS AND SITUATIONS.

ALL rational improvement of grounds is, necessarily, founded on a due attention to the CHARACTER and SITUATION of the place to be improved: the *former* teaches what is advisable, the *latter* what is *possible*, to be done; while the extent of the premises has less influence than is generally imagined; as, however large or small it may be, one of the fundamental principles of landscape gardening is to disguise the real boundary.

In deciding on the *character* of any place, some attention must be given to its situation with respect to other places; to the natural shape of the ground on which the house is, or may be, built; to the size and style of the house, and even to the rank of its possessor; together with the use which he intends to make of it, whether as a mansion or constant residence, a sporting seat, or a villa; which particular objects require distinct and opposite treatment. To give some idea of the variety that abounds in the *characters* and *situations* of different places, it will be proper to insert a few specimens from different subjects: I shall begin this work, therefore, by a remarkable instance of situation, only two miles distant from the capital.

'BRANDSBURY. Is situated on a broad swelling hill,
'the ground gently falling from the house (which looks on

' rich distances) in almost every direction. Except a very
' narrow slip of plantation to the north, two large elms near
' the house, and a few in hedge rows at a distance, the spot is
' destitute of trees: the first object, therefore, must be to
' shelter the house by home shrubberies; as on land of such
' value extensive plantations would be an unpardonable want
' of economy.

' No *general plan* of embellishment can, perhaps, be de-
' vised which is more eligible than that so often adopted by
' Mr. Brown, viz., to surround a paddock with a fence, inclos-
' ing a shrubbery and gravel walk round the premises: this
' idea was happily executed by him at Mr. Drummond's de-
' lightful place near Stanmore; but as an attempt has been
' made* to follow the same plan at Brandsbury, without con-
' sidering the difference of the two situations, I shall beg leave
' to explain myself by the following sections and remarks.'

SECTION OF STANMORE.

' Where the natural shape of the ground is *concave*, as
' that at Stanmore, see [fig. 2,] nothing can be more desirable

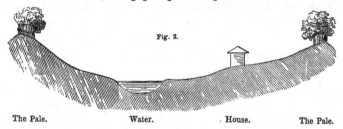

Fig. 2.

The Pale.   Water.   House.   The Pale.

' than to enrich the horizon by plantations on the highest
' ground, and to flood the lowest by a lake or river: in such
' a situation, the most pleasing scenes will be, *within the pale,*
' looking on the opposite rising bank fringed with trees, or
' occasionally catching distant views over or beyond the fence.'

* The house was altered under the direction of a gentleman whose long
experience in building has deservedly placed him high as an architect, and
for whose abilities I have the greatest respect; although, in this instance, I
did not adopt his ideas.   Every one seems to imagine that the art of laying
out ground is within a very small compass, and indeed I once thought so
myself; but I have found by long experience, that the closest application,
and, I may add, the enthusiastic partiality of a whole life spent in the pur-

SECTION OF BRANDSBURY.

' On the contrary, if the natural shape be *convex* [see
' fig. 3], any fence crossing the declivity must intercept those

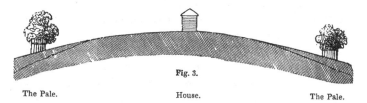

Fig. 3.

The Pale.   House.   The Pale.

' distant views which an eminence should command, and which
' at Brandsbury are so rich and varied that nothing can jus-
' tify their total exclusion. A walk round a paddock in such
' a situation, inclosed by a lofty fence, would be a continual
' source of mortification; as every step would excite a wish
' either to peep through, or look over, the pale of confine-
' ment.'—[See fig. 4.]

[Fig. 4. A scene in the garden at Brandsbury, where a fence of pales is used, instead of a sunk
fence, or, ha! ha!]

suit, is barely sufficient to qualify the artist for this profession. It is, therefore, no more an impeachment of a person's taste, to suppose him incompe-
tent to the embellishment of ground, without having previously studied the
art, than to suppose him unable to build a house without having studied
architecture.

G

'Where all the surrounding country presents the most
'beautiful pasture ground, instead of excluding the vast herds
'of cattle which enliven the scene, I recommend that only a
'sufficient quantity of land round the house be inclosed, to
'shelter and screen the barns, stables, kitchen garden, offices,
'and other useful, but unpleasing, objects; and within this
'inclosure, though not containing more than ten or twelve
'acres, I propose to conduct walks through shrubberies, plan-
'tations, and small sequestered lawns, sometimes winding into
'rich internal scenery, and sometimes breaking out upon the
'most pleasing points for commanding distant prospects: at
'such places the pale may be sunk and concealed, while in
'others it will be so hid by plantation, that the twelve acres
'thus enclosed will appear considerably larger than the sixty
'acres originally intended to be surrounded by a park pale.'*—
[See fig. 5.]

[Fig. 5. Garden scene at Brandsbury, with the sunk fence substituted for the pales.]

'RIVENHALL PLACE. The present *character* of Riven-

---

* When I had first the honour of being consulted on this subject, in
1789, the property annexed to the house consisted of little more than sixty
acres: it has since been augmented, by several purchases, to so great an
amount, that my plan, and indeed the house itself, are on too small a scale
for the present size of the estate; which extends two miles in length from
the toll gate of Kilburn turnpike, and is therefore one of the largest landed
properties within so short a distance of London.

' hall Place is evidently *gloomy* and *sequestered*, with the

[Fig. 6. Rivenhall Place, in its gloomy and sequestered state.]

' appearance of being *low* and *damp*. [See fig. 6.] The
' interference of *art*, in former days, has, indeed, rendered the
' improvement and restoration of its natural beauties a work
' of some labour; yet, by availing ourselves of those *natural*
' beauties, and displacing some of the encumbrances of *art*
' [see fig. 7,] the character of the place may be made *pic-*

[Fig. 7. Another view of Rivenhall Place, shewing its formal canal and bridge.]

' *turesque* and *cheerful*, and the situation, which is *not really*
' *damp*, may be so managed as *to lose that appearance*. The
' first object is to remove the stables, and all the trees and

' bushes in the low meadow, which may then with ease be
' converted into a pleasing piece of water, in the front of the
' house.

[Fig. 8. Rivenhall Place rendered picturesque and cheerful, by the removal of the tall trees and
bushes which encumber the house, as shewn in fig. 6 ; and by continuing the water along the valley, and
altering the colour of the house from a brick-red to a stone colour.]

' The effect of this alteration is shewn by plate No. II.
' [our figs. 6, 7, and 8.] In its present state, two tall elms
' are the first objects that attract our notice [see fig. 6:] from
' the tops of these trees the eye measures downwards to the
' house, that is very indistinctly seen amidst the confusion of
' bushes and buildings with which it is encumbered; and the
' present water appearing above the house, we necessarily
' conclude that the house stands low: but instead of this con-
' fusion, let water be the leading object [see fig. 8,] and the
' eye will naturally measure upwards to the house, and we
' shall then pronounce that it no longer appears in a low
' situation.'

' LIVERMERE PARK. However delightful a romantic
' or mountainous country may appear to a traveller, the more
' solid advantages of a flat one to live in, are universally
' allowed; and in such a country, if the gentle swell of the
' ground occasionally presents the eye with hanging woods,
' dipping their foliage in an expanse of silvery lake, or softly
' gliding river, we no longer ask for the abrupt precipice or
' foaming cataract.

' Livermere Park possesses ample lawns, rich woods, and
' an excellent supply of good-coloured water: its greatest

' defect is a want of clothing near the house, and round that
' part of the water where the banks are flat; yet, in other
' parts, the wood and water are most beautifully connected
' with each other.'

' MILTON PARK. Where the ground naturally pre-
' sents very little inequality of surface, a great appearance of
' extent is rather disgusting than pleasing, and little advan-
' tage is gained by attempts to let in distant objects; yet there
' is such infinite beauty to be produced by judicious manage-
' ment of the home scenery, as may well compensate the want
' of prospect. There is always great cheerfulness in a view on
' a flat lawn, well stocked with cattle, if it be properly bounded
' by a wood at a distance, neither too far off to lessen its im-
' portance, nor too near to act as a confinement to the scene;
' and which contributes also to break those straight lines that
' are the only causes of disgust in a flat situation. Uneven
' ground may be more striking as a picture, and more interest-
' ing to the stranger's eye, it may be more bold, or magnificent,
' or romantic, but the *character of cheerfulness* is peculiar to
' the plain. Whether this effect be produced by the apparent
' ease of communication, or by the larger proportion of sky
' which enters into the landscape, or by the different manner
' in which cattle form themselves into groups on a plain, or on
' a sloping bank, I confess I am at a loss to decide: all three
' causes may, perhaps, contribute to produce that degree of
' cheerfulness which every one must have observed in the
' scenery of Milton.'

' HASELLS HALL. There has hardly been proposed
' to my consideration a spot in which both situation and cha-
' racter have undergone a greater change than at Hasells Hall.
' From the former mode of approaching the house, especially
' from the Cambridge side, a stranger could hardly suppose
' there was any unequal ground in the park: even to the
' south, where the ground naturally falls towards a deep valley,
' the mistaken interference of art, in former days, had bol-
' stered it up by flat bowling greens, and formal terraces;
' while the declivity was so thickly planted as entirely to
' choke up the lowest ground, and shut out all idea of inequa-
' lity. The first object of improvement is to point out those
' beautiful shapes in the ground which so copiously prevail in

'several parts of this park; the second, is to change its cha-
'racter of gloom and sombre dampness, to a more cheerful
'shade; and the third, is to mark the whole with that degree
'of importance and extent, which the size of the house, and
'the surrounding territory demand.'

'CULFORD. The house stands on the side of a hill,
'gently sloping towards the south; but nearly one-half of the
'natural depth of the valley has been destroyed to obtain an
'expanse of water, which, in so flat a situation, I think ought
'not to have been attempted; and I am certain, by proper
'management of the water, the house would appear to stand
'on a sufficient eminence above it, and not so low as the pre-
'sent surface of the water seems to indicate; since the eye
'is always disposed to measure from the surface of neighbour-
'ing water, in forming a judgment of the height of any
'situation.'

'CREWE HALL. In judging the character of any
'place to which I am a stranger, I very minutely observe the
'first impression it makes upon my mind, and, comparing it
'with subsequent impressions, I inquire into the causes which
'may have rendered my first judgment erroneous. I confess
'there has hardly occurred to me an instance where I have
'experienced so great a fluctuation of opinion as in this place.
'I was led, from a consideration of the antiquity of the Crewe
'family in Cheshire, to expect a certain degree of magni-
'ficence; but my first view of the house being from an unfa-
'vourable point, and at too great a distance to judge of
'its real magnitude, I conceived it to be very small; and,
'measuring the surrounding objects by this false standard,
'the whole place lost that importance which I afterwards
'found it assume on a closer examination.

'In former days, the dignity of a house was supposed to
'increase in proportion to the quantity of walls and buildings
'with which it was surrounded: to these were sometimes
'added tall ranks of trees, whose shade contributed to the
'gloom at that time held essential to magnificence.

'Modern taste has discovered, that greatness and cheerful-
'ness are not incompatible; it has thrown down the ancient
'palisade and lofty walls, because it is aware that liberty is
'the true portal of happiness; yet, while it encourages more

' cheerful freedom, it must not lay aside becoming dignity.
' When we formerly approached the mansion through a village
' of its poor dependants, we were not offended at their prox-
' imity, because the massy gates and numerous courts suffi-
' ciently marked the distance betwixt the palace and the
' cottage: these being removed, other expedients must be
' adopted to restore the native character of Crewe Hall.'

' TATTON PARK. The situation of Tatton may be
' justly described as too splendid to be called interesting, and
' too vast to be deemed picturesque; yet it is altogether beau-
' tiful, in spite of that greatness which is rather the attribute
' of sublimity than of beauty.

' The mind is astonished and pleased at very extensive
' prospect, but it cannot be interested, except by those objects
' which strike the eye distinctly; and the scenery of Tatton is
' at present of a kind much beyond the pencil's power to imi-
' tate with effect: it is like the attempt to paint a giant by
' himself in a miniature picture.

' Perfection in landscape may be derived from various
' sources: if it is sublime, it may be wild, romantic, or greatly
' extensive: if beautiful, it may be comfortable, interesting,
' and graceful in all its parts; but there is no incongruity in
' blending these attributes, provided the natural situation con-
' tinues to prevail; for this reason, no violation will be offered
' to the genius of Tatton Park, if we add to its splendour the
' amenity of interesting objects, and give to its vastness the
' elegance of comfort.

' It is not from the *situation* only that the *character* of
' Tatton derives its greatness. The command of adjoining
' property, the style and magnitude of the mansion (from the
' elegant design of Samuel Wyat, Esq.), and all its appendages,
' contribute to confer that degree of importance which ought
' here to be the leading object in every plan of improvement.

' Vastness of extent will no more constitute greatness of
' character in a park, than a vast pile of differently coloured
' building will constitute greatness of *character* in a house. A
' park, from its vast extent, may perhaps surprise, but it will
' not impress us with the *character* of greatness and impor-
' tance, unless we are led to those parts where beauty is shewn
' to exist, with all its interest, amidst the boundless range of
' undivided property.'

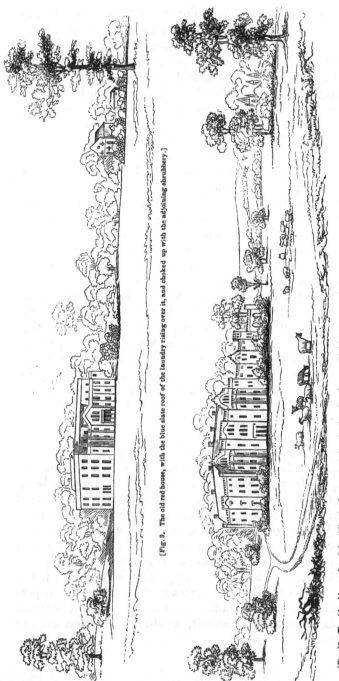

[Fig. 9. The old red house, with the blue slate roof of the laundry rising over it, and choked up with the adjoining shrubbery.]

[Fig. 10. The old red house altered, by adding battlements, and changing its colour from a red brick to a cream-coloured stone; the offices, before at a distance, being brought near, so as to join the house and add to its effect; and the shrubbery removed, to shew more extent of park and prospect.]

' WEMBLY. In the vicinity of the metropolis there
' are few places so free from interruption as the grounds at
' Wembly ; and, indeed, in the course of my experience, I
' I have seen no spot within so short a distance of London,
' more perfectly secluded from those interferences which are
' the common effects of divided property, and a populous
' neighbourhood. Wembly is as quiet and retired at seven
' miles distance, as it could have been at seventy.

' The fatal experience of some, who begin improvements
' by building a house too sumptuous for the grounds, has
' occasionally induced others to consider the ground indepen-
' dent of the house; but this, I conceive, will unavoidably lead
' to error. It is not necessary that the house and grounds
' should correspond with each other in point of size, but the
' *characters* of each should be in strict harmony, since it is
' hardly less incongruous to see a palace by the side of a
' neglected common, than an ugly ill-designed mansion, whether
' large or small, in the midst of highly-improved scenery;
' to every part of which it must be considered as a disgrace.

' Plate No. III. [our figs. 9 and 10], presents the general
' view of the house, offices, and stables, as they appear in the
' approach. In the present state [see fig. 9] there is a gloomi-
' ness and confinement about the house, proceeding from the
' plantation, necessary to hide the vast quantity of unsightly
' buildings with which it was encumbered; yet one of those
' buildings, *viz.* the laundry, is so large and lofty [see the
' sloping roof, rising over the square mass of the house, in fig.
' 9], that it divides the interest with the mansion, or, rather,
' takes the lead of the house itself, by its colour [being covered
' with blue slates] and more extravagant form. I have sup-
' posed an opening made betwixt the house and the mass of
' wood, surrounding the stables [on the right-hand side of the
' landscape], to detach them from each other, and to give an
' extent and cheerfulness ; which is the more advisable on
' that side, as, from the shape of the ground on the other,
' there is some confinement: though I confess, if the house
' were Gothic, that shape would rather be a circumstance of
' picturesque beauty, since we are accustomed to see elegant
' Gothic structures at the foot, or on the sloping side of a hill.
' The stables, without being too conspicuous, may be just seen

H

' to rise above the shrubbery, so that while they give impor-
' tance to the mansion, they will possess only a subordinate
' place in the general scenery ; still contributing to that unity
' of design which makes a composition perfect.' [See fig. 10.]

  ' WELBECK. The house appears to stand much lower
' than it really does, by the entrance in the basement story
' [see fig. 11] ; which, being carried up to the principal floor,

[Fig. 11. The entrance front of Welbeck, with the entrance porch on the ground floor, which makes the house
appear to be situated much lower than it really is, from the ground having a slope towards it.]

' will not only be of great advantage to the inside, by remov-
' ing all necessity for ascending the present staircase, but the
' effect on the outside will be much greater than may at first
' be imagined ; since, by giving an opportunity of altering the
' shape of the ground, it will take the house out of an hollow,
' and set it on a pleasing eminence.

  ' The ground, at present, slopes gradually towards the
' house, with a flat hanging level, which is evidently artificial ;
' and, from the north-west corner of the projecting wing there
' is a ridge of earth which divides this platform from the
' adjoining valley : the superfluous earth from this ridge will
' be sufficient to answer every purpose of raising the lawn to
' the house ; and I propose to slope the ground with a gra-
' dual fall from the riding-house to the valley, and to cross
' this fall by an additional steep from the west front, making
' both to wind naturally towards the low ground of the valley.

  ' The earth may be raised just above the tops of the win-
' dows in the basement story, which may still be sufficiently
' lighted by an area ; but when the lower row of windows is

'totally hid, the house will appear too long for its height, and
'the depth of roof will be still more conspicuous. Having
'hinted this objection to Mr. Carr, he immediately assented
'to it; and, after various attempts to counteract this awkward
'effect, without any great operation, the following appeared
'the most simple: *viz.* that the present pediment (which is
'incongruous to the battlements) should be raised as a square
'tower, and that the parapets, also, at the ends of the build-
'ing, should be raised to unite with the chimneys in the gabels.

[Fig. 12. The west or entrance front of Welbeck altered, by raising the earth above the lower story, and thus
placing the house upon an eminence. The roof is also partly hid by turrets, and farther improved by
changing its colour from that of red tiles to that of blue slates.]

'This will serve not only to hide more of the roof, but will
'give that importance to the whole fabric, which, in a large
'mass of Gothic building, is always increased by the irregu-
'larity of its outline. [See fig. 12.]

'The following drawing, plate IV. [our figs. 11 and 12],
'may serve to shew this effect. I have also changed the
'colour of the roof and chimneys: for, though such minutiæ
'are apt to pass unnoticed in the great outline of improvement,
'I consider the mention of them as a duty of my profession;
'as the motley appearance of red brick with white stone, by
'breaking the unity of effect, will often destroy the magnifi-
'cence of the most splendid composition.'

CHAP. II.

CONCERNING BUILDINGS.

THE perfection of landscape gardening depends on a concealment of those operations of *art* by which *nature* is embellished; but where buildings are introduced, *art* declares herself openly, and should, therefore, be very careful, lest she have cause to blush at her interference.

It is this circumstance that renders it absolutely necessary for the *landscape gardener* to have a competent knowledge of *architecture:* I am, however, well aware that no art is more difficult to be acquired; and although every inferior workman pretends to give plans for building, yet perfection in that art is confined to a very few gentlemen, who, with native genius, and a liberal education, have acquired good taste by travel and observation. This remark proceeds from the frequent instances I continually see of good houses built without any taste, and attempts to embellish scenery by ornamental buildings, that are totally incongruous to their respective situations. The country carpenter or bricklayer is only accustomed to consider detached parts; the architect, on the contrary, finds it his office to consider the whole. There is some degree of merit in building good rooms, but there is more in connecting these rooms together: however, it is the regular bred architect alone who can add to these an outside according to the established rules of art: and where these rules are grossly violated, the eye of genuine taste will instantly be offended, although it may not always be able to explain the cause of its disgust.

To my profession belongs chiefly the external part of

architecture, * or a knowledge of the effect of buildings on the surrounding scenery.

' WELBECK. As every conspicuous building in a park ' should derive its character from that of the house, it is very ' essential to fix, with some precision, what that character ' ought to be ; yet the various tastes of successive ages have so ' blended opposite styles of architecture, that it is often diffi- ' cult, in an old house, to determine the date to which its true ' character belongs. I venture to deliver it as my opinion, ' that there are only two characters of buildings ; the one may ' be called *perpendicular*, and the other *horizontal*. Under ' the first, I class all buildings erected in England before, and ' during the early part of, Queen Elizabeth's reign, whether ' deemed Saracenic, Saxon, Norman, or the Gothic of the ' thirteenth and fourteenth centuries; and even that peculiar ' kind called Queen Elizabeth's Gothic, in which turrets pre- ' vailed, though battlements were discarded, and Grecian ' columns occasionally introduced. Under the *horizontal* ' character I include all edifices built since the introduction of ' a more regular architecture, whether it copies the remains of ' Grecian, or Roman models. There is, indeed a third kind, ' in which neither the horizontal nor perpendicular lines pre-

* I am happy to defend my predecessor, as well as myself, from the imputation of blending *architecture* with *gardening*, by the following extract of a letter from the celebrated author of the ENGLISH GARDEN :—

"I have lately had some correspondence with Mr. Penn, concerning " the intended monument you mention (to Gray, the poet, who is buried in the " churchyard adjoining to Stoke Park,) and finding that he means to con- " sult you on the subject, I have presumed to tell him that he will do well, " if he gives you the absolute choice of the spot, as well as the size of the " building which he means to erect to my excellent friend's memory : for, " although I hold the architectural taste of Mr. Wyat in supreme estimation, " I also am uniformly of opinion, that where a place is to be formed, he who " disposes the ground, and arranges the plantations, ought to fix the situation, " at least, if not to determine the shape and size of the ornamental buildings. " Brown, I know, was ridiculed for turning architect, but I always thought " he did it from a kind of necessity, having found the great difficulty which " must frequently have occurred to him in forming a picturesque whole, " where the previous building had been ill placed, or of improper dimensions.
" I am, Sir, your most obedient servant,
" ASTON, April 24, 1792. W. MASON."
[" Gray's tomb is at the end of the chancel of Stoke Pogeis church. Not far from the churchyard is the cenotaph erected by Mr. Penn, to the memory of Gray, from a design, I believe, by the late Mr. Wyat."— *Mitford's Life of Gray*. First Edition. p. 79.—J. C. L.]

'vail, but which consists of a confused mixture of both : this
' is called CHINESE.

'The two characters of architecture might, perhaps, be
' distinguished by merely calling the one GOTHIC, or *of old
' date*, and the other GRECIAN, or *modern :* but it is not the
' style or date that necessarily determines the character, as
' will appear from plate V. [our figs. 13 and 14] ; which re-
' presents a view of a house at such a distance that none of
' its parts can be distinguished, yet the prevalence of horizon-
' tal or perpendicular lines at once fixes and determines the
' character.   The first [fig. 13] we should call a Grecian, or

[Fig. 13.   View of the water at Welbeck ; introduced to shew the effect of a Grecian or Roman building, or one
in which horizontal lines prevail.]

' modern house ; the latter [fig. 14,] a Gothic one : and there

[Fig. 14.   View of the water at Welbeck ; introduced to shew the effect of Gothic architecture, or buildings of
old date, in which perpendicular lines prevail.]

' can be little doubt, in such a situation, which ought to be
' preferred. I may here observe, that it is unnecessary to
' retain the Gothic character within the mansion, at least not
' farther than the hall, as it would subject such buildings to
' much inconvenience; for since modern improvement has
' added glass-sashed windows to the ancient Grecian and
' Roman architecture, in like manner the inside of a Gothic
' building may, with the same propriety, avail itself of modern
' comforts and convenience.

' The character of the house should, of course, prevail in
' all such buildings as are very conspicuous, or in any degree
' intended as ornaments * to the general scenery; such as

[Fig. 15. The gate used at Welbeck, in which the perpendicular lines prevail; and which, therefore, is better
adapted to scenes of Gothic than to those of Grecian architecture.]

' lodges, pavilions, temples, belvederes, and the like. Yet, in
' adapting the Gothic style to buildings of small extent, there
' may be some reasonable objection : the fastidiousness even of
' good taste will, perhaps, observe, that we always see vast
' piles of buildings in ancient Gothic remains, and that it is a
' modern, or false Gothic only, which can be adapted to so
' small a building as a keeper's lodge, a reposoir, or a pavilion.
' There may be some force in this objection, but there is always
' so much picturesque effect in the small fragments of those

* In consequence of the general observation, respecting the prevalence
of perpendicular lines in the Gothic, at plate VI. [our fig. 15], is introduced a design of a gate, which is everywhere used at Welbeck, but would
be utterly incongruous to Grecian architecture.

'great piles, that, without representing them as ruins, it is
'surely allowable to copy them for the purposes of ornament:
'and, with respect to the mixture of different styles, in Gothic
'edifices, I think there is no incongruity, provided the same
'character of perpendicular architecture be studiously re-
'tained; because there is hardly a cathedral in England in
'which such mixture may not be observed: and while the
'antiquary only can discover the Saxon and Norman styles
'from the Gothic of later date, the eye of taste will never be
'offended, except by the occasional introduction of some
'Grecian or Roman ornaments.'

'WEMBLY. The characters of *Grecian* and *Gothic archi-*
'*tecture* are better distinguished by an attention to their
'general effects, than to the minute parts peculiar to each.
'It is in architecture as in painting, beauty depends on light
'and shade, and these are caused by the openings or pro-
'jections in the surface: if these tend to produce horizontal
'lines, the building must be deemed Grecian, however whim-
'sically the doors or windows may be constructed: if, on the
'contrary, the shadows give a prevalence of perpendicular
'lines, the general character of the building will be Gothic:
'and this is evident from the large houses built in Queen
'Elizabeth's reign, where Grecian columns are introduced;
'nevertheless, we always consider them as Gothic buildings.

'In Grecian architecture, we expect large cornices, windows
'ranged perfectly on the same line, and that line often more
'strongly marked by an horizontal fascia; but there are few
'breaks of any great depth; and if there be a portico, the
'shadow made by the columns is very trifling, compared with
'that broad horizontal shadow proceeding from the soffit;
'and the only ornament its roof will admit, is either a flat
'pediment, departing very little from the horizontal tendency,
'or a dome, still rising from an horizontal base. With such
'buildings it may often be observed, that trees of a pointed or
'conic shape have a beautiful effect, I believe chiefly from the
'circumstances of contrast; though an association with the
'ideas of Italian paintings, where we often see Grecian edi-
'fices blended with firs and cypresses, may also have some
'influence on the mind.

'Trees of a conic shape mixed with Gothic buildings dis-

' please, from their affinity with the prevalent lines of the
' architecture; since the play of light and shadow in Gothic
' structures must proceed from those bold projections, either
' of towers or buttresses, which cause strong shadows in a per-
' pendicular direction : at the same time, the horizontal line
' of roof is broken into an irregular surface, by the pinnacles,
' turrets, and battlements, that form the principal enrichment
' of Gothic architecture; which becomes, therefore, peculiarly
' adapted to those situations, where the shape of the ground
' occasionally hides the lower part of the building, while its
' roof is relieved by trees, whose forms contrast with those of
' the Gothic outline.

' As this observation is new, and may, perhaps, be thought
' too fanciful, I must appeal to the eye, by the help of the
' plate No. VII. [our figs. 16, 17, 18, and 19], which I hope

[Fig. 16. Grecian Architecture, or Architecture on which horizontal lines prevail, contrasted with
round-headed trees.]

' will find that my observation is not wholly chimerical; and
' will, consequently, lay the foundation for this general prin-
' ciple; *viz.* that the lines of Gothic buildings are contrasted
' with round-headed trees; or, as Milton observes,—

" Towers and battlements he sees,
" Embosom'd high in *tufted trees;*"

I

' and that those of the Grecian will accord either with round
' or conic trees; but, if the base be hid, the contrast of the
' latter will be most pleasing.

 ' The Gothic style of architecture being the most calcu-
' lated for additions or repairs to an old house, I might here
' venture to recommend it on the score of mere utility ; but
' when we take into the account that picturesque effect which
' is always produced by the mixture of Gothic buildings with
' *round-headed* trees, I confess myself to be rather sanguine in
' my hopes of producing such beauty at Wembly, as will ren-
' der that house, which has hitherto been a reproach to the
' place, the leading feature of the scenery.

[Fig. 17. Gothic Architecture, or Architecture in which perpendicular lines prevail, contrasted with
round-headed trees.]

 ' Instead of clogging all the improvements with the dread
' of shewing the house, I conceive it possible, without any
' very great expense, to convert the house itself into the most
' pleasing object throughout every part of the grounds from
' whence it may be visible.*

 * I confess there is much danger in adopting the Gothic, where it is
not executed under the direction of architects who have had great experience in that style of building ; nor does it always happen that the gentlemen who have studied their profession in Italy are competent to the task.
The most correct specimens of true Gothic recently built, in places where
I have the honour to be concerned, are Sheffield Place and Nacton, both

' Having stated some arguments for adopting the Gothic
style, I shall now proceed to consider the objections that
' may be urged against it.

' The first objection will arise from the expense of altering
' the outside, without any addition to the internal comfort
' of the mansion.

' The same objection may, indeed, be made to every species
' of external ornament in dress, furniture, equipage, or any
' other object of taste or elegance : the outside case of an harp-
' sichord does not improve the tone of the instrument, but it

[Fig. 18. Grecian Architecture, or the Architecture of horizontal lines, contrasted with spiry-topped trees]

' decorates the room in which it is placed: thus it is as an or-
' nament to the beautiful grounds at Wembly, that I contend
' for the external improvement of the house.

' But in altering the house, we may add a room to any
' part of the building without injuring the picturesque out-
' side, because an exact symmetry, so far from being neces-
' sary, is rather to be avoided in a Gothic building.

old houses, altered by James Wyat, Esq.; and Donington, a new house,
building from the designs of W. Wilkins, Esq. I have never yet seen Mr.
Barrat's house in Kent.

Many other observations respecting Gothic houses have occurred in my
red books for Cobham, Lamer, Little Warley, Nacton, Gayhurst, Tyrring-
ham, Wansley Park, Port Eliot, and Cotchfrench.

'Another objection may arise from the smallness of the
'house, as Gothic structures are in general of considerable
'magnitude; but the character of great or small is not go-

[Fig. 19.　Gothic Architecture, or that in which perpendicular lines prevail, contrasted with spiry-topped
trees.]

'verned by measurement: a great building may be made to
'appear small; and it is from the quantity of windows, and
'not their size, that we should pronounce the house at
'Wembly to be a very considerable edifice.'

CHAP. III.

CONCERNING PROPER SITUATIONS FOR A HOUSE.

' WELBECK. However various opinions may be on the
' choice of a situation for a house, yet there appear to be cer-
' tain principles on which such choice ought to be founded;
' and these may be deduced from the following considerations:
' First. The natural character of the surrounding country.
' Secondly. The style, character, and size of the house.
' Thirdly. The aspects of exposure, both with regard to
' the sun and the prevalent winds of the country.
' Fourthly. The shape of the ground near the house.
' Fifthly. The views from the several apartments; and,
' Sixthly. The numerous objects of comfort:—such as a
' dry soil, a supply of good water, proper space for offices,
' with various other conveniences essential to a mansion in the
' country; and which in a town may sometimes be dispensed
' with, or at least very differently disposed.
' It is hardly possible to arrange these six considerations
' according to their respective weight or influence, which
' must depend on a comparison of one with the other, under
' a variety of circumstances; and even on the partiality of
' individuals in affixing different degrees of importance to each
' consideration. Hence it is obvious, that there can be no
' danger of sameness in any two designs conducted on princi-
' ples thus established; since in every different situation some
' one or more of these considerations must preponderate; and
' the most rational decision will result from a combined view
' of all the separate advantages or disadvantages to be fore-
' seen from each.*

* Having always had these considerations in view whenever I have
been consulted in the site of a new house, or on the preservation of the old
one, I shall take the liberty of mentioning several instances, in some of
which the original red books may possibly be consulted, to shew the variety
of manner in which these general rules have been applied to particular
purposes :—Sunninghill, Sundridge, Courteen Hall, Whersted, Waresley
Park, Ouston, Bessacre Manor, Northrepps, Buckminster, Little Green,
Holme Park, Purley.

'It was the custom of former times, in the choice of
' domestic situations, to let comfort and convenience prevail
' over every other consideration : thus the ancient baronial
' castles were built on the summit of hills, in times when de-
' fence and security suggested the necessity of placing them
' there; and difficulty of access was a recommendation which,
' in our happier days, exists no more.   But when this neces-
' sity no longer operated (as mankind are always apt to fly
' from one extreme to the other), houses were universally
' erected in the lowest situations, with a probable design to
' avoid those inconveniences to which the lofty positions had
' been subject; hence the frequent sites of many large man-
' sions, and particularly abbeys and monasteries, the residence
' of persons who were willing to sacrifice the beauty of prospect
' for the more solid and permanent advantages of habitable
' convenience : amongst which, shelter from wind, and a sup-
' ply of water, were predominant considerations.   Nor shall I
' withhold the following conjecture, which I hope will not be
' considered as a mere suggestion of fancy.—When such
' buildings were surrounded by trees, for the comfort of shade,
' might not the occasional want of circulation in the air, have
' given the first idea of cutting long narrow glades through the
' woods, to admit a current of wind ? and is it not possible
' that this was the origin of those avenues which we frequently
' see pointing, from every direction, towards the most respect-
' able habitations of the two last centuries ?'

'LANGLEY.   It seems to have been as much the fashion
' of the present century to condemn avenues, as it was in
' the last to plant them ; and yet the subject is so little
' understood, that most people think they sufficiently justify
' their opinion, in either case, by merely saying, " I like an
' avenue," or, " I hate an avenue:" it is my business to
' analyze this approbation or disgust.

' The several degrees of pleasure which the mind derives
' from the love of order, of unity, antiquity, greatness of
' parts, and continuity, are all in some measure gratified by
' the long perspective view of a stately avenue: for the truth
' of this assertion, I appeal to the sensations that every one
' must have felt who has visited the lofty avenues of Wind-
' sor, Hatfield, Burleigh, &c. &c. before experience had

' pointed out that tedious sameness, and the many inconveni-
' ences which have deservedly brought avenues into disrepute.
' This sameness is so obvious, that, by the effect of avenues,
' all novelty or diversity of situation is done away; and the
' views from every house in the kingdom may be reduced to
' the same *landscape*, if looking up or down a straight line,
' betwixt two green walls, deserves the name of landscape.

   ' Among the inconveniences of long, straight avenues, may
' very properly be reckoned that of their acting as wind-spouts
' to direct cold blasts with more violence upon the dwelling,
' as driven through a long tube. But I propose rather to
' consider the objections in point of beauty. If at the end of
' a long avenue be placed an obelisk, or temple, or any other
' eye-trap, ignorance or childhood alone will be caught or
' pleased by it: the eye of taste or experience hates compul-
' sion, and turns away with disgust from every artificial means
' of attracting its notice: for this reason an avenue is most
' pleasing, which, like that at Langley Park, climbs up a hill,
' and, passing over its summit, leaves the fancy to conceive its
' termination.

   ' One great mischief of an avenue is, that it divides a
' park, and cuts it into separate parts, destroying that *unity*
' of lawn or wood which is necessary to please in every com-
' position: this is so obvious, that where a long avenue runs
' through a park from east to west, it would be hardly pos-
' sible to avoid distinguishing it into the north and south lawn,
' or north and south division of the park.

   ' But the greatest objection to an avenue is, that (espe-
' cially in uneven ground) it will often act as a curtain drawn
' across to exclude what is infinitely more interesting than
' any row of trees, however venerable or beautiful in them-
' selves; and it is in undrawing this curtain at proper places,
' that the utility of what is called *breaking an avenue* consists:
' for it is in vain we shall endeavour, by removing nine-tenths
' of the trees in rows, to prevent its having the effect of an
' avenue when seen from either end.   The drawing No. VIII.
' [our figs. 20 and 21] may serve to shew the effect of cutting
' down some chestnut trees in the avenue at Langley, to let in
' the hill, richly covered with oaks, and that majestic tree,
' which steps out before its brethren like the leader of a host.

' Such openings may be made in several parts of this avenue
' with wonderful effect; and yet its venerable appearance from
' the windows of the saloon will not be injured, because the
' trees removed from the rows will hardly be missed in the

[Fig. 20. Oblique view of the avenue at Langley Park.]

' general perspective view from the house. And though I
' should not advise the planting such an avenue, yet there will
' always be so much of ancient grandeur in the front trees, and
' in looking up this long vista at Langley, that I do not wish
' it should be further disturbed, especially as the views on

[Fig. 21. Effect of cutting down some of the trees in the avenue at Langley Park.]

' each side are sufficiently capable of yielding beauty; and,
' when seen from the end rooms of the house, the avenue will
' act as a foreground to either landscape.'

' HANSLOPE. Most of the large trees at Hanslope
' stand in avenues, yet their pleasant shade forbids the cutting
' down many of them, merely because the false taste of former
' times has planted them in rows; at least till those planta-
' tions which are now made shall better replace the shelter,
' which the avenues in some measure afford. The following
' sketch [our figs. 22 and 23] gives an idea of breaking the

[Fig. 22. Avenue at Hanslope Park.]

' avenue to the north, which is not to be done by merely
' taking away certain trees, but also by planting a thicket
' before the trunks of those at a distance; as we may be thus

[Fig 23. Part of the avenue at Hanslope Park cut down. This view shews that, in looking along an avenue
its effect will not be destroyed by cutting down a number of trees, unless the trunks of a portion of those
that remain be disguised by bushes, such as thorns, hollies, &c.

' induced to forget that they stand in rows. The addition of a
' few single trees, guarded by cradles, though often used as an
' expedient to break a row, never produces the desired effect:
' the original lines are for ever visible.*

* It is of little consequence from what spot the drawing, No. IX. [our
figs. 22 and 23] was taken, since all avenues bear so great a resemblance
to each other. I shall here enumerate a few instances in which avenues
have been submitted to my consideration. At Cobham Park I give reasons
for preserving one, and destroying the rest; at Prestwood, for retaining the

K

' WELBECK. Besides the character which the style
' and size of the house will confer on a place, there is a *natural*
' *character of country*, which must influence the site and dis-
' position of a house ; and though, in the country, there is not
' the same occasion, as in towns, for placing offices under
' ground, or for setting the principal apartments on a base-
' ment story, as it is far more desirable to walk from the house
' on the same level with the ground, yet there are situations
' which require to be raised above the natural surface : this is
' the case at Welbeck, where the park not only abounds with
' bold and conspicuous inequalities, but in many places there
' are almost imperceptible swellings in the ground, which art
' would in vain attempt to remedy, from their vast breadth ;
' though they are evident defects whenever they appear to cut
' across the stems of trees and hide only half their trunks ; for
' if the whole trunk were perfectly hid by such a swell, the in-
' jury would be less, because the imagination is always ready
' to sink the valley and raise the hill, if not checked in its
' efforts by some actual standard of measurement.   In such
' cases the best expedient is to view the ground from a gentle
' eminence, that the eye may look over and, of course, lose
' these trifling inequalities.

' The family apartments are to the south, the principal
' suite of rooms to the east, and the hall and some rooms of
' less importance to the west ; when, therefore, the eating-
' room and kitchen offices shall be removed to the north, it is
' impossible to make a better disposition of the whole, with
' regard to aspect.   I shall therefore proceed to the fourth
' general head proposed for consideration, *viz.*, the *shape of*
' *the ground* near the house : and as the improvement at Wel-
' beck, originally suggested by his Grace the Duke of Port-

avenue ; at Tatton Park, for quitting the avenue, and planting it up ; at
Trewarthenick, an avenue was very easily broken, from its having been
planted on uneven ground ; and at Brookmans, I elucidate the necessity
of fixing on proper trees to form the outline in breaking an avenue ; or,
if the trees have stood so long near each other that no good outline can be
formed, then the tops of some neighbouring trees may be so introduced
as in some degree to supply the defect.

An avenue of firs is the most obstinate to break, because they leave no
lateral branches ; and, therefore, in the stupendous double row of large silver
firs, which the false taste of the last century has planted at Herriard's house,
I have advised the destruction of one half, leaving the other as a magnificent specimen of the ancient style in gardening.

' land, has, I confess, far exceeded even my own expectations,
' I shall take the liberty of drawing some general conclusions
' on the subject, from the success of this bold experiment. At
' the time I had the honour to deliver my former opinion, my
' idea of raising the ground near the house was confined to the
' west front alone; and, till it had been exemplified and exe-
' cuted, few could comprehend the seeming paradox of burying
' the bottom of the house, as the means of elevating the whole
' structure ; or, as it was very wittily expressed, " moulding
' up the roots of the venerable pile, that it might shoot up in
' fresh towers from its top."

' All natural shapes of ground must necessarily fall under
' one of these descriptions, *viz.*, *convex, concave, plane*, or *in-*
' *clined plane*, as represented in the following sections, plate
' No. X. [our fig. 24.] I will suppose it granted that, except
' in very romantic situations, all the rooms on the principal
' floor ought to range on the same level; and that there must
' be a platform, or certain space of ground, with a gentle

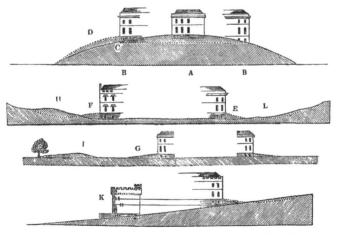

[Fig 24. Sections to shew the manner of adapting houses to different natural shapes of ground.]

' descent from the house every way. If the ground be natu-
' rally convex, or what is generally called a knoll, the size of
' the house must be adapted to the size of the knoll :* this is

* There is a recent instance of a house adapted to the shape of a beautiful knoll at Courteen Hall, where an elegant mansion, with three fronts,
has been lately built, under the direction of S. Saxon, Esq.

'shewn by the small building A, supposed to be only one
'hundred feet in front, which may be placed upon such an
'hillock, with a sufficient platform round it; but if a building
'of three hundred feet long, as B, B, should be required, it is
'evident that the crown of the hill must be taken off, and then
'the shape of the ground becomes very different from its
'original form: for although the small house would have a
'sufficient platform, the large one will be on the brink of a
'very steep bank at C; and this difficulty would be increased
'by raising the ground to the dotted line D, to set the large
'house on the same level with the smaller one. It therefore
'follows, that if the house must stand on a natural hillock, the
'building should not be larger than its situation will admit;
'and where such hillocks do not exist in places proper for a
'house in every other respect, it is sometimes possible for art
'to supply what nature seems to have denied: but it is not
'possible in all cases; a circumstance which proves the absur-
'dity of those architects who design and plan a house, with-
'out any previous knowledge of the situation or shape of the
'ground on which it is to be built.—Such errors I have had
'too frequent occasion to observe.

    'When the shape is naturally either *concave*, or perfectly
'flat, the house would not be habitable, unless the ground
'sloped sufficiently to throw the water from it: this is often
'effected, in a slight degree, merely by the earth that is dug
'from the cellars and foundations: but if, instead of sinking
'the cellars, they were to be built upon the level of the
'ground, they may afterwards be so covered with earth, as to
'give all the appearance of a natural knoll, the ground falling
'from the house to any distance where it may best unite with
'the natural shape, as shewn at E, F, and G: or, as it fre-
'quently happens that there may be small hillocks, H and I,
'near the house, one of them may be removed to effect this
'purpose.* This expedient can also be used in an *inclined*
'*plane*, falling towards the house, where the inclination is not
'very great, as shewn at L; but it may be observed of the
'*inclined plane*, that the size of the house must be governed

    * As at Donington, a seat of Earl Moira, where the house forms a
quadrangle, inclosing an inner court, a whole story lower than appears
externally.

' in some measure by the fall of the ground ; since it is evi-
' dent, that although a house of a hundred feet deep might
' stand at K, yet it would require an artificial terrace on that
' side ; because neither of the dotted lines shewn there would
' connect with the natural shape ; and where the ground can-
' not be made to look natural, it is better, at all times, to avow
' the interference of art, than to attempt an ineffectual con-
' cealment of it.   Such situations are peculiarly applicable to
' the Gothic style, in which horizontal lines are unnecessary.

 ' These sections can only describe the shape of the ground
' as it cuts across in any one direction : but another shape is
' also to be considered : thus it generally happens that a knoll
' is longer one way than the other, or it may even extend to a
' natural ridge, of sufficient length for a long and narrow
' house ; but such a house must be fitted to the ground, for
' it would be absurd in the architect to place it either diago-
' nally or directly across such a ridge : the same holds good of
' the *inclined plane*, which is, in fact, always the side of a val-
' ley, whose general inclination must be consulted in the posi-
' tion of the building.   A square house would appear awry,
' unless its fronts were made to correspond with the shape of
' the adjacent ground.

 ' I shall conclude this digression by observing, that, on a
' dead flat or plain, the principal apartments ought to be ele-
' vated, as the only means of shewing the landscape to advan-
' tage.   Where there is no inequality, it will be very difficult
' to unite any artificial ground with the natural shape : it will,
' in this case, be advisable either to raise it only a very few
' feet, or to set the house on a basement story.   But where-
' ever a park abounds in natural inequalities, even though the
' ground near the house should be flat, we may boldly venture
' to create an artificial knoll, as it has been executed at
' Welbeck.'

CHAP. IV.

CONCERNING WATER.

THERE being no part of my profession so captivating in its effect, and oftentimes so readily executed, as making a large pieee of artificial water, it may be proper, in this volume, to give a few specimens of different improvements presumed to have been produced by it:—though, if all that I have written to explain and elucidate this subject were to be inserted, the whole of the volume would be engrossed by it. I must, therefore, for the present, only mention a few places where artificial pieces of water have been ornamented under my directions:* *viz.*, at *Holkham*, the magnificent lake has been dressed by walks on its banks, and a peculiar ferry-boat invented to unite the opposite shores.

Sheffield Place. A very beautiful lake has been added to the scenery of a place, which abounds in the most perfect specimens of the picturesque effect produced by Gothic architecture.

Sunning Hill. This large piece of water, which consists of a lake, with a river flowing into it, is nearly completed, and will be one of the most pleasing objects that can be produced by art.

Milton in Cambridgeshire. A small river has been made, with great effect, in proportion to its quantity.

Gayhurst. The water in the park, though it consists of several pieces of different levels, has the effect of being in one single sheet when seen from the house: this was very ingeniously executed by Mr. Brown; but I have also connected the neighbouring river with the park, by means of a dressed walk which passes under the turnpike road; and the banks of this river are worthy of every effort to make them a part of the beautiful scenery of the place.

---

* This subject has also been mentioned in the following red books, *viz.*, Ferney Hall, Rudding Hall, Widdial Hall, Babworth, Scrielsby, Milton, Livermere, Garnons, Crewe Hall, Brocklesby, Thoresby, Stoneaston, Nacton, &c.

[Fig. 25. View of the lake and some of the large oaks at Welbeck, before the ground in front of the house was raised, and the elevation of the house altered, or the lake lengthened so as to assume the character of a river.]

[Fig. 26. The lake at Welbeck altered so as to assume the character of a river; and the ground being raised near the house, the situation of the building is, in appearance, considerably elevated.]

'WELBECK. From the number of small promontories
'and bays, together with its termination full in view of the
'house, the water at Welbeck had acquired the character,
'and indeed the name, of a lake: but as a large river is always
'more beautiful than a small lake, the character has been
'changed, not only by continuing it beyond the house, but
'also by altering its line, and taking off those projections
'which were inconsistent with the course of a natural river.
'This is discovered in fig. 26, as compared with fig. 25; the
'former figure also shews, in some degree, the effect of raising
'the earth towards the house; though it appears, in the
'reality, much more striking, from the difference of the scale
'on which it is presented. In this view [fig. 26], only a very
'small part of the house is exhibited, merely to shew its situ-
'ation; the design for the proposed additions to this front not
'being finally settled.'

'TATTON PARK. It has often been asserted by authors
'on gardening, that all pieces of fresh water must come
'under one of these descriptions,—a *lake*, a *pool*, a *river*,
'or a *rivulet:* but since my acquaintance with Cheshire, I am
'inclined to add the *meer*, as an intermediate term between
'the lake and the pool; it being, frequently, too large to be
'deemed a pool, and too small, as well as too round in its
'form, to deserve the name of a lake: for the beauty of a
'lake consists not so much in its size, as in those deep bays
'and bold promontories which prevent the eye from ranging
'over its whole surface. What is best respecting the two
'large *meers* in Tatton Park, is a question of some difficulty,
'and on which there has been a variety of opinions. I shall
'now proceed to deliver mine, and endeavour to explain the
'reasons on which it is founded.

'An unity of design in all compositions is, confessedly,
'one of the first principles in each of the polite arts; and
'nothing, perhaps, evinces more strongly the love of unity
'acting on the mind in landscape gardening, than the follow-
'ing fact,—*viz.*, that the most superficial observer of any park
'scene will be displeased by the view of two separate pieces
'of water; and he will probably ask, without reflecting on
'the difference of levels, why they are not formed into one?
'The first opinion seems, therefore, that these two waters

' should be united: but if the union is not clearly possible, it
' certainly ought not to be attempted. The second opinion is,
' that the upper pool ought to be destroyed; or, as some
' express themselves, should be filled up: but the latter would
' be an Herculean labour to very little purpose; and the
' former, though practicable, would not be advisable, because
' so deep a hollow immediately in front of the house, would
' be a yawning chasm, very difficult to convert into an object
' of beauty. My opinion, therefore, is, that the two waters
' should, from the house, appear to be connected with each
' other, although in reality they are very far asunder; and the
' means of effecting such a deception will require some theo-
' retical reasoning to explain.

' The deception at present operates to the disadvantage of
' the waters; for I was myself greatly deceived in the size of
' this pool when I looked at it from the house; and as it pro-
' duces a similar effect on every person who first sees it, I
' must explain the causes of the *deceptio visus.*

' First. The net fence, through which the water appears,
' is so near the windows, that, by the laws of perspective (of
' which I will explain some general rules in the sequel), it acts
' as a false standard, and by it we measure the size of the pool.
' It was for this reason that I desired some cattle might be
' driven on the banks, which, as I have elsewhere shewn,* are
' the best standard for assisting the judgment with respect to
' the distance, and, of course, the dimensions of other objects.

' Secondly. The pool is almost circular, and the eye darts
' round its border with such instantaneous imperceptible velo-
' city, that it is impossible to suppose its circumference to be
' nearly a mile, unless we can see cattle on the opposite shores;
' and then, by their respective dimensions, we judge of the
' comparative distance. This effect, the drawing, No. XII.
' [our figs. 27 and 28], will elucidate, in which the sheep on
' one side the water appear to be larger than the cows on the
' other. The bay or creek may be hid by shrubs, which will
' give the eye a check in its circuitous progress.

' To explain the uses of the other bay [which seems to
' connect the water in the fore-ground with the water in

---

* Castle Hill, a villa of H. Beaufoy, Esq.

L

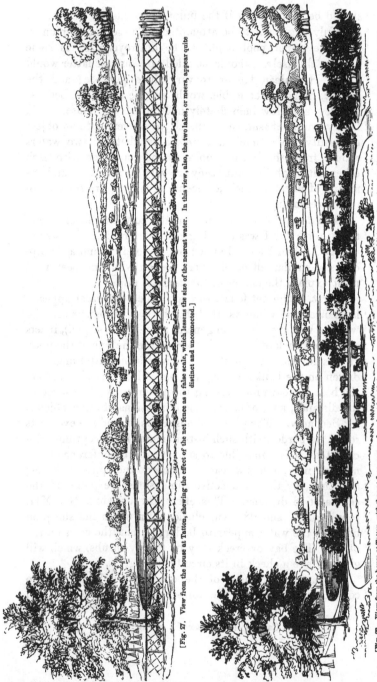

[Fig. 27. View from the house at Tatton, shewing the effect of the net fence as a false scale, which lessens the size of the nearest water. In this view, also, the two lakes, or meers, appear quite distinct and unconnected.]

[Fig. 28. View from the house at Tatton, with the net fence removed, and castle added, to serve as a scale ; and, also, with an addition made to the more distant water, so as to give it an apparent connexion with that in the fore-ground.]

' the distance], I must take the liberty to describe some
' effects in *perspective,* not, I believe, generally attended to in
' gardening. PERSPECTIVE, in painting, is known to be of
' two kinds; the first is called *linear perspective,* and is that
' by which objects appear to diminish in proportion to the
' distance at which they are viewed: this I have here already
' mentioned, in referring to the use of cattle as a scale of mea-
' surement: a horse, a cow, or a sheep, is very nearly of the
' same size, and with this size the mind is perfectly acquainted;
' but trees, bushes, hills, or pools of water, are so various in
' their dimensions, that we are never able to judge exactly of
' their size, or at what distance they appear to us.

   ' The second kind of *perspective* is *aerial,* as it depends
' on the atmosphere; since we observe that objects not only
' diminish in their size, but in their distinctness, in proportion
' to the body of air betwixt the eye and the objects: those
' nearest are strongly represented, while other parts, as they
' recede, become less distinct, till at last the outline of a
' distant hill seems melting into the air itself. Such are the
' laws of *aerial perspective* on all objects, but not on all alike;
' since it is the peculiar property of *light,* and the reflection of
' light, unmixed by colour, to suffer much less by comparison
' than any other object. It is for this reason that we are so
' much deceived in the distance of *perfectly white* objects: the
' light reflected from a white-washed house, makes it appear
' out of its place; snow, at many miles distance, appears to be
' in the next field; indeed, so totally are we unable to judge
' of light, that a meteor within our atmosphere is some-
' times mistaken for a lantern; at others, for a falling star.

   ' Water, like a mirror, reflecting the light, becomes equally
' uncertain in its real distance; and, therefore, an apparent
' union of the two meers in Tatton Park, may be effected by
' attending to this circumstance.

   ' The large piece of water crosses the eye in the view from
' the house; consequently it looks much less considerable
' than it really is, and its effect is of little advantage to the
' scene, being too distant, and too widely separated by the
' vast tract of low ground betwixt the pool and the lake. I
' propose that this water should be rendered more interesting,

'by making it appear as if the arm of a river proceeded from
'the lake; and its termination will easily be hid in the [dis-
'tant] valley.   From the drawing [fig. 28], I hope it will
'appear that the ideal connexion of the two waters may be
'accomplished, although the actual junction is impracticable.
'The facility of deception arises from the causes already
'stated, *viz.*, that water is a mirror from which light is
'strongly reflected, and that of the distance betwixt any light
'and the eye we form a very inaccurate judgment: it is,
'therefore, impossible to know, by looking on the surfaces of
'two distinct waters, whether they are of the same level, un-
'less some ground betwixt them assists the measurement.
'We have, therefore, only to bring the two meers nearer to
'each other, and give their forms such curvature as I have
'described, to produce that effect of apparent unity, which is
'all that is necessary in this instance.

'I am aware of the common objection to all efforts that
'may be deemed *deceptions;* but it is the business of taste, in
'all the polite arts, to avail itself of stratagems, by which the
'imagination may be deceived.   The images of poetry and of
'painting are then most interesting, when they seduce the
'mind to believe their fictions; and, in landscape gardening,
'everything may be called a deception by which we endea-
'vour to conceal the agency of art, and make our works
'appear the sole product of nature.   The most common
'attempts to improve may, indeed, be called deceptions: we
'plant a hill, to make it appear higher than it is; we open
'the banks of a brook, to give it the appearance of a river;
'or stop its current, to produce an expanse of surface; we
'sink the fence betwixt one lawn and another, to give ima-
'ginary extent, without inconvenience or confinement; and
'every piece of artificial water, whether it take the shape of
'a lake, a river, or a pool, must look natural, or it will fail to
'be agreeable.   Nor is the imagination so fastidious as to
'take offence at any well supported deception, even after the
'want of reality is discovered.   When we are interested at a
'tragedy, we do not inquire whence the characters are copied:
'on the contrary, we forget that we see a Garrick or a
'Siddons, and join in the sorrows of a Belvidere or a Beverley,

'though we know that no such persons ever existed: it is
'enough, if so much as we are shewn of the character appears
'to be a just resemblance of nature. In the same manner,
'the magnificent water at Blenheim strikes with wonder and
'delight, while we neither see its beginning nor end; and we
'do not view it with less pleasure after we are told, that it
'was not originally a natural lake, but that Mr. Brown, stop-
'ping the current of a small river, collected this vast body of
'water into the beautiful shape we now admire.'

Mr. Burke very justly observes, " that a true artist should
" put a generous deceit on the spectators, and effect the
" noblest designs by easy methods. Designs that are vast
" only by their dimensions, are always the sign of a common
" and low imagination. No work of art can be great, but as
" it deceives;* to be otherwise is the prerogative of nature
" only."—*Essay on the Sublime, Part II, Section* 10.

[* This is unquestionably a false principle, though laid down by so great
a master. It is possible, indeed, that Burke may have intended it to be
taken in some sense which we do not clearly perceive; but whether we consider it as an isolated sentence, or take it in connexion with what goes before and after in the " *Essay on the Sublime,*" &c., it appears to us alike
false. A marble statue is a work of art, and one which all allow to be great;
but in what respect does it deceive? If coloured so as to resemble nature,
it might possibly " deceive," and be mistaken at a distance for a living being; but it would cease immediately to be admired as a work of art, and be
looked on as an attempt to deceive the spectator, by making him believe it
a work of nature. If, by the word " great," the idea of magnitude is intended, the principle appears to us to be equally false. The Doric columns
at the London entrance to the Birmingham Railway, and St. Paul's Cathedral, are, undoubtedly, great objects—that is, objects of great magnitude;
the columns are great as compared with other columns, and St. Paul's is
great as compared with other churches; but surely they are not the less
great to those who know their real height, than to those who are ignorant
of it, or who imagine it to be greater than it is. J. C. L.]

CHAP. V.

CONCERNING PARK SCENERY.

'Welbeck. The view from the principal apartments
'should bear some proportion to the importance of the house
'itself; not so much in the quantity or extent of the prospect
'as in the nature of the objects which compose the scenery;
'an extensive prospect being only applicable to a castle, a
'villa, or a belvedere. The landscape from a palace should
'everywhere appear appropriate to the magnificence or plea-
'sure of its inhabitants: the whole should be, or, at least,
'*appear to be*, a park, unlimited and unconfined by those
'lines of division or boundary which characterize the large
'grass fields of a dairy farm. Yet a *park* has a character dis-
'tinct from a *forest;* for while we admire, and even imitate,
'the romantic wildness of nature, we ought never to forget
'that a park is the habitation of men, and not solely devoted
'to beasts of the forest. I am convinced that some enthusi-
'astic admirers of uncultivated nature are too apt to overlook
'this distinction. Park scenery compared with forest scenery,
'is like an historical picture compared with a landscape;
'nature must alike prevail in both, but that which relates to
'man should have a higher place in the scale of arts.

'The objects which nature has furnished at Welbeck are
'of the most beautiful kind, and truly in character with the
'dignity of the place. The vast range of woods, the extensive
'lawns, the broad expanse of river, and the astonishing oaks
'scattered about the park, seem to require but a little atten-
'tion from art to mark the residence of a noble possessor;
'yet, as there are a few instances in which the interference of
'art can openly be acknowledged, those few should not be
'neglected. Buildings, however simple, if in character, and
'not too numerous, will more than anything contribute to
'display magnificence.

'Woods enriched by buildings, and water enlivened by a
'number of pleasure-boats, alike contribute to mark a visible
'difference betwixt the magnificent scenery of a park, and

'that of a sequestered forest: the trees, the water, the lawns,
'and the deer, are alike common to both.

'There is another distinction betwixt park and forest
'scenery on which I shall beg leave to state my opinion, as
'it has been a topic of some doubt and difficulty amongst the
'admirers of my profession, *viz.*, *How far gravel roads are*
'*admissible across the lawns of a park:* yet surely very little
'doubt will remain on this subject, when we consider a park
'as a place of *residence;* and see the great inconvenience to
'which grass roads are continually liable.

'I have endeavoured to discover two reasons which may have
'given rise to the common technical objection, that a gravel road
'*cuts up a lawn;* the first arises from the effect observed after
'an avenue has been destroyed, where the straight line of
'gravel, which formerly was less offensive, while accompanied
'by trees, becomes intolerable when it divides a small lawn
'directly through the middle. The other arises from the effect
'which even a winding turnpike road has in destroying the
'sequestered and solemn dignity of forest scenery: but in a
'park, a road of convenience, and of breadth proportioned to
'its intention, as an approach to the house for visitors, will
'often be a circumstance of great beauty ; and is a character-
'istic ornament of art, allowable in the finest inhabited scenes
'of nature.'

'WEMBLY. The park * at Wembly is only defective in
'two circumstances ; the first is the common defect of all places
'where hedges have been recently removed, and too many
'single trees are left; the natural reluctance felt by every man of
'taste and experience to cut down large trees, at the same
'time that he sees the unpleasant effect of artificial rows, is
'very apt to suggest the idea of breaking those rows by plant-
'ing many young trees; and thus the whole composition
'becomes frittered into small parts, which are neither compa-
'tible with the ideas of the sublime or beautiful. The masses
'of light and shade, whether in a natural landscape or a pic-

* There is at present no word by which we express that sort of territory adjacent to a country mansion, which being too large for a garden, too
wild for pleasure ground, and too neat for a farm, is yet often denied the
name of a park, because it is not fed by deer. I generally waive this distinction, and call the wood and lawns, near every house, a park, whether
fed by deer, by sheep, or heavy cattle.

'ture, must be broad and unbroken, or the eye will be
'distracted by the flutter of the scene; and the mind will be
'rather employed in retracing the former lines of hedge-rows,
'than in admiring the ample extent of lawn, and continuity of
'wood which alone distinguishes the park from the grass or
'dairy farm. This defect will, of course, easily be remedied
'when the new plantations shall have acquired a few years'
'growth, and many of the old trees shall be either taken down
'or blended into closer groups by young ones planted very
'near them: but there can be little occasion for dotting young
'trees with such profusion; and I do not hesitate to affirm,
'that of several hundred such trees now scattered upon the
'lawn, not more than twenty can be absolutely necessary.

'The other defect of Wembly arises from a sameness of
'objects; and this is a defect common to all the countries
'where the grass land is more generally mowed than fed. It
'proves what no landscape painter ever doubted, that a scene
'consisting of vegetable productions only can seldom make a
'pleasing picture. The contrasted greens of wood and lawn
'are not sufficient to gratify the eye; it requires other objects,
'and those of different colours, such as rocks, water, and cattle;
'but where these natural objects cannot easily be had, the
'variety may be obtained by artificial means, such as a build-
'ing, a tent, or a road; and, perhaps, there is no object more
'useful in such countries than a good coloured gravel road,
'gracefully winding, and, of course, describing those gentle
'swells of the ground, which are hardly perceptible from the
'uniform colour of grass land.

'The approach-road to the house will be a feature on the
'lawn, both as seen from thence, and also from the high ground
'about the park. Cattle might be more frequently introduced
'than seems to be the custom of this country, especially sheep,
'than which nothing contributes more to enliven a lawn, and
'even to improve and fertilize its verdure; and though some
'objections may arise, from the nature of the soil, they are by
'no means insurmountable.'

'CASTLE HILL. A scene, however beautiful in itself,
'will soon lose its interest unless it is enlivened by moving
'objects. This may be effected by sunk fences; and, from the
'shape of the ground, there is another material use in having

'cattle to feed the lawn before the windows. The eye forms
'a very inaccurate judgment of extent, unless there be
'some standard by which it can be measured; bushes and
'trees are of such various sizes, that it is impossible to use
'them as a measure of distance; but the size of a horse, a
'sheep, or a cow, varies so little, that we immediately judge
'of their distance from their apparent diminution, according
'to the distance at which they are placed; and as they occa-
'sionally change their situation, they break that surface over
'which the eye passes, without observing it, to the first object
'it meets to rest upon. This doctrine will, I hope, be ex-
'plained by a reference to plate No. XIII. [Our figs. 29 and
'30.] It has been objected to the slides with which I eluci-

[Fig. 29. A view at Castle Hill, seen disadvantageously, for the want of cattle or other moving objects in the
foreground, as well as by the flatness of the surface by which it is foreshortened.]

[Fig. 30. The view at Castle Hill, with cattle introduced in the foreground.]

'date my proposed alterations, that I generally introduce, in
'the improved view, boats on the water, and cattle on
'the lawns. To this I answer, that both are real objects
'of improvement, and give animation to the scene; indeed it
'cannot be too often inculcated, that a large lake without
'boats, is a dreary waste of water, and a large lawn without

M

' cattle, is one of the melancholy appendages of solitary gran-
' deur observable in the pleasure-grounds of the past century.'

   ' WEMBLY. The expedient of producing variety at
' Wembly, by buildings, is perhaps the most difficult, and
' requires the greatest attention; because one source of our
' admiration is, that in the neighbourhood of the metropolis a
' place should exist so perfectly secluded and detached from
' the " busy haunts of men:" we must, therefore, be particu-
' larly cautious that every building should appear to be an
' appendage or inmate of the place, and not a neighbour
' intruding on its privacy. From hence arose some difficulty
' in the style of building proper for the prospect on the hill :
' a very small one would have been inadequate to the purpose
' of containing such companies as may resort thither; as well
' as forming a dwelling house for those who should have the
' care of the prospect rooms, and the dairy; yet in building
' a large house, there was danger of making it appear to
' belong to some other person. A design has at length been
' made for such a building as is worthy of the situation, from
' whence a view is presented, of which it is very difficult for
' the pencil to give any just idea; yet it is here inserted, No.
' XIV. [Our figs. 31 and 32], for the sake of shewing the im-
' provement of which it is capable, on the principles already
' enumerated, viz.—

   ' First. By collecting the wood into larger masses, and
' distinguishing the lawns in a broad masterly manner, without
' the confused frittering of too many single trees.

   ' Secondly. By the interesting line of road winding through
' the lawn.

   ' Thirdly. By the introduction of cattle, to enliven the
' scene; and,

   ' Lastly. By the appearance of a seat on the knoll; and a
' part of the house, with its proposed alterations, displaying
' its turrets and pinnacles amongst the trees.

   ' To the common observer, the beauties of Wembly may
' appear to need no improvement; but it is the duty of my
' profession to discover how native charms may be heightened
' by the assistance of taste : and that even beauty itself may
' be rendered more beautiful, this place will furnish a striking
' example.'

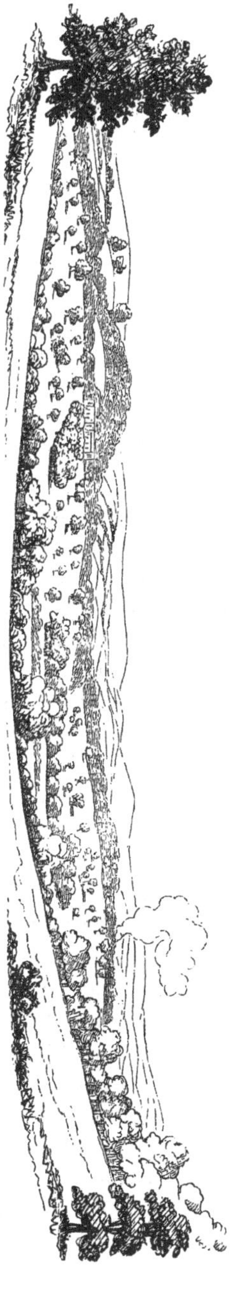

[Fig. 31. View from the tower at Wembly, in which there is an evident confusion; and the chief circumstance attracting notice, is the smoke of a distant limekiln seen in the horizon.]

[Fig. 32. View from the tower at Wembly, in which it is attempted to shew how breadth of light and shade is produced, and that flutter corrected which had been the consequence of too many trees dotted on the lawn. The attention of the spectator is no longer attracted by the smoke of the limekiln, in consequence of introducing objects within the park, by which the view becomes more appropriate and concentrated; and the distance rendered more subordinate in the general composition.]

CHAP. VI.

ON THE ANCIENT STYLE OF GARDENING; OF SYMMETRY
AND UNIFORMITY.

' FINEDON.   There is no part of my profession more diffi-
' cult and troublesome, than the attempt to modernise, *in part*
' *only*, those places which have been formerly decorated by
' the line and square of GEOMETRIC TASTE.   To explain this
' difficulty, I will briefly state the difference between the prin-
' ciples on which improvements are now conducted, and those
' which governed the style of former periods.

' The perfection of *Landscape Gardening* consists in the
' four following requisites: First, it must display the natural
' beauties, and hide the natural defects of every situation.
' Secondly, it should give the appearance of extent and free-
' dom, by carefully disguising or hiding the boundary.
' Thirdly, it must studiously conceal every interference of art,
' however expensive, by which the scenery is improved; mak-
' ing the whole appear the production of nature only; and,
' fourthly, all objects of mere convenience or comfort, if inca-
' pable of being made ornamental, or of becoming proper
' parts of the general scenery, must be removed or concealed.
' Convenience and comfort, I confess, have occasionally misled
' modern improvers into the absurdity of not only banishing
' the appearance, but the reality, of all comfort and conveni-
' ence to a distance; as I have frequently found in the bad
' choice of a spot for the kitchen-garden.

' Each of the four objects here enumerated, are directly
' opposite to the principles of ancient gardening, which may
' thus be stated.   First, the natural beauties or defects of a
' situation had no influence, when it was the fashion to exclude,
' by lofty walls, every surrounding object.   Secondly, these
' walls were never considered as defects; but, on the con-

' trary, were ornamented with vases, expensive iron gates,
' and palisades, to render them more conspicuous. Thirdly,
' so far from making gardens appear natural, every expedient
' was used to display the expensive efforts of art, by which
' nature had been subdued:—the ground was levelled by a
' line; the water was squared, or scollopped into regular
' basins; the trees, if not clipped into artificial shape, were at
' least so planted by line and measurement, that the formal
' hand of art could no where be mistaken. And, lastly, with
' respect to objects of convenience, they were placed as near
' the house as possible:—the stables, the barns, and the
' kitchen-garden, were among the ornaments of a place ; while
' the village, the almshouse, the parish school and churchyard,
' were not attempted to be concealed by the walls or palisades
' that divided them from the embellished pleasure-ground.'

    ' LATHOM. Congruity of style, uniformity of character,
' and harmony of parts with the whole, are different modes
' of expressing that *unity*, without which no composition can
' be perfect : yet there are few principles in gardening which
' seem to be so little understood. This essential unity has
' often been mistaken for symmetry, or the correspondence of
' similar parts ; as where

    " Grove nods at grove, each alley has a brother,
    And half the platform just reflects the other."
                                                    POPE.

' Indeed, this symmetry in the works of art was perfectly jus-
' tifiable under that style of gardening, which confined, within
' lofty walls, the narrow enclosure appropriated to ancient
' grandeur.

    ' When the whole design is meant to be surveyed at a
' single glance, the eye is assisted in its office by making its
' divisions counterparts of each other ; and as it was con-
' fessedly the object of the artist to display his labour, and the
' greatness of the effort by which he had subdued nature, it
' could not possibly be more conspicuous than in such shapes
' of land and water as were most unnatural and violent.
' Hence arose the flat terrace, the square and octagon pool,
' and all those geometric figures which were intended to con-
' trast, and not to assimilate with any scenes in nature. Yet
' within this small enclosure, an *unity* of design was strictly

' preserved, and few attempts made to extend it farther than
' the garden wall.'

From the prodigious difference of taste in gardening betwixt the last and the present century, it seems, at first sight,
almost impossible to lay down any fixed principles; but, on
duly considering the subject, it will be found that in this
instance, as well as in many others, mankind are apt to fly
from one extreme to the other; thus, because straight lines,
and highly finished and correspondent parts prevailed in the
ancient style, some modern improvers have mistaken crooked-
ness for the line of beauty, and slovenly carelessness for natural
ease; they call every species of regularity formal, and, with
the hackneyed assertion, that " *nature abhors a straight line*,"
they fatigue the eye with continual curvatures.

There appears to be in the human mind a natural love of
order and symmetry. Children who at first draw a house
upon a slate, generally represent it with correspondent parts:
it is so with the infancy of taste; those who, during the early
part of life, have given little attention to objects of taste,
are captivated with the regularity and symmetry of correspondent parts, without any knowledge of congruity, or a harmony of parts with the whole: this accounts for those
numerous specimens of bad taste, which are too commonly
observable in the neighbourhood of great towns, where we see
Grecian villas spreading their little Gothic wings, and red-
brick castles supported by Grecian pavilions; but though
congruity may be banished, symmetry is never forgotten. If
such be the love of symmetry in the human mind, it surely
becomes a fair object of inquiry, how far it ought to be
admitted or rejected in modern gardening. The following
observation from Montesquieu, on Taste,* seems to set the
matter in a fair light.

" Wherever symmetry is useful to the soul, and may
" assist her functions, it is agreeable to her; but wherever it
" is useless, it becomes distasteful, because it takes away
" variety. Therefore, things that we see in succession ought
" to have variety, for our soul has no difficulty in seeing them;
" those, on the contrary, that we see at one glance, ought to

[* See " *An Essay on Taste.*" By A. Gerard, D.D. To which are prefixed three Dissertations on the same subject:—viz., by M. de Voltaire,
M. d'Alembert, and M. de Montesquieu. Edin. 1764. 12mo.]

" have symmetry: thus, at one glance we see the front of a
" building, a parterre, a temple; in such things there is
" always a symmetry, which pleases the soul by the facility it
" gives her of taking the whole object at once."

It is upon this principle that I have frequently advised the
most perfect symmetry in those small flower-gardens which
are generally placed in the front of a green-house, or orangery,
in some inner part of the grounds; where, being secluded
from the general scenery, they become a kind of episode to the
great and more conspicuous parts of the place. In such small
enclosures, irregularity would appear like affectation. Symmetry is also allowable, and indeed necessary, at or near the
front of a regular building; because, where that displays correspondent parts, if the lines in contact do not also correspond,
the house itself will appear twisted and awry. Yet this degree of symmetry ought to go no farther than a small distance
from the house, and should be confined merely to such objects
as are confessedly works of art for the uses of man; such as
a road, a walk, or an ornamental fence, whether of wood or
iron; but it is not necessary that it should extend to plantations, canals, or over the natural shape of the ground. " In
" forming plans for embellishing a field, an artist without
" taste employs straight lines, circles, and squares, because
" these look best upon paper. He perceives not, that to
" humour and adorn nature is the perfection of his art; and
" that nature, neglecting regularity, distributes her objects in
" great variety, with a bold hand. (Some old gardens were
" disposed like the human frame; alleys, like legs and arms,
" answering each other; the great walk in the middle repre-
" senting the trunk of the body.) Nature, indeed, in organ-
" ized bodies comprehended under one view, studies regularity;
" which, for the same reason, ought to be studied in architec-
" ture; but in large objects, which cannot be surveyed but in
" parts, and by succession, regularity and uniformity would be
" useless properties, because they cannot be discovered by the
" eye. Nature, therefore, in her large works, neglects these
" properties; and, in copying nature, the artist ought to neg-
" lect them."*—*Lord Kaims's Elements of Criticism.*

[* This extract would require to be analyzed, and some erroneous principles in it pointed out. To say that straight lines, circles, &c. look best on

'LATHOM. It is hardly to be conceived how much this
'view to the north, No. XV. [our figs. 33 and 34], will be

[Fig. 33. View from the house at Lathom, in which a pond is so near the eye, that its glare prevents the lawn
beyond it from being seen to advantage]

[Fig. 34. View from the house at Lathom, shewing the effect of removing the pond, and the introduction of
moving objects.]

'improved by the removal of the large square pond. Water
'reflecting only the sky (which is the case with this and every
'other pond raised above the level of the natural ground), acts
'like a mass of light placed betwixt the eye and the more dis-
'tant objects. Every one knows the effect that a lantern or a
'torch has, to prevent our seeing what is beyond it; and this
'same cause operates in all cases in proportion to the quantity

paper, is, as a general principle, not true; for they can only look better than
other lines by being better suited for the particular purpose for which they
are introduced. In a country where all is uncultivated nature, and consequently all the lines are irregular, geometrical lines and forms will unquestionably be admired, as indications of art and refinement; and this is the
principle on which the architectural style of gardening was founded. "To
humour and adorn nature," would not, in such a case, be the perfection of
a landscape gardener's art; on the contrary, it would, in such a case, consist
in controlling nature, and subjecting her to those forms and dispositions
which indicated the wealth, the power, and the civilization and taste of
man.—J. C. L.]

' of rays reflected, whether from water, from snow, from white
' paling, or any other luminous object. This accounts for the
' pleasure we derive from seeing water at a proper distance,
' and of a natural shape. Water is said to attract our notice
' with irresistible power; but the pond at Lathom, placed in
' the foreground, engrosses too much of the landscape, and is
' neither sufficiently pleasing in its shape, nor natural in its
' situation, to deserve the place it holds, as the leading feature
' of the scene.

' The management of the view to the north will further
' serve to elucidate another general principle in gardening,
' viz., that although we do not require a strict symmetry in
' the two sides of the landscape, yet there is a certain *balance*
' *of composition,** without which the eye is not perfectly satis-
' fied.

' The two screens of wood beyond the pond may be varied
' and contrasted; that to the west may be left as a thick and
' impenetrable mass of trees and underwood, while great part
' of that to the east should be converted into an open grove;
' thus destroying the *formality*, while the *balance of composi-*
' *tion* may still be preserved.'

* The subject has been more fully treated, in my Remarks on Holwood,
in Kent, a seat of the Right Hon. Wm. Pitt; and Stoke, in Herefordshire,
a seat of the Hon. Edw. Foley.

CHAP. VII.

CONCERNING APPROACHES; WITH SOME REMARKS ON THE AFFINITY BETWIXT PAINTING AND GARDENING.

It was not my original intention to have treated of *Approaches* in this volume, as it is a subject that requires to be elucidated by many plates; but the publication of a didactic poem,* where much is said on that subject, under the sanction and authority of two gentlemen of acknowledged taste, obliges me to defend not only my own principles, and the reputation of my late predecessor, Mr. Brown, but also the *art* itself, from attacks, which are the more dangerous, from the manner in which they are conveyed; and because they are accompanied by some doctrines, to which every person of true taste must give his assent.   Yet, while I pay this tribute due to the merit of a work containing many things worthy of admiration, and while I acknowledge my personal obligation for being the only individual, in my profession, to whom any degree of merit is allowed by the author of it, I feel it a kind of duty to watch, with a jealous eye, every innovation on the principles of taste in Landscape Gardening; since I have been honoured with the care of so many of the finest places in the kingdom.

The road by which a stranger is supposed to pass through the park or lawn to the house, is called an approach; and there seems the same relation betwixt the approach and the house externally, that there is internally betwixt the hall or entrance, and the several apartments to which it leads.   If the hall be too large or too small, too mean or too much ornamented for the style of the house, there is a manifest incongruity in the architecture, by which good taste will be offended; but if the hall be so situated as not to connect well with the several apartments to which it ought to lead, it will then be defective in point of convenience.   So it is with

---

* The Landscape, a Poem, by R. P. Knight, Esq.; addressed to Uvedale Price, Esq.

respect to an approach:—it ought to be convenient, interesting, and in strict harmony with the character and situation of the mansion to which it belongs.

' COBHAM HALL. There seems to be as much absurdity
' in carrying an approach round, to include those objects which
' do not naturally fall within its reach, as there was formerly
' in cutting through a hill, to obtain a straight line pointing
' to the hall door. A line of red gravel across a lawn, is apt to
' offend, by cutting it into parts, and destroying the unity of
' verdure, so pleasing to the eye: but I have in some places
' seen the aversion *of showing a road* carried to such a length,
' that a gap has been dug in the lawn, by way of road; and,
' in order to hide it, the approach to a palace must be made
' along a ditch. In other places, I have seen what is called a
' *grass approach*, which is a broad, hard road, thinly covered
' with bad verdure, or even moss, to hide it from the sight;
' and thus, in a dusky evening, after wandering about the park
' in search of a road, we suddenly find ourselves upon grass, at
' the door of the mansion, without any appearance of mortals
' ever having before approached its solitary entrance.

' Thus do improvers seem to have mistaken the most
' obvious meaning of an *approach*, which is simply this—
' A ROAD TO THE HOUSE. If that road be greatly circuitous,
' no one will use it when a much nearer is discovered: but if
' there be two roads of nearly the same length, and one be
' more beautiful than the other, the man of taste will certainly
' prefer it; while, perhaps, the clown, insensible to every
' object around him, will indifferently use either.'

' TATTON. The requisites to a good approach may be
' thus enumerated:—

' First. An approach is *a road to the house;* and to that
' principally.

' Secondly. If it is not naturally the nearest road pos-
' sible, it ought artificially to be made impossible to go a
' nearer.

' Thirdly. The artificial obstacles which make this road
' the nearest, ought to appear natural.

' Fourthly. Where an approach quits the high road, it
' ought not to break from it at right angles, or in such a man-
' ner as robs the entrance of importance; but rather at some
' bend of the public road, from whence a lodge, or gate, may

' be more conspicuous; and where the high road may appear
' to branch from the approach, rather than the approach from
' the high road.

' Fifthly. After the approach enters the park, it should
' avoid skirting along its boundary, which betrays the want of
' extent, or unity of property.

' Sixthly. The house, unless very large and magnificent,
' should not be seen at so great a distance as to make it
' appear much less than it really is.

' Seventhly. The house should be at first presented in a
' pleasing point of view.

' Eighthly. As soon as the house is visible from the
' approach, there should be no temptation to quit it; which
' will ever be the case, if the road be at all circuitous; unless
' sufficient obstacles, such as water, or inaccessible ground,
' appear to justify its course.

' I shall not here speak of the convenience or inconve-
' nience of a large town situated very near a park; but of the
' influence that the proximity of a large town has on the cha-
' racter of a park, which is very considerable; because it must
' either serve to increase or to diminish its importance: the
' latter is at present the case with respect to Tatton and
' Knutsford.

' The first essential of greatness in a place, is the appear-
' ance of united and uninterrupted property; and it is in vain
' that this is studied within the pale, if it is too visibly contra-
' dicted without it. It is not to be expected that a large
' manufacturing town, like Knutsford, can be the entire pro-
' perty of one individual; but the proportion of interest
' belonging to the adjoining family, should impress the mind
' with a sense of its influence. There are various ways by
' which this effect is occasionally produced, and I will mention
' some of them,—viz., the church, and churchyard, may be
' decorated in a style that shall in some degree correspond
' with that of the mansion;—the market-house, or other pub-
' lic edifice, an obelisk, or even a *mere stone*,* with distances,

---

* This passage having excited a very severe attack from Mr. Knight,
I must beg leave to transcribe the following note from his poem, entitled
" THE LANDSCAPE:"—

" That I may not be supposed to deal unfairly with the modern im-
" provers of places, or landscape gardeners, I must inform the reader, that
" I have taken this passage from one, who will be readily and universally

' may be made an ornament to the town, and bear the arms of
' the family; or the same arms may be the sign of the princi-
' pal inn of the place; but there are no means so effectual as
' that which presents itself at Knutsford, of which I have given
' a hint in the *slide* [our fig. 36] of the following sketch:—

[Fig. 35. View of the Town of Knutsford, from the background of which the road proceeds to Tatton Park

[Fig. 36. View of the Town of Knutsford, as proposed to be improved. and shewing, in the background, a
new entrance-gate and lodges to Tatton Park.]

' By taking down a few miserable cottages, and rebuilding
' them as tenements, in a plain, uniform manner, the end
' of the street will be opened, to shew the entrance of the park
' through a simple, handsome arch. The arch should be of

" allowed to be the most skilful and eminent among them. Mr. Repton, in
" his plan for improving Tatton Park, in Cheshire, with which he means to
" favour the public in the general collection of his works, and in which he
" has confessedly detailed the principles of his art, suggests many expedients
" for shewing the extent of property; and, among others, that of placing
" the family arms upon the neighbouring *mile-stones;* but as difficulties
" might arise among the trustees of the turnpikes, who might each wish to
" have his own arms on some particular stone, I flatter myself that the more
" direct and explicit means of gratifying purse-proud vanity, which I here
" propose, may not be thought unworthy of the attention of those improvers
" who make this gratification the object of their labours."
    The expedient proposed, is to hang up a map of every estate at the por-
ter's lodge. This introduces a sarcasm on WEALTH and RANK.—But whatever reasons Mr. Knight may be able to assign for indulging his spleen on

' stone colour; but the tenements of red brick, as according
' better with the other houses in the town.'

' ANTONY HOUSE. In this country there will, I hope,
' for ever exist different orders and degrees of society, which
' must generally depend on the proportion of property, either
' inherited or acquired by different individuals; and so long
' as such distinctions remain, it will be proper that the resi-
' dence of each should be marked by such distinct characters
' as may not easily be mistaken.

' Before the introduction of modern gardening, there
' always existed a marked difference betwixt the residence of
' the landlord, and that of his tenant; not only in the size and
' style of the house itself, but in that also of the land imme-
' diately adjoining. The importance of the mansion was sup-
' ported by a display of convenience, rather than of beauty;
' and thus the *Hall-house* was distinguished from the neigh-
' bouring cottage, not by the extent of lawn, or the variety of
' landscape, but the quantity of barns, stables, and offices,
' with which it was surrounded : and, as our ancestors thought
' a certain degree of gloom and confinement necessary to great-
' ness, the views from the windows were confined by lofty

these subjects, all his ingenuity will not qualify him to gloss over the injustice, to say no more, of misrepresenting my sentiments, and mistaking my
expressions.

" But in your grand approach (the critic cries),
" Magnificence requires some sacrifice :—
" As you advance unto the palace gate,
" Each object should announce the owner's state ;
" His vast possessions, and his wide domains,
" His waving woods, and rich unbounded plains.
" He therefore leads you many a tedious round,
" To show th' extent of his employer's ground ;
" Climbs o'er the hills, and to the vales descends ;
" Then mounts again, through lawn that never ends."

How far the poet's licence may have been used with fairness and discretion, will appear, by comparing the sentiments conveyed in my observations on Tatton, and his poem. But it seems to be the opinion of this
writer, that any approach is a defective part of modern gardening; because,
in some instances, it has been injudiciously made to display the whole beauties of the place at the first entrance. I perfectly agree with him, that
those ostentatious approaches, from whence the whole scenery is spread before
the stranger's eye, as upon a map, are not to be justified, because they rob
the mind of that pleasure which arises from novelty and variety, from
expectation and surprise; but surely there is no more incongruity in marking the entrance of a park with some distinction, and displaying some of its
beauties in the course of a road that must run through it, than in shewing,
by the external appearance of a house, that it is the residence of great
wealth or exalted station.

' walls, surrounding quadrangular courts, or the kitchen-gar-
' den; which, being felt as an object of the greatest conve-
' nience, was deemed the properest object of sight from the
' principal apartments. This taste in gardening continued
' long after the vaulted kitchen, the buttery-hatch, the carved
' cellar door, and other internal marks of hospitable splendour,
' had been banished by modern improvements in architecture.

' It is now acknowledged that gloom is not necessary to
' magnificence, that liberty is not incompatible with greatness,
' and that convenience is not the sole object of ornament; for
' though such things as are useful may occasionally be orna-
' mental, it does not follow that ornaments must always be
' useful; on the contrary, many of those productions of the
' polite arts which are most admired, are now merely consi-
' dered ornaments, without any reference to their original
' uses. This is confessedly the case with works of painting
' and sculpture (except in that inferior branch of each which
' relates to portraits); for whatever might be the original uses
' of pictures or statues, they are now only considered as orna-
' ments, which, by their number and excellence, distinguish
' the taste, the wealth, and dignity of their possessors. To
' use these internal marks of distinction only, might be pru-
' dent in those countries where it would be dangerous to dis-
' play any external ornaments of grandeur; but rank and
' affluence are not crimes in England; on the contrary, we
' expect to see a marked difference in the style, the equipage,
' and the mansions of wealthy individuals; and this differ-
' ence must also be extended to the grounds in the neighbour-
' hood of their mansions; since congruity of style, and unity
' of character, are amongst the first principles of good taste.'

It has already been remarked in this volume, that there
ought to be some difference betwixt a park and a forest; and
as the whole of that false and mistaken theory, which Mr.
Knight endeavours to introduce, by confounding the two
ideas, proceeds from not duly considering the degree of affinity betwixt painting and gardening, I shall transcribe a few
passages from manuscripts, written long before I saw his poem;
although the inquiry was originally suggested by conversations
I have occasionally had, both with Mr. Knight and Mr. Price,
at their respective seats in the county of Hereford.

'HOLME PARK. A great difference betwixt a scene in
'nature, and a picture on canvas, will arise from the following
'considerations:—

'First. The spot from whence the view is taken, is in a
'fixed state to the painter; but the gardener surveys his
'scenery while in motion; and, from different windows in the
'same front, he sees objects in different situations; therefore,
'to give an accurate portrait of the gardener's improvement,
'would require pictures from each separate window, and even
'a different drawing at the most trifling change of situation,
'either in the approach, the walks, or the drives, about each
'place.

'Secondly. The quantity of view, or field of vision, is
'much greater than any picture will admit.

'Thirdly. The view from an eminence down a steep hill
'is not to be represented in painting, although it is often one
'of the most pleasing circumstances of natural landscape.

'Fourthly. The light which the painter may bring from
'any point of the compass, must, in real scenery, depend on
'the time of day. It must also be remembered, that the light
'of a picture can only be made strong by contrast of shade;
'while in nature, every object may be strongly illumined,
'without destroying the composition, or disturbing the keep-
'ing. And,

'Lastly. The foreground, which, by framing the view,
'is absolutely necessary to the picture, is often totally defi-
'cient, or seldom such as a painter chooses to represent;
'since the neat gravel walk, or close mown lawn, would ill
'supply the place, in painting, of a rotten tree, a bunch
'of docks, or a broken road, passing under a steep bank,
'covered with briers, nettles, and ragged thorns.'

'STOKE POGIES.] Real landscape, or that which my art
'professes to improve, is not always capable of being repre-
'sented on paper or canvas: for although the rules for good
'natural landscape may be found in the best painters' works,
'in which

———" we ne'er shall find
" Dull uniformity, contrivance quaint,
" Or labour'd littleness; but contrasts broad,
" And careless lines, whose undulating forms
" Play through the varied canvas;"

MASON.

' yet Monsieur Gerardin* is greatly mistaken, when he directs,
' that no scene in nature should be attempted till it has first
' been painted : and I apprehend the cause of his mistake to
' be this,—In an artificial landscape, the foreground is the
' most important object ; indeed some of the most beautiful
' pictures of Claude de Lorraine, consist of a dark foreground,
' with a very small opening to distant country.   But this
' ought not to be copied in the principal view from the windows
' of a large house, because it can only have its effect from one
' window out of many ; and, consequently, the others must all
' be sacrificed to this sole object.   In a picture, the eye is con-
' fined within certain limits, and unity is preserved by artificial
' means, incapable of being applied to real landscape, in all the
' extent which Mons. Gerardin recommends.'

    ' HOLWOOD.   By LANDSCAPE, I mean a view capable of
' being represented in painting.   It consists of two, three, or
' more, well marked distances, each separated from the other
' by an unseen space, which the imagination delights to fill up
' with fancied beauties, that may not perhaps exist in reality.

> " Of Nature's various scenes, the painter culls
> " That for his favourite theme, where the fair whole
> " Is broken into ample parts, and bold ;
> " Where, to the eye, three well mark'd distances
> " Spread their peculiar colouring."———

MASON.

    ' Here Mr. Mason supposes an affinity between painting
' and gardening, which will be found, on a more minute exa-
' mination, not strictly to exist.

    ' The landscape painter considers all these three distances
' as objects equally within the power of his art ; but his com-
' position must have a foreground ; and though it may only
' consist of a single tree, a rail, or a piece of broken road, it is
' absolutely necessary to the painter's landscape.

    ' The subjects of the landscape gardener are very different ;
' though his scenery requires, also, to be broken into distinct

---

    * Gerardin Visconte d'Ermenonville sur le Paysage.   A work contain-
ing many just observations ; but often mixed with whimsical conceits, and
impracticable theories of gardening.   [The work alluded to is translated
under the title of " *An Essay on Landscape ; or, on the means of Improving
and Embellishing the Country round our Habitations.*"   Translated from the
French of R. L. Gerardin Visconte d'Ermenonville.   London, 1783. 12mo.]

O

' parts or distances, because the eye is never long delighted,
' unless the imagination has some share in its pleasure: an
' intricacy and entanglement of parts heightens the satis-
' faction. The landscape gardener may also class his distances
' under three distinct characters, but very different from those
' of the painter. The first includes that part of the scene
' which it is in his power to improve; the second, that which
' it is not in his power to prevent being injured; and the
' third, that which it is not in the power of himself, or any
' other, either to injure or improve: of this last kind, is the
' distant line of the horizon in the views from Holwood. The
' part which the painter calls his middle distance, is often that
' which the landscape gardener finds under the control of
' others; and the foreground of the painter can seldom be
' introduced into the composition of the gardener's landscape,
' from the whole front of a house, because the best landscapes
' of Claude will be found to owe their beauty to that kind of
' foreground, which could only be applied to one particular
' window of a house, and would exclude all view from that
' adjoining.'

' RUDDING HALL.   Having frequently been asked,
' whether my drawings were made upon such a scale, as not
' to deceive, I shall take this opportunity of answering that
' question, by discussing its possibility.

' That a rural scene in reality, and a rural scene upon
' canvas, are not precisely one and the same thing, Dr. Burgh,
' in his Commentary on Mason, says, is a self-evident propo-
' sition: and Mr. Gilpin has very ingeniously shewn, that a
' picture can hardly be an exact imitation of nature, without
' producing disgust as a picture; but the question, whether
' landscape is reducible to a scale, can only proceed from a
' total inexperience of the art of painting.   A scale can only
' be applied to a diagram, representing parts on the same
' plane, whether horizontal, as in a map, or perpendicular, as
' in the elevation of a building; but even in these cases the
' scale is erroneous, if the surface of the ground plot be
' uneven; or if the elevation presents parts in perspective:
' how then shall any scale be applied to a landscape which
' presents parts innumerable, and those at various distances
' from the eye? my sketches, therefore, do not attempt to

' describe the minutiæ of a scene, but the general effects; and
' all the accuracy of portraiture to which I pretend, is, never
' to insert objects that do not exist, although I cannot repre-
' sent all that do. The large single trees shewn in the sketch
' [in the *Red Book* of Rudding Hall], are all nearly in the situ-
' ations of their prototypes; but it may be possible to leave
' in reality more small trees and bushes than I have shewn on
' paper; because such actual groups will cause no confusion
' to the eye on the spot, although it would be impossible to
' separate them in the picture, even if it were finished with
' the laboured accuracy of Paul Bril, or Velvet Breugel.'

The enthusiasm for picturesque effect, seems to have so
completely bewildered the author of the *poem* already mentioned, that he not only mistakes the essential difference
between the landscape painter and the landscape gardener;
but appears even to forget that a dwelling-house is an object
of comfort and convenience, for the purposes of habitation;
and not merely the frame to a landscape, or the foreground of
a rural picture. The want of duly considering the affinity
between painting and gardening, is the source of those errors
and false principles, which I find too frequently prevailing in
the admirers of, or connoisseurs in, painting: and I do not
hesitate to acknowledge, that I once supposed the two arts to
be more intimately connected, than my practice and experience have since confirmed. I am not less an admirer of
those scenes which painting represents; but I have discovered
that *utility* must often take the lead of beauty, and *convenience* be preferred to picturesque effect, in the neighbourhood
of man's habitation. From Mr. Knight's poem, which is not
without ingenious observations, and beautiful images, I will
enrich my work with the following quotations:—

    " The *quarry long neglected,* and o'ergrown
    " With thorns, that hang o'er mould'ring beds of stone,
    " May oft the place of nat'ral rocks supply,
    " And frame the verdant picture to the eye;
    " Or, closing round the solitary seat,
    " Charm with the simple scene of calm retreat."

    " Bless'd is the man, in whose sequester'd glade
    " Some *ancient abbey's* walls diffuse their shade;
    " With mould'ring windows pierc'd, and turrets crown'd,
    " And pinnacles with clinging ivy bound.

" Bless'd, too, is he, who, 'midst his tufted trees,
" Some *ruin'd castle's* lofty towers sees,
" Imbosom'd high upon the mountain's brow,
" Or nodding o'er the stream that glides below.
" Nor yet unenvied, to whose humbler lot
" Falls the retired, *antiquated cot :—*
" Its roof with weeds and mosses cover'd o'er,
" And honeysuckles climbing round the door;
" While mantling vines along its walls are spread,
" And clust'ring ivy decks the chimney's head."

Insensible, and tasteless, indeed, must that mind be, which cannot admire such scenes as these, whether in reality, in poetry, or in painting : they are precious relics, which deserve the utmost care and preservation ;—pictures worthy the study of the connoisseur; but not tea-boards for common use.   They are objects to be visited with admiration, and protected amidst all their wild and native charms; but they are situations ill adapted to the residence of man.   The *quarry long neglected,* may supply a home for swallows and martens; the *mouldering abbey,* for ravens and jackdaws; the *ruined castle,* for bats and owls; and the *antiquated cot,* whose chimney is choked up with ivy, may perhaps yield a residence for squalid misery and want.—But is affluence to be denied a suitable habitation, because

—" Harsh and cold the builder's work appears,
" Till soften'd down by long revolving years;
" Till time and weather have conjointly spread
" Their mould'ring hues and mosses o'er its head" ?

or because, in some wild and romantic scenery, the appearance of art would offend the eye of taste, are we to banish all convenience from close-mown grass, or firm gravel-walks, and to bear with weeds, and briers, and docks, and thistles, in compliment to the slovenly mountain nymphs, who exclaim with this author :—

" Break their fell scythes, that would these beauties shave,
" And sink their iron rollers in the wave" ?

And again, in the bitterness of prejudice against all that is neat and cleanly,—

" Curse on the shrubbery's insipid scenes
" Of tawdry fringe, encircling vapid greens! "

By those who do not know the author's situation, such a curse may perhaps be attributed to the same spirit of discontent, which laments that,

> " Vain is the pomp of wealth, its splendid halls,
> " And vaulted roofs, sustain'd by marble walls; "

but it is evident to me, that the only source of disgust excited in this gentleman's mind, on viewing the scenes improved by Mr. Brown, proceeds from their not being fit objects for the representation of the pencil.—The painter turns with indignation from the trim-mown grass, and swept gravel-walk; but the gardener, who knows his duty, will remove such unsightly weeds as offend the view from a drawing-room window, although perfectly in harmony with the savage pride and dignity of the forest;

> " Where every shaggy shrub, and spreading tree,
> " Proclaims the seat of native liberty."

It would have been far more grateful to my feelings and inclination, to have pointed out those passages in which I concur with the author of the *Landscape;* but I am compelled, by the duties of my profession, to notice those parts only, which tend to vitiate the taste of the nation, by introducing false principles; by recommending negligence for ease, and slovenly weeds for native beauty. Extremes are equally to be avoided; and I trust that the taste of this country will neither insipidly slide into the trammels of that smooth-shaven " *genius* of the bare and bald," which he so justly ridicules, nor enlist under the banners of that shaggy and harsh-featured *spirit,* which knows no delight but in the scenes of Salvator Rosa;—scenes of horror, well calculated for the residence of banditti,

> " Breathing blood, calamity, and strife."

Thus I have been led to consider the theory* of this inge-

* In Mr. Knight's work, there are two etchings from the masterly pencil of Mr. Hearne, which, though intended as examples of good and bad taste, serve rather to exemplify bad taste in the two extremes of artificial neatness and wild neglect. I can hardly suppose any humble follower of Brown, or any admirer of the " bare and bald," to shave, and smooth, and serpentine a scene like this caricature of modern improvement; nor would any architect of common taste suggest such a house, instead of the venerable

nious author; or rather, to analyze and examine what he
deems

" Harmless drugs, roll'd in a gilded pill,"

lest the subtle poison they contain should not only influence
the art of gardening, but infuse itself into the other polite
arts. In *Sculpture*, we ought to admire the graces of the
Venus de Medicis, as well as the majestic Apollo, the brawny
Hercules, or the agonizing Laocoon. In *Architecture*, there
is not less beauty in the Grecian columns, than in the Gothic
spires, pinnacles and turrets. In *Music*, it is not only the
bravura, the march, or allegro furioso, that ought to be
permitted; we must sometimes be charmed by the soft plain-
tive movement of the Siciliano, or the tender graces of an
amoroso. In like manner, *Gardening* must include the two
opposite characters of native wildness, and artificial comfort,
each adapted to the genius and character of the place; yet,
ever mindful, that, near the residence of man, convenience,
and not picturesque effect, must have the preference, wherever
they are placed in competition with each other.

I flatter myself that no part of this chapter will be deemed
irrelevant to the subject of my work, which is an attempt to
explain and elucidate certain general principles in the art I
profess: especially as those principles have been formally
attacked and misrepresented, by one who has given such con-
summate proof of good taste in the improvement of his own
place, Downton Vale, near Ludlow, one of the most beautiful
and romantic valleys that the imagination can conceive. It is
impossible, by description, to convey an idea of its natural
charms, or to do justice to that taste which has displayed
these charms to the greatest advantage,

" With art clandestine, and conceal'd design."

A narrow, wild, and natural path sometimes creeps under the

---

pile in the other drawing. At the same time, there is a concomitant absurdity in the other view, unless we are to consider it as the forsaken mansion
of a noble family gone to decay: for if it be allowable to approach the
house by any road, and if that road must cross the river, there are architects in this country, who would suggest designs for a bridge in unison with
the situation, without either copying fantastic Chinese models, or the no
less fantastic wooden bridge here introduced; which, though perfectly picturesque in its form, and applicable to the steep banks of the Teme, yet, in
this flat situation, looks like the miserable expedient of poverty, or a ridiculous affectation of rural simplicity.

beetling rock, close by the margin of a mountain stream. It sometimes ascends to an awful precipice, from whence the foaming waters are heard roaring in the dark abyss below, or seen wildly dashing against its opposite banks; while, in other places, the course of the river Teme being impeded by natural ledges of rock, the vale presents a calm, glassy mirror, that reflects the surrounding foliage. The path, in various places, crosses the water by bridges of the most romantic and contrasted forms; and, branching in various directions, including some miles in length, is occasionally varied and enriched by caves and cells, hovels, and covered seats, or other buildings, in perfect harmony with the wild but pleasing horrors of the scene. Yet, if the same picturesque objects were introduced in the gardens of a villa near the capital, or in the more tame, yet interesting, pleasure-grounds which I am frequently called upon to decorate, they would be as absurd, incongruous, and out of character, as a Chinese temple from Vauxhall transplanted into the Vale of Downton.

" Whate'er its essence, or whate'er its name,
" Whate'er its modes, 'tis still in all the same ,
" 'Tis *just congruity* of parts combined,
" Must please the sense, and satisfy the mind."

APPENDIX.

THE delay occasioned by the want of punctuality in the several artists who had undertaken to etch, engrave, and colour the plates for this Volume,[*] has exposed me to anticipation in several parts of my work.—This is the case in Mr. Price's ingenious *Essay on the Picturesque,* which its author calls "a direct and undisguised attack on the art;" and which, in fact, is also a "caustic" satire upon the taste of the present century; and particularly on those gentlemen, who not having been so fortunate as to consider "drawing and painting" to be the first requisites in a polite education, have never been taught to refer each object of sight to its effect upon canvas. The attack on the *Art* itself, I have already answered in a Letter, which is here reprinted as a note.†—

[* This passage, of course, refers to the original edition. J. C. L.]

† *A Letter to* UVEDALE PRICE, *Esq., of Foxley in Herefordshire.*

SIR,

I am much obliged by your attention, in having directed your bookseller to send me an early copy of your ingenious work. It has been my companion during a long journey, and has furnished me with entertainment, similar to that which I have occasionally had the honour to experience, from your animated conversation on the subject. In the general principles and theory of the art, which you have considered with so much attention, I flatter myself that we agree; and that our difference of opinion relates only to the *propriety,* or, perhaps, *possibility,* of reducing them to practice.

I must thank both Mr. Knight, and yourself, for mentioning my name as an exception to the tasteless herd of Mr. Brown's followers. But while you are pleased to allow me some of the qualities necessary to my profession, you suppose me deficient in others; and therefore strongly recommend the study of "what the higher artists have done, both in their pictures and drawings:" a branch of knowledge which I have always considered to be not less essential to my profession than hydraulics or surveying; and without which I should never have presumed to arrogate to myself, the title of "LANDSCAPE GARDENER," which (in allusion to my having adopted it) you observe is, "*a title of no small pretension.*"

It is difficult to define GOOD TASTE in any of the polite arts; and among the respective professors of them, I am sorry to observe that it is seldom allowed in a rival; while those who are not professors, but, being free from the business or dissipation of life, find leisure to excel in any of these arts, generally find time also to cultivate the others; and because there really does exist some affinity betwixt them, they are apt to suppose it still greater. Thus *Music* and *Poetry* are often coupled together, although very few instances occur in which they are made to assimilate; because the melody

That letter was written under the immediate impression of surprise, on my first perusing the work, of which I had not the most distant idea; or I should certainly have been more guarded in my conversations with its author, who has frequently adopted my ideas; and has, in some instances, robbed me of originality; particularly in that observation concerning the prevalence of lines in architecture; on which subject the Right Honourable Mr. Burke, in a letter to me, says, " I " have no sort of doubt that you are in the right; your

of an air is seldom adapted either to the rhyme or measure of the verse. In like manner, *Poetry* and *Painting* are often joined; but the canvas rarely embodies those figurative personages to advantage, which the poet's enthusiasm presents to the reader's imagination.

During the pleasant hours we passed together amidst the romantic scenery of the Wye, I do remember my acknowledging that an enthusiasm for the picturesque, had originally led me to fancy greater affinity betwixt *Painting* and *Gardening,* than I found to exist after more mature consideration, and more practical experience; because, *in whatever relates to man, propriety and convenience are not less objects of good taste, than picturesque effect;* and a beautiful garden scene is not more defective because it would not look well on canvas, than a didactic poem because it neither furnishes a subject for the painter or the musician.   There are a thousand scenes in nature to delight the eye, besides those which may be copied as pictures; and, indeed, one of the keenest observers of picturesque scenery (Mr. Gilpin), has often regretted, that few are capable of being so represented, without considerable license and alteration.

If, therefore, the painter's landscape be indispensable to the perfection of gardening, it would surely be far better to paint it on canvas at the end of an avenue, as they do in Holland, than to sacrifice the health, cheerfulness, and comfort of a country residence, to the wild but pleasing scenery of a painter's imagination.

There is no exercise so delightful to the inquisitive mind, as that of deducing theories and systems from favourite opinions: I was therefore peculiarly interested and gratified by your ingenious distinction betwixt the *beautiful* and the *picturesque;* but I cannot admit the propriety of its application to landscape gardening; because *beauty,* and not *"picturesqueness,"* is the chief object of modern improvement: for although some nurserymen, or labourers in the kitchen garden, may have badly copied Mr. Brown's manner, the unprejudiced eye will discover innumerable beauties in the works of that great self-taught master: and since you have so judiciously marked the distinction betwixt the *beautiful* and the *picturesque,* they will perhaps discover, that where the habitation and convenience of man can be improved by *beauty,* *"picturesqueness"* may be transferred to the ragged gipsy, with whom " the wild ass, the Pomeranian dog, and shaggy goat," are more in harmony, than " the sleek-coated horse," or the dappled deer.   The continual motion and lively agitation observable in herds of deer, is one of the circumstances which painting cannot represent; but it is not less an object of beauty and cheerfulness in park scenery.

Amidst the severity of your satire on Mr. Brown and his followers, I cannot be ignorant that many pages are directly pointed at my opinions; although with more delicacy than your friend Mr. Knight has shewn, in

" observation seems not more acute and ingenious than solid;
" and I believe it is quite new, at least I do not recollect to
" have seen it anywhere else: nor has it, in my thoughts on
" the subject, ever occurred to myself."

I had the honour of knowing Mr. Price, as a gentleman,
long before he became an author; and I trust he knew me as
such before I entered into a profession, which he must have
known I was endeavouring to render liberal, rational, and
respectable, at the very time which he selects for loading its

the attempt to make me an object of ridicule, by misquoting my un-
published MSS.

It is the misfortune of every liberal art, to find, among its professors,
some men of uncouth manners; and since my profession has more frequently been practised by mere day labourers, and persons of no education, it
is the more difficult to give it that rank in the polite arts, which I conceive
it ought to hold. But I am now more particularly called upon to support
its respectability, since you attack the very existence of that profession at
the head of which both you and Mr. Knight have the goodness to say that
I am deservedly placed.

Your new theory of deducing *landscape gardening* from *painting* is so
plausible, that, like many other philosophic theories, it may captivate and
mislead, unless duly examined by the test of experience and practice. I
cannot help seeing great affinity betwixt deducing gardening from the
painter's studies of wild nature, and deducing government from the
uncontrolled opinions of man in a savage state. The neatness, simplicity,
and elegance of English gardening, have acquired the approbation of the
present century, as the happy medium betwixt the wildness of nature and
the stiffness of art; in the same manner as the English constitution is the
happy medium between the liberty of savages, and the restraint of despotic
government; and so long as we enjoy the benefit of these middle degrees,
between the extremes of each, let experiments of untried theoretical
improvement be made in some other country.

So far I have endeavoured to defend Mr. Brown, with respect to the
general principle of improvement; but it is necessary to enter something
farther into the detail of his practice of what has been ludicrously called
clumping and *belting*. No man of taste can hesitate betwixt the natural
group of trees, composed of various growths, and those formal patches of
firs which too often disfigure a lawn, under the name of a clump: but the
most certain method of producing a group of five or six trees, is to plant
fifty or sixty within the same fence; and this Mr. Brown frequently advised,
with a mixture of firs, to protect and shelter the young trees during their
infancy: unfortunately, the neglect or bad taste of his employers would
occasionally suffer the firs to remain, long after they had completed their
office as nurses; while others had actually planted *firs only* in such clumps,
totally misconceiving Mr. Brown's original intention. Nor is it uncommon
to see these black patches surrounded by a painted rail, a quick hedge, or
even a stone wall, instead of that temporary fence which is always an
object of necessity, and not of choice.

If a large expanse of lawn happens unfortunately to have no single
trees or groups to diversify its surface, it is sometimes necessary to plant

professors with contempt and ridicule, as belt-makers—
deformers—shavers of Nature—dealers in ready-made taste,
and such like opprobrious epithets.   However, amidst this
despicable herd, Mr. Price has the goodness to distinguish me
in the following note : " Mr. Repton (who is deservedly at
" the head of his profession) might effectually correct the
" errors of his predecessors, if to his taste and facility in
" drawing (an advantage they did not possess), to his quick-
" ness of observation, and to his experience in the practical

them ; and if the size and quantity of these clumps or masses bear a just
proportion to the extent of lawn, or shape of the ground, they are surely
less offensive than a multitude of starving single trees, surrounded by heavy
cradle fences, which are often dotted over the whole surface of a park.   I
will grant, that where a few old trees can be preserved of former hedgerows, the clump is seldom necessary, except in a flat country.   The clump,
therefore, is never to be considered as an object of present beauty, but as a
more certain expedient for producing future beauties, than young trees,
which very seldom grow when exposed singly to the wind and sun.
    I shall now proceed to defend my predecessor's *belt*, on the same
principle of expedience.   Although I perfectly agree, that, in certain
situations, it has been executed in a manner to be tiresome in itself, and
highly injurious to the general scenery ; yet there are many places in which
no method could be more fortunately devised, than a belt or boundary of
plantation to encompass the park or lawn.   It is often too long, and always
too narrow ; but from my own experience I am convinced, that notwithstanding the obstinacy and presumption of which Mr. Brown is accused, he
had equal difficulties to surmount from the profusion and the parsimony of
his employers, or he would never have consented to those meagre girdles of
plantation, which are extended for many miles in length, although not above
twenty or thirty yards in breadth.
    Let me briefly trace the origin, intention, and uses of a belt.   The
comfort and pleasure of a country residence requires, that some ground, in
proportion to the size of the house, should be separated from the adjoining
ploughed fields ; this enclosure, call it park, or lawn, or pleasure-ground,
must have the air of being appropriated to the peculiar use and pleasure of
the proprietor.   The love of seclusion and safety is not less natural to man
than that of liberty, and I conceive it would be almost as painful to live in
a house without the power of shutting any door, as in one with all the doors
locked : the mind is equally displeased with the excess of liberty, or
of restraint, when either is too apparent.   From hence proceeds the
necessity of enclosing a park, and also of hiding the boundary by which it
is enclosed : now a plantation being the most natural means of hiding a park
pale, nothing can be more obvious than a drive or walk in such a plantation.
If this belt be made of one uniform breadth, with a drive as uniformly
serpentining through the middle of it, I am ready to allow that the way
can only be interesting to him who wishes to examine the growth of his
young trees : to every one else it must be tedious, and its dulness will
increase in proportion to its length.   On the contrary, if the plantation be
judiciously made of various breadth, if its outline be adapted to the natural
shape of the ground, and if the drive be conducted irregularly through its
course, sometimes totally within the dark shade, sometimes skirting so near

" part, he were to add an attentive study of what the higher
" artists have done, both in their pictures and drawings.
" Their selections and arrangements would point out many
" beautiful compositions and effects in nature, which, without
" such a study, may escape the most experienced observer.
" The fatal rock on which all professed improvers are likely to
" split, is system: they become mannerists, both from getting
" fond of what they have done before, and from the ease of
" repeating what they have so often practised: but to be

its edge as to show the different scenes betwixt the trees, and sometimes
quitting the wood entirely, to enjoy the unconfined view of distant prospects,
—it will surely be allowed that such a plantation is the best possible means
of connecting and displaying the various pleasing points of view, at a
distance from each other, within the limits of the park;—and the only just
objection that can be urged, is—where such points do not occur often
enough, and where the *length* of a drive is substituted for its *variety*.

This letter, which has been written, at various opportunities, during my
journey into Derbyshire, has insensibly grown to a bulk which I little
expected when I began it; I shall therefore cause a few copies to be printed,
to serve as a general defence of an art, which, I trust, will not be totally
suppressed, although you so earnestly recommend every gentleman to
become his own landscape gardener. With equal propriety might every
gentleman become his own architect, or even his own physician: in short,
there is nothing that a man of abilities may not do for himself, if he will
dedicate his whole attention to that subject only. But the life of man
is not sufficient to excel in all things; and as "a little knowledge is a
dangerous thing," so the professors of every art, as well as that of medicine,
will often find that the most difficult cases are those, where the patient has
begun by *quacking himself*.

The general rules of art are to be acquired by study, but the manner of
applying them can only be learned by practice; yet there are certain good
plans, which, like certain good medicines, may be proper in almost every
case; it was therefore no greater impeachment of Mr. Brown's taste, to
anticipate his belt in a naked country, than it would be to a physician to
guess, before he saw the patient, that he would prescribe James's powders in
a fever.

In the volume of my works now in the press, I have endeavoured to
trace the difference betwixt *painting* and *gardening*, as well as to make a
distinction betwixt a *landscape* and a *prospect;* supposing the former to be
the proper subject for a painter, while the latter is that in which everybody
delights; and, in spite of the fastidiousness of connoisseurship, we must
allow something to the general voice of mankind. I am led to this remark
from observing the effect of picturesque scenery on the visitors of Matlock
Bath (where this part of my letter has been written). In the valley, a
thousand delightful subjects present themselves to the painter, yet the
visitors of this place are seldom satisfied till they have climbed the
neighbouring hills, to take a bird's-eye view of the whole spot, which no
painting can represent:—the love of prospect seems a natural propensity,
an inherent passion of the human mind, if I may use so strong an expression.

" reckoned a mannerist, is at least as great a reproach to the
" improver as to the painter.   I have never seen any piece of
" water that Mr. Repton had both planned and finished
" himself.   Mr. Brown seems to have been perfectly satisfied
" when he had made a natural river look like an artificial one :
" I hope Mr. Repton will have a nobler ambition ;—that of
" having his artificial rivers and lakes mistaken for real
" ones."

This advice concerning the study of pictures I have already
answered : and with respect to artificial waters, I must only
observe, that for some years the banks of a new-made lake

This consideration confirms my opinion, that *painting* and *gardening*
are nearly connected, but not so intimately related as you imagine : they
are not sister arts, proceeding from the same stock, but rather congenial
natures, brought together like man and wife ; while, therefore, you exult
in the office of mediator betwixt these two "imaginary personages," you
should recollect the danger of interfering in their occasional differences, and
especially how you advise them both to wear the same articles of dress.

I shall conclude this long letter, by an allusion to a work, which it is
impossible for you to admire more than I do.—Mr. Burke, in his *Essay on
the Sublime and Beautiful*, observes, that habit will make a man prefer the
taste of tobacco to that of sugar ; yet the world will never be brought to say
that sugar is not sweet.   In like manner, both Mr. Knight and you are in
the habit of admiring fine pictures, and both live amidst bold and picturesque
scenery : this may have rendered you insensible to the beauty of those
milder scenes that have charms for common observers.   I will not arraign
your taste, or call it vitiated, but your palate certainly requires a degree of
"irritation" rarely to be expected in garden scenery ; and, I trust, the
good sense and good taste of this country will never be led to despise the
comfort of a gravel walk, the delicious fragrance of a shrubbery, the soul
expanding delight of a wide extended prospect,* or a view down a steep
hill, because they are all subjects incapable of being painted.

Notwithstanding the occasional asperity of your remarks on my
opinions, and the unprovoked sally of Mr. Knight's wit, I esteem it a very
pleasant circumstance of my life to have been personally known to you
both, and to have witnessed your good taste in many situations.   I shall beg
leave, therefore, to subscribe myself, with much regard and esteem,

Sir,
Your most obedient,
humble servant,
H. REPTON.

Hare-street, near Romford,
    July 1, 1794.

* An extensive *prospect* is here mentioned as one of the subjects that may be delightful,
although not picturesque.—But I have repeatedly given my opinion, that however desirable a
prospect may be from a tower or belvidere, it is seldom advisable from the windows of a
constant residence.

will generally appear bald and naked: for this reason, I have myself ridiculed the absurd *fashion* of cutting down trees in rich valleys, to make a vast sheet of water, without any accompaniment of wood.—The following lines are extracted from the *Red Book* of Babworth, written in the beginning of the year 1790:

> Despotic FASHION, in fantastic garb,
> Oft, by her vot'ries, for the magic-robe
> Of TASTE mistaken, with ill guiding step
> Directs our path.   Perchance among the roots
> Of shadowy alders (by entangled grass
> Half veiled), the shining face of some clear brook,
> That winding gurgles o'er its pebbly bed,
> Her prying glance discerns:——" A lake," she cries,
> " A lake shall fill this undulating vale! "
> Nor heeds she that the naked banks, alas!
> Shall many a tedious year be naked still.
> Slow is the progress of great NATURE's work;
> While ART, by raising high the *puddled* mound,
> Suddenly drowns a country, spreading wide
> The watery desolation.——Here, perhaps,
> Some venerable trees, by grandsires rear'd,
> " From storms their shelter, and from heat their shade,"
> With stubborn, knotty roots impede the plan
> Of FASHION's deluge.   Then aloud she calls
> ART, and her ruthless myrmidons, to rear
> The sacrilegious axe—See, it descends!
> Too late the shaggy *Genius* of the place
> Bewails his comforts gone.   The deed is done!
> The lake expands—the guardian trees are fell'd—
> And chilling *Eurus* howls along the vale.

The author of the Essay has very unfairly attributed to Mr. Brown all the bad taste of the day-labourers who became his successors; but of his own good taste, there is surely one lasting monument in the first entrance of Blenheim park, the pride of this country, and the astonishment of foreigners. It was this part of the water that Mr. Brown viewed with exultation, and not the serpentine river below the cascade, which I believe he never saw finished. There is another misrepresentation concerning that self-taught genius: so far from his being insensible to the wild scenery of nature, he frequently passed whole days in studying the sequestered haunts of Needwood forest, as I have done those in the forest of Hainault; and, I trust, from these studies, we have both acquired not only picturesque ideas, but this useful lesson; *that the landscape ought to be adapted to the beings which are to inhabit it*—to men, and not to beasts. The landscape painter may consider men subordinate objects in his scenery, and place them merely as "*figures*, to adorn his picture." The landscape gardener does more :—he undertakes to study their comfort and convenience.

I will allow that there is a shade of difference betwixt the opinions of Mr. *Price* and Mr. *Knight*, which seems to have arisen from the different characters of their respective places; *Foxley* is less romantic than *Downton*, and therefore Mr. Price is less extravagant in his ideas, and more willing to allow some little sacrifice of picturesque beauty to neatness, near the house ; but by this very concession he acknowledges, that real *comfort*, and his ideas of *picturesqueness*, are incompatible. In short, the mistake of both these gentlemen arises from their not having gone deep enough in the inquiry, and not having carefully traced, to all its sources, that pleasure which the mind receives from landscape gardening ; for although picturesque effect is a very copious source of our delight, it is far from being the only one.

After sedulously endeavouring to discover other causes of this pleasure, I think it may occasionally be attributed to each of the following different heads; which I have enumerated in my *Red Book* of Warley, near Birmingham, a seat of Samuel Galton, Esquire.

SOURCES OF PLEASURE IN LANDSCAPE GARDENING.

I. *Congruity ;* or a proper adaptation of the several parts to the whole ; and that whole to the character, situation, and circumstances of the place and its possessor.

II. *Utility.* This includes convenience, comfort, neatness, and everything that conduces to the purposes of habitation with elegance.

III. *Order.* Including correctness and finishing ; the cultivated mind is shocked by such things as would not be

visible to the clown : thus, an awkward bend in a walk, or
lines which ought to be parallel, and are not so, give pain ;
as a serpentine walk through an avenue, or along the course
of a straight wall or building.

IV. *Symmetry ;* or that correspondence of parts expected
in the front of buildings, particularly Grecian ; which, however formal in a painting, require similarity and uniformity of
parts to please the eye, even of children. So natural is the
love of order and of symmetry to the human mind, that it is
not surprising it should have extended itself into our gardens,
where nature itself was made subservient, by cutting trees
into regular shapes, planting them in rows, or at exact equal
distances, and frequently of different kinds in alternate
order.

These first four heads may be considered as generally
adverse to picturesque beauty ; yet they are not, therefore, to
be discarded : there are situations in which the ancient style
of gardening is very properly preserved : witness the academic
groves and classic walks in our universities; and I should
doubt the taste of any improver, who could despise the
congruity, the utility, the order, and the symmetry of the
small garden at Trinity college, Oxford, because the clipped
hedges and straight walks would not look well in a picture.

V. *Picturesque Effect.* This head, which has been so fully
and ably considered by Mr. Price, furnishes the gardener
with breadth of light and shade, forms of groups, outline,
colouring, balance of composition, and occasional advantage
from roughness and decay, the effect of time and age.

VI. *Intricacy.* A word frequently used by me in my *Red
Books*, which Mr. Price has very correctly defined to be,
" that disposition of objects, which, by a partial and uncertain
" concealment, excites and nourishes curiosity."

VII. *Simplicity ;* or that disposition of objects which,
without exposing all of them equally to view at once, may
lead the eye to each by an easy gradation, without flutter,
confusion, or perplexity.

VIII. *Variety.* This may be gratified by natural landscape,
in a thousand ways that painting cannot imitate ; since it
is observed of the best painters' works, that there is a sameness
in their compositions, and even their trees are all of one

general kind, while the variety of nature's productions is endless, and ought to be duly studied.

IX. *Novelty.* Although a great source of pleasure, this is the most difficult and most dangerous for an artist to attempt; it is apt to lead him into conceits and whims, which lose their novelty after the first surprise.

X. *Contrast* supplies the place of novelty, by a sudden and unexpected change of scenery, provided the transitions are neither too frequent nor too violent.

XI. *Continuity.* This seems evidently to be a source of pleasure, from the delight expressed in a long avenue, and the disgust at an abrupt break between objects that look as if they ought to be united; as in the chasm betwixt two large woods, or the separation betwixt two pieces of water; and even a walk, which terminates without affording a continued line of communication, is always unsatisfactory.

XII. *Association.* This is one of the most impressive sources of delight; whether excited by local accident, as the spot on which some public character performed his part; by the remains of antiquity, as the ruin of a cloister or a castle; but more particularly by that personal attachment to long known objects, perhaps indifferent in themselves, as the favourite seat, the tree, the walk, or the spot endeared by the remembrance of past events: objects of this kind, however trifling in themselves, are often preferred to the most beautiful scenes that painting can represent, or gardening create: such partialities should be respected and indulged, since true taste, which is generally attended by great sensibility, ought to be the guardian of it in others.

XIII. *Grandeur.* This is rarely picturesque, whether it consists in greatness of dimension, extent of prospect, or in splendid and numerous objects of magnificence; but it is a source of pleasure mixed with the sublime: there is, however, no error so common as an attempt to substitute extent for beauty in park scenery, which proves the partiality of the human mind to admire whatever is vast or great.

XIV. *Appropriation.* A word ridiculed by Mr. Price as lately coined by me, to describe extent of property; yet the appearance and display of such extent is a source of pleasure not to be disregarded; since every individual who possesses anything, whether it be mental endowments, or power, or property, obtains respect in proportion as his possessions are known, provided he does not too vainly boast of them; and it is the sordid miser only who enjoys for himself alone, wishing the world to be ignorant of his wealth. The pleasure of appropriation is gratified in viewing a landscape which cannot be injured by the malice or bad taste of a neighbouring intruder: thus an ugly barn, a ploughed field, or any obtrusive object which disgraces the scenery of a park, looks as if it belonged to another, and therefore robs the mind of the pleasure derived from appropriation, or the unity and continuity of unmixed property.

XV. *Animation;* or that pleasure experienced from seeing life and motion; whether the gliding or dashing of water, the sportive play of animals, or the wavy motion of trees; and particularly the playsomeness peculiar to youth, in the two last instances, affords additional delight.

XVI. And lastly, the *seasons,* and times of day, which are very different to the gardener and the painter. The noontide hour has its charms; though the shadows are neither long nor broad, and none but a painter or a sportsman will prefer the sear and yellow leaves of autumn to the fragrant blossoms and reviving delights of spring, "the youth of the year."

I cannot better conclude my remarks on this new theory of Landscape Gardening (though in fact it ought rather to be called *Picture Gardening),* than by the following abstract of a letter, which I received from a Right Honourable Friend, whose name, * were I permitted to mention, would confer lustre on this work, as it does on every cause to which he gives his support.

"DEAR SIR,
"I must not delay to thank you at once for your obliging offer of the "use of your house, and for the very agreeable present of your printed "letter to Mr. Price. I read it the moment that I received it, and read it "in the way most flattering to the writer, by taking it up without any "settled purpose, and being carried on by approbation of what I found "there. You know of old that I am quite of your side in the question "between you, and am certain that the farther you go in this controversy,

[* William Wyndham, Esq., of Fellbrig, Norfolk. See Biographical Notice, p. 11. J. C. L.]

" the more you will have the advantage. Nothing indeed can be so absurd,
" nor so unphilosophical, as the system which Mr. Knight and Mr. Price
" seem to set up. It not only is not true in practice, that men should
" expose themselves to agues and rheumatisms, by removing from their
" habitations every convenience that may not happen to fall in with the
" ideas of picturesque beauty; but it is not true that what is adverse to
" comfort and convenience, is in situations of that sort the most beautiful.
" The writers of this school, with all their affectation of superior sensibility,
" shew evidently that they *do not trace with any success the causes of their*
" *pleasure.* Does the pleasure that we receive from the view of parks and
" gardens result from their affording in their several parts subjects that
" would appear to advantage in a picture? In the first place, what is most
" beautiful in nature is not always capable of being represented most
" advantageously by painting; the instance of an extensive prospect, the
" most affecting sight that the eye can bring before us, is quite conclusive.
" I do not know anything that does, and naturally should, so strongly
" affect the mind, as the sudden transition from such a portion of space as
" we commonly have in our minds, to such a view of the habitable globe as
" may be exhibited in the case of some extensive prospects. Many things
" too, as you illustrate well in the instance of deer, are not capable of
" representation in a picture at all; and of this sort must everything be
" that depends on motion and succession. But, in the next place, the
" beauties of nature itself, and which painting *can* exhibit, are many, and
" most of them, probably, of a sort which have nothing to do with the
" purposes of habitation, and are even wholly inconsistent with them. A
" scene of a cavern, with banditti sitting by it, is the favourite subject of
" Salvator Rosa; but are we therefore to live in caves, or encourage the
" neighbourhood of banditti?—Gainsborough's country girl is a more pic-
" turesque object than a child neatly dressed in a white frock; but is that a
" reason why our children are to go in rags? Yet this is just the proposition
" which Mr. Knight maintains, in the contrast which he exhibits of the
" same place, dressed in the modern style, and left, as he thinks, it ought
" to be. The whole doctrine is so absurd, that when set forth in its true
" shape, no one will be hardy enough to stand by it, and accordingly they
" never do set it forth, nor exhibit it in any distinct shape at all; but only
" take a general credit for their attachment to principles which everybody
" is attached to as well as they; and where the only question is of the
" application which they afford you no means of making. They are lovers
" of picturesque beauty, so is everybody else; but is it contended that in
" laying out a place, whatever is most picturesque, is most conformable to
" true taste? If they say so, as they seem to do in many passages, they
" must be led to consequences which they can never venture to avow: if
" they do not say so, the whole is a question of how much, or how little; which,
" without the instances before you, can never be decided; and all that they
" do is to lay down a system as depending on one principle, which they
" themselves are obliged to confess afterwards, depends upon many. They
" either say what is false, or what turns out upon examination to be nothing
" at all.

" I hope, therefore, that you will pursue the system which I conceive
" you to have adopted, and vindicate to the art of laying out ground its true
" principles, which are wholly different from those which these wild im-
" provers would wish to introduce. Places are not to be laid out with a view
" to their appearance in a picture, but to their uses, and the enjoyment of
" them in real life ; and their conformity to those purposes is that which con-
" stitutes their beauty : with this view, gravel walks, and neat mown lawns,
" and in some situations, straight alleys, fountains, terraces, and, for aught
" I know, parterres and cut hedges, are in perfect good taste, and infinitely
" more conformable to the principles which form the basis of our pleasure
" in these instances, than the docks and thistles, and litter and disorder,
" that may make a much better figure in a picture."

OBSERVATIONS

ON

THE THEORY AND PRACTICE

OF

LANDSCAPE GARDENING.

INCLUDING

SOME REMARKS ON GRECIAN AND GOTHIC

ARCHITECTURE,

Collected from Various Manuscripts,

IN THE

POSSESSION OF THE DIFFERENT NOBLEMEN AND GENTLEMEN,

FOR WHOSE USE THEY WERE ORIGINALLY WRITTEN;

THE WHOLE TENDING TO ESTABLISH FIXED PRINCIPLES IN
THE RESPECTIVE ARTS.

BY H. REPTON, ESQ.

[*Published in* 1803.]

TO

THE KING,

WITH

HIS MAJESTY'S

MOST GRACIOUS PERMISSION,

THIS WORK

IS HUMBLY INSCRIBED, BY

HIS MAJESTY'S

MOST FAITHFUL, OBEDIENT,

AND HUMBLE SUBJECT,

HUMPHRY REPTON.

Hare Street, near Romford,
Dec. 31, 1802.

ADVERTISEMENT,

EXPLAINING THE NATURE OF THIS WORK.

SEVEN years have now elapsed since the publication of my
" *Sketches and Hints on Landscape Gardening*," during which,
by the continued duties of my profession, it is reasonable to
suppose much experience has been gained and many principles established. Yet so difficult is the application of any
rules of ART to the works of NATURE, that I do not presume
to give this Book any higher title, than " *Observations tending
to establish fixed Principles in the Art of Landscape Gardening.*"

After various attempts to arrange, systematically, the
matter of this Volume, I found the difficulties increase with
the number of the subjects; and although each was originally
treated with order and method in a separate state, yet, in
combining many of these subjects, the same order and method
could not easily be preserved. I have, however, with as
much attention to arrangement as my professional duties
would admit, collected such observations as may best vindicate
the Art of Landscape Gardening from the imputation of being
founded on caprice and fashion: occasionally adding such
matter as I thought might suit the various taste or inclinations
of various readers. Some delight in speculative opinions, some
in experimental facts; others prefer description, others look
for novelty, and some, perhaps, for what I hope will not be
found in this Work, impracticable theories.

The present Volume neither supersedes nor contradicts
my former Work, neither is it a repetition nor a continuation;
but to avoid the oblong and inconvenient shape of that book,

the present Volume is printed under a different form and title, because I am less ambitious of publishing a book of beautiful prints, than a book of precepts: I must therefore entreat that the plates be rather considered as necessary than ornamental; they are introduced to illustrate the arguments, rather than to attract the attention. I wish to make my appeal less to the eye than to the understanding.

In excuse for the frequent use of the first personal pronoun, it should be remembered, that when an author relates his own theory, and records his own practice, it is hardly possible to avoid the language of egotism.

When called upon for my opinion concerning the improvement of a place, I have generally delivered it in writing, bound in a small book, containing maps and sketches, to explain the alterations proposed: this is called the *Red Book* of the place; and thus my opinions have been diffused over the kingdom in nearly two hundred such manuscript volumes. From many of these, with the permission of their respective proprietors, this Volume has been composed; sometimes adopting the substance, and sometimes quoting the words of the *Red Book*.

The severity of criticism is seldom abated in consideration of the circumstances under which a work is produced; yet should it be objected that some parts of this Volume are unequal, the author can plead in excuse, that the whole has been written in a carriage during his professional journeys from one place to another, and being seldom more than three days together in the same place, the difficulty of producing this Volume, such as it is, can hardly be conceived by those who enjoy the blessings of stationary retirement, or a permanent home.

The Plates are fac-similes of my sketches in the original *Red Books*, and have been executed by various artists, whose names are affixed to each; to whom I thus publicly express my acknowledgments, and when tempted to complain of delay, disappointment, and want of punctuality in artists, I am checked by the consideration that works of genius cannot be restricted by time, like the productions of daily labour.

The necessity of blending Architecture with Landscape Gardening, mentioned in my former Work, induced me to turn the studies of one of my sons to that auxiliary part of my profession; it is, therefore, to the assistance of *Mr. John Adey Repton* that I am indebted for many valuable ornaments to this Volume. His name has hitherto been little known as an architect, because it was suppressed in many works begun in that of another person, to whom I freely, unreservedly, and confidentially gave my advice and assistance, while my son aided, with his architectural knowledge and his pencil, to form plans and designs, from which we have derived neither fame nor profit; but amongst the melancholy evils to which human life is subject, the most excruciating to a man of sensibility, is the remembrance of disappointed hope from misplaced confidence.

PREFACE,

CONTAINING SOME OBSERVATIONS ON TASTE.

———

IN every other polite Art, there are certain established rules or general principles, to which the professor may appeal in support of his opinions; but in Landscape Gardening every one delivers his sentiments, or displays his taste, as whim or caprice may dictate, without having studied the subject, or even thought it capable of being reduced to any fixed rules. Hence it has been doubted, whether each proprietor of his own estate, may not be the most proper person to plan its improvement.

Had the art still continued under the direction of working gardeners, or nurserymen, the proprietor might supersede the necessity of such landscape gardeners, provided he had previously made this art his study; but not, (as it is frequently asserted) because the gentleman who constantly resides at his place, must be a better judge of the means of improving it, than the professor whose visits are only occasional: for if this reason for a preference were granted, we might with equal truth assert, that the constant companion of a sick man has an advantage over his physician.

Improvements may be suggested by any one, but the *professor only* acquires a knowledge of effects before they are produced, and a facility in producing them by various methods, expedients, and resources, the result of study, observation, and experience. He knows what can, and what can not be accomplished within certain limits. He ought to know what to adopt, and what to reject; he must endeavour to accommodate his plans to the wishes of the person who consults him,* although, in some cases, they may not strictly accord with his own taste.

Good sense may exist without *good taste*,† yet, from their intimate connexion, many persons are as much offended at having their taste, as their understanding, disputed; hence the most ignorant being generally the most obstinate, I have occasionally found that, as " a little learning is a dangerous thing," a little taste is a troublesome one.

Both taste and understanding require cultivation and improvement. *Natural taste*, like *natural genius*, may exist to a certain degree, but without study, observation, and experience, they lead to error: there is, perhaps, no circumstance which so strongly marks the decline of public taste, as the extravagant applause bestowed on early efforts of unlettered and uncultivated genius: extraordinary instances of prematurity deserve to be patronised, fostered, and encouraged, provided

* Thus before a house is planned, the proprietor must describe the kind of house he wishes to build. The architect is to consider what must be had, and what may be dispensed with. He ought to keep his plan as scrupulously within the expense proposed, as within the limits of the ground he is to build upon: he is, in short, to enter into the views, the wishes, and the ideas of the gentleman who will inhabit the house proposed.

† The requisites of taste are well described by Dr. Beattie, under five distinct heads. " 1. A lively and correct imagination ; 2. the power of distinct apprehension ; 3. the capacity of being easily, strongly, and agreeably affected with sublimity, beauty, harmony, correct imitation, &c. ; 4. sympathy, or sensibility of heart; and, 5. judgment or good sense, which is the principal thing, and may not very improperly be said to comprehend all the rest."

they excite admiration from excellence, independent of peculiar circumstances; but the public taste is endangered by the circulation of such crude productions as are curious only from the youth or ignorance of their authors. Such an apology to the learned will not compensate for the defects of grammar in Poetry, nor to the scientific artist for the defects of proportion and design in Architecture; while the incorrectness of such efforts is hardly visible to the bulk of mankind, incapable of comparing their excellence with works of established reputation. Thus in poetry, in painting, and in architecture, *false taste is propagated by the sanction given to mediocrity.*

Its dangerous tendency, added to its frequency, must plead my excuse for taking notice of the following vulgar mode of expression: "I do not profess to understand these matters, but I know what pleases me." This may be the standard of perfection with those who are content to gratify their own taste without inquiring how it may affect others; but the man of good taste endeavours to investigate the causes of the pleasure he receives, and to inquire whether others receive pleasure also. He knows that the same principles which direct taste in the polite arts, direct the judgment in morality; in short, that a knowledge of what is good, what is bad, and what is indifferent, whether in actions, in manners, in language, in arts, or science, constitutes the basis of good taste, and marks the distinction between the higher ranks of polished society, and the inferior orders of mankind, whose daily labours allow no leisure for other enjoyments than those of mere sensual, individual, and personal gratification.

" In most countries, novelty, in every form of extrava-
" gance, broad humour, and caricature, affords the greatest
" delight to the populace. This preference is congenial with

" their love of coarse pleasures, and distinguishes the multi-
" tude from the more polite classes of every nation.   The
" inferior orders of society are therefore disqualified from de-
" ciding upon the merits of the fine arts ; and the department
" of taste is consequently confined to persons enlightened by
" education and conversant with the world, whose views of
" nature, of art, and of mankind, are enlarged and elevated by
" an extensive range of observation." *Kett's Elements of
General Knowledge.*

Those who delight in depreciating the present by com-
parisons with former times, may, perhaps, observe a decline
of taste in many of the polite arts; but surely in architecture
and gardening, the present era furnishes more examples of
attention to comfort and convenience than are to be found in
the plans of Palladio, Vitruvius, or Le Nôtre, who, in the
display of useless symmetry, often forgot the requisites of
habitation.   The leading feature in the good taste of modern
times, is the just sense of GENERAL UTILITY.

So difficult is the task of giving general satisfaction, that
I am aware I shall cause offence to some by mentioning their
places ; to others, by not mentioning them : to some, by having
said too much ; to others, by having said too little.   Yet to
establish principles from experience, and theory from practice,
it was necessary to quote examples ; I have therefore prefixed a list of those places only to which I refer in the course
of the work.

It will, perhaps, be observed, that some of these places are
of great extent and importance, whilst others are so inconsiderable that they might have been omitted.   But to the proprietor his own place is always important; and to the professor

a small place may serve to illustrate the principles of his art; and his whole attention and abilities should be exerted, whether he is to build a palace or a cottage, to improve a forest or a single field. Well knowing that every situation has its facilities and its difficulties, I have never considered how many acres I was called upon to improve, but how much I could improve the subject before me, and have occasionally experienced more pleasure and more difficulties in a small flower-garden, than amidst the wildest scenery of rocks and mountains.

Some of the places here enumerated are subjects which I have visited only once; others, from the death of the proprietors, the change of property, the difference of opinions, or a variety of other causes, may not, perhaps, have been finished according to my suggestions. It would be endless to point out the circumstances in each place where my plans have been partially adopted or partially rejected. To claim as my own, and to arrogate to myself all that I approve at each place, would be doing injustice to the taste of the several proprietors who may have suggested improvements. On the other hand, I should be sorry, that to *my taste* should be attributed all the absurdities which fashion, or custom, or whim, may have occasionally introduced in some of these places. I can only advise, I do not pretend to dictate, and, in many cases, must rather conform to what has been ill begun, than attempt to pull to pieces and remodel the whole Work.

"Non mihi res sed me rebus subjungere conor."
[Circumstances do not yield to me, I am forced to yield to them.]

To avoid the imputation of having fully *approved*, where I have found it necessary merely to *assent*, I shall here beg leave to subjoin my opinion negatively, as the only means of doing so without giving offence to those from whom I may differ; at the same time, with the humility of experience, I am conscious my opinion may, in some cases, be deemed wrong. The same motives which induce me to mention what I recommend, will also justify me in mentioning what I disapprove; a few observations, therefore, are subjoined to mark those errors, or absurdities in modern gardening and architecture, to which I have never willingly subscribed, and from which it will easily be ascertained how much of what is called the improvement of any place in the list, may properly be attributed to my advice. It is rather upon my opinions in writing, than on the partial and imperfect manner in which my plans have sometimes been executed, that I wish my fame to be established.

OBJECTION No. 1.

There is no error more prevalent in modern gardening, or more frequently carried to excess, than taking away hedges to unite many small fields into one extensive and naked lawn, before plantations are made to give it the appearance of a park; and where ground is subdivided by sunk fences, imaginary freedom is dearly purchased at the expense of actual confinement.

No. 2.

The baldness and nakedness round the house is part of the same mistaken system, of concealing fences to gain extent. A palace, or even an elegant villa, in a grass field, appears to me incongruous; yet I have seldom had sufficient influence to correct this common error.

No. 3.

An approach which does not evidently lead to the house,

or which does not take the shortest course, cannot be right. [This rule must be taken with certain limitations. The shortest road across a lawn to a house will seldom be found graceful, and often vulgar. A road bordered by trees in the form of an avenue, may be straight without being vulgar; and grandeur, not grace or elegance, is the expression expected to be produced.]

No. 4.

A poor man's cottage, divided into what is called *a pair of lodges*, is a mistaken expedient to mark importance in the entrance to a Park.

No. 5.

The entrance gate should not be visible from the mansion, unless it opens into a court yard.

No. 6.

The plantation surrounding a place, called a *Belt*, I have never advised; nor have I ever willingly marked a drive, or walk, completely round the verge of a park, except in small villas, where a dry path round a person's own field is always more interesting to him than any other walk.

No. 7.

Small plantations of trees, surrounded by a fence, are the best expedients to form groups, because trees planted singly seldom grow well; neglect of thinning and removing the fence, has produced that ugly deformity called a *Clump*.

No. 8.

Water on an eminence, or on the side of a hill, is among

the most common errors of Mr. Brown's followers : in numerous instances I have been allowed to remove such pieces of
water from the hills to the valleys; but in many my advice has
not prevailed.

No. 9.

Deception may be allowable in imitating the works of
NATURE; thus artificial rivers, lakes, and rock scenery, can
only be great by deception, and the mind acquiesces in the
fraud, after it is detected: but in works of ART every trick
ought to be avoided. Sham churches, sham ruins, sham
bridges, and everything which appears what it is not, disgusts
when the trick is discovered.

No. 10.

In buildings of every kind the *character* should be strictly
observed. No incongruous mixture can be justified. To add
Grecian to Gothic, or Gothic to Grecian, is equally absurd ;
and a sharp pointed arch to a garden gate or a dairy window,
however frequently it occurs, is not less offensive than Grecian
architecture, in which the standard rules of relative proportion are neglected or violated.

The perfection of landscape gardening consists in the full-
est attention to these principles, *Utility*, *Proportion*, and *Unity*,
or harmony of parts to the whole.

LIST OF THE PLACES

REFERRED TO AS EXAMPLES.

Abington Hall ......Cambridgeshire......John Mortlock, Esq.
Adlestrop ....Gloucestershire......J. H. Leigh, Esq.
Antony ....Cornwall............R. P. Carew, Esq. M.P.
Ashton Court........Somersetshire........Sir Hugh Smyte, Bart.
Aston ....Cheshire ....Hon. Mrs. Harvey Aston
Attingham ....Shropshire ....Right Hon. Lord Berwick
Babworth ....Nottinghamshire ....Hon. J. B. Simpson, M.P.
Bank Farm ....Surry ....Hon. Gen. St. John
Bayham ....Kent................Earl Camden
Betchworth ....Surry ....Hon. W. H. Bouverie, M.P.
Blaize Castle.......Gloucestershire......J. S. Harford, Esq.
Bowood ....Wiltshire ....Marquis Lansdown
Brandsbury ....Middlesex ....Hon. Lady Salusbury
Bracondale.........Norfolk ....P. Martineau, Esq.
Brentrey Hill ......Gloucestershire .. ..Wm. Payne, Esq.
Buckminster ....Leicestershire ......Sir Wm. Manners, Bart.
Bulstrode ....Buckinghamshire....His Grace the Duke of Portland
Burleigh on the Hill Rutlandshire ....Earl Winchelsea
Catton ....Norfolk ....Jer. Ives, Esq.
Cashiobury ....Hertfordshire........Earl of Essex
Catchfrench ....Cornwall............Francis Glanville, Esq. M.P.
Chilton Lodge ......Berkshire ....John Pearse, Esq.
Clayberry Hall ......Essex ....James Hatch, Esq.
Cobham Hall........Kent................Earl Darnley
Courteen Hall ......Northamptonshire ..Sir Wm. Wake, Bart.
Corsham House ....Wiltshire ....Paul Cob. Methuen, Esq.
Condover Park.....Shropshire ....Owen Smyth Owen, Esq.
Coombe Lodge .....Berks & Oxfordshire Samuel Gardener, Esq.
Cote Bank ....Gloucestershire......Wm. Broderip, Esq.
Crewe ....Cheshire ....John Crewe, Esq. M.P.
Culford ....Suffolk..............Marquis Cornwallis
Donington Park ....Leicestershire ......Earl Moira
Dulwich Casina ....Surry ....Richard Shawe, Esq.
Dullingham House ..Cambridgeshire......Colonel Jeaffreson
Dyrham Park ......Gloucestershire......Wm. Blathwayte, Esq.
Fort ....Bristol..............T. Tyndall, Esq.
Garnons ....Herefordshire ......J. G. Cotterel, Esq. M.P.
Gayhurst...........Buckinghamshire....George Wright, Esq.
Glemham ....Suffolk..............Dudley North, Esq. M.P.
The Grove ....Southgate ....Walker Gray, Esq.
Hasells ....Bedfordshire ....Francis Pym, Esq.
Harewood House....Yorkshire ....Right Hon. Lord Harewood
Heathfield ....Sussex ....Francis Newberry, Esq.
High Legh.........Cheshire ..,........G. J. Legh, Esq.
Higham Hills........Essex ....John Harman, Esq.
Highlands ....Essex ....C. H. Kortright, Esq.
Hill Hall ....Essex ....Sir Wm. Smyth, Bart.

Holkham........... Norfolk .... T. W. Coke, Esq. M.P.
Holwood........... Kent................ Right Hon. Wm. Pitt.
Holme Park .... Berkshire .... Richard Palmer, Esq.
Hooton.............. Cheshire .... Sir Thomas Stanley, Bart.
Hurlingham .... in Fulham .... John Ellis, Esq.
Kenwood .... Middlesex .... Earl Mansfield
Langley Park ...... Kent................ Right Hon. Lord Gwydr
Lathom House ..... Lancashire .... Wilbraham Bootle, Esq. M.P.
Langleys........... Essex .... W. Tuffnel, Esq.
Livermere .... Suffolk.............. N. Lee Acton, Esq.
Luscombe .... Devonshise.......... Ch. Hoare, Esq.
Maiden Early ..... Berkshire .... E. Golding, Esq. M.P.
Magdalen College ..Oxford............. President and Fellows
Merly House........ Dorsetshire.......... W. Willet Willet, Esq.
Milton House ..... Cambridgeshire...... Sam. Knight, Esq.
Milton Abbey........ Northamptonshire . Earl Wentworth Fitzwilliam
Michel Grove........ Sussex .... Richard Walker, Esq.
Moccas Court ...... Herefordshire ...... Sir George Cornewall, Bart. M.P.
Mulgrave .... Yorkshire .... Right Hon. Lord Mulgrave
Newton Park........ Somersetshire ...... W. Gore Langton, Esq. M.P.
Normanton.......... Rutlandshire........ Sir Gilbert Heathcote, Bart. M.P.
Oldbury Court ...... Gloucestershire ...... T. Græme, Esq.
Organ Hall.......... Hertfordshire........ Wm. Togwood, Esq.
Panshanger .... Hertfordshire........ Earl Cowper
Port Eliot .... Cornwall............ Right Hon. Lord Crags Eliot
Prestwood .... Staffordshire .... Hon. Edw. Foley, M.P.
Plas Newyd .... Anglesea............ Earl of Uxbridge
Purley.............. Berkshire .... J. Ant. Storer, Esq.
Rendlesham .... Suffolk............. P. Thellusson, Esq. M.P.
Rüg .... North Wales .... Col. E. V. W. Salesbury
Sarsden .... Oxfordshire .... J. Langston, Esq. M.P.
Scarrisbrick .... Lancashire .... T. Scarrisbrick Eccleston, Esq.
Sheffield Place ..... Sussex .... Right Hon. Lord Sheffield
Shardeloes .... Buckinghamshire.... Wm. Drake, Esq. M.P.
Stoke Park.......... Herefordshire ...... Hon. E. Foley, M.P.
Stoke Pogies .... Berkshire .... John Penn, Esq.
Stoneaston .... Somersetshire ...... Hippesley Coxe, Esq. M.P.
St. John's .... Isle of Wight........ Edw. Simeon, Esq.
Stapleton............ Gloucestershire...... Dr. Lovell, M.D.
Stratton Park ...... Hampshire .... Sir Francis Baring, Bart. M.P.
Streatham Villa ....Surry .... Robert Brown, Esq.
Sufton Court .... Herefordshire ...... James Hereford Esq.
Sundridge Park .... Kent................ Claude Scott, Esq. M.P.
Suttons .... Essex .... Charles Smith, Esq. M.P.
Taplow .... Buckinghamshire.... J. Fryer, Esq.
Tendring............ Suffolk.............. Sir Wm. Rowley, Bart.
Thoresby........... Nottinghamshire ... Lord Viscount Newark
Valleyfield .... Perthshire .... Sir Robert Preston, Bart. M.P.
Wall Hall .... Hertfordshire........ G. W. Thellusson, Esq. M.P.
West Wycombe...... Buckinghamshire.... Sir J. Dashwood King, Bart.
Wentworth House ..Yorkshire .... Earl Wentworth Fitzwilliam
Welbeck............ Nottinghamshire .... His Grace the Duke of Portland
Whitton Park ...... Middlesex .... Samuel Prime, Esq.
Wimpole............ Cambridgeshire...... Earl Hardwicke
Woodley............ Berkshire .... Right Hon. H. Addington, M.P.
Wycombe .... Buckinghamshire.... Right Hon. Lord Carrington.

[Fig. 37. Diagram to shew the use of the human figure as a scale for measuring objects.]

CHAPTER I.

Introduction—General Principles—Utility—Scale—Various Examples of comparative Proportion—Use of Perspective—Example from THE FORT—Ground—Several Examples of removing Earth—The great Hill at WENTWORTH.

THE Theory and practice of Landscape Gardening have seldom fallen under the consideration of the same author; because those who have delivered their opinions in writing on this art have had little practical experience, and few of its professors have been able to deduce their rules from theoretical principles. To such persons indeed had its practice been committed, that it required no common degree of fortitude and perseverance to elevate the art of landscape gardening to its proper rank, and amongst those which distinguish the pleasures of civilized society from the pursuits of savage and barbarous nations.

Not deterred by the sneer of ignorance,* the contradiction of obstinacy, the nonsense of vanity, or the prevalence of false

---

* The ignorance and obstinacy here alluded to, relate to the frequent opposition I have experienced from gardeners, bailiffs, and land stewards, who either wilfully mar my plans, or ignorantly mistake my instructions.

taste, I made the attempt; and with the counsels and advice of men of science, and the countenance of some of the first characters in the kingdom, a very large portion of its scenery has been committed to my care for improvement.   Hence it might be expected, that with some degree of confidence I now should deliver the result of my observations; yet, from the difficulties continually increasing with my knowledge of the subject, I submit this work to the public with far more diffidence than I did my former volume: because in this, as in every other study, reflection and observation on those things which we do know, teach us to regret our circumscribed knowledge, and the difficulty of reducing to fixed principles the boundless variety of the works of Nature.

If any general principles could be established in this art, I think that they might be deduced from the joint consideration of *relative fitness* or UTILITY, and *comparative proportion* or SCALE; the former may be referred to the mind, the latter to the eye, yet these two must be inseparable.

Under relative fitness I include the comfort, the convenience, the character, and every circumstance of a place, that renders it the desirable habitation of man, and adapts it to the uses of each individual proprietor; for it has occasionally happened to me to have been consulted on the same subject by two different proprietors, when my advice has been materially varied, to accord with the respective circumstances or intentions of each.

The second is that leading principle which depends on sight, and which I call comparative proportion; because all objects appear great or small by comparison only, or as they have a reference to other objects with which they are liable to be compared.

As this will be more clearly explained by an example, the vignette* at the beginning of the chapter presents two

---

* Besides the obelisks in the vignette, are several other emblems relating to landscape gardening : *the proportional compasses* are often necessary to fix the exact comparative dimensions on paper, to reduce or enlarge the scale, and the flowing lines of ribbon or linen cloth are frequently necessary to mark the outline of a piece of water, when its effect is to be judged of at a distance ; but, above all, the *eye* to observe and the *hand* to delineate, are always *necessary*, and will often supersede the use of every instrument; because the judicious artist must rather consider things as they appear than

obelisks, of exactly the same size, yet by the figures placed
near each, they appear to be of very different dimensions.
The height of a man we know to be generally from five to six
feet, but an obelisk may be from ten to a hundred feet high;
we therefore compare the unknown with the known object,
and immediately pronounce one of these obelisks to be twice
the size of the other.   Yet without some such scale to assist
the eye, it would be equally difficult either in nature, or in a
picture, to form a correct judgment concerning objects of
uncertain dimensions.

    At HOLKHAM, about twenty years ago, the lofty obelisk
seen from the portico, appeared to be surrounded by shrubbery, but on a nearer approach, I found that these apparent
shrubs were really large trees, and only depressed by the
greater height of the obelisk.   A similar instance occurs at
WELBECK; the large grove of oaks seen from the house across
the water, consists of trees most remarkable for their straight
and lofty stems; yet, to a stranger, their magnitude is apparently lessened by an enormous large and flourishing ash,
which rises like a single tree out of a bank of brushwood.
When I was first consulted respecting WENTWORTH HOUSE,
the lawn behind it appeared circumscribed, and the large
trees which surrounded that lawn appeared depressed by four
tall obelisks: these have since been removed, the stately trees
have assumed their true magnitude, and the effect of confinement is done away.

    I have illustrated these observations by the example of an
obelisk [fig. 37], because its height being indeterminate, it
may mislead the eye as a scale ; since, according to its size and
situation, the very same design may serve for a lamp-post, a
mile-stone in the market-place of a city, an ornament to a
public square, or it may be raised on the summit of a hill, a
monument to a nation's glory.

    The necessity of observing scale or comparative proportion,
may be further elucidated by a reference to WEST WYCOMBE,
a place generally known, from its vicinity to the road to
Oxford.   Amongst the profusion of buildings and ornament

as they really exist, by which he may unite distant objects, and separate
those in contact; his effects must be studied with the eye of the painter,
and reduced to proper scale with the measurement of the land surveyor.

which the false taste of the last age lavished upon this spot, many were correct in design, and, considered separately, in proportion; but even many of the designs, although perfect in themselves, were rendered absurd, from inattention either to the scale or situation of the surrounding objects. The summit of a hill is covered by a large mass of Grecian architecture, out of which apparently rises a small square projection, with a ball at the top, not unlike the kind of cupolas misplaced over stables; but in reality this building is the tower of a church,* and the ball a room sufficiently large to contain eight or ten people.

This comparative proportion, or, in other words, this attention to scale or measurement, is not only necessary with regard to objects near each other, but it forms the basis of all improvement depending on perspective, by the laws of which it is well known that objects diminish in apparent size in proportion to their distance: yet the application of this principle may not, perhaps, have been so universally considered. I shall, therefore, mention a few instances in which I have availed myself of its effects.

At HURLINGHAM, on the banks of the Thames, the lawn in front of the house was necessarily contracted by the vicinity of the river, yet being too large to be kept under the scythe and roller, and too small to be fed by a flock of sheep, I recommended the introduction of Alderney cows only; and the effect is that of giving imaginary extent to the place, which is thus measured below a true standard; because if

---

* On the summit of another building, viz. a saw-mill in the park, was a figure of a man in a brown coat and a broad-brimmed hat, representing the great Penn, of Pensylvania, which being much larger than the natural proportion of a man, yet having the appearance of a man upon the roof of the building, diminished the size of every other object by which it was surrounded.   It has since been removed, and is now in the possession of Mr. Penn, at Stoke Pogies, where, placed in a room, it seems a colossal figure. Another instance of false scale at this place, was the diminutive building with a spire at the end of the park, which, perhaps, when the neighbouring trees were small, might have been placed there with a view of extending the perspective.   This artifice may be allowable in certain cases, and to a certain degree, yet a cathedral in miniature must in itself be absurd; and when we know that it was only the residence of a shoemaker, and actually dedicated to St. Crispin, it becomes truly ridiculous.
    I have drawn these examples of defects from West Wycombe, because they are obvious to every passenger on a very public road, and because I shall, in the course of this volume, have occasion to mention the many beauties of this place.

distance will make a large animal appear small, so the distance will be apparently extended by the smallness of the animal.

The same reasoning induced me to prefer, at STOKE POGIES, a bridge of more arches than one over a river which is the work of art, whilst in natural rivers a single arch is often preferable, because in the latter we wish to increase the magnitude of the bridge, whilst in the former we endeavour to give importance to the artificial river.

Another instance of the necessity of attending to comparative scale, occurred near the metropolis, where a gentleman wished to purchase a distant field for the purpose of planting out a tile-kiln, but I convinced him, that during the life of man the nuisance could never be hid from his windows by planting near the kiln, whilst a few trees, judiciously placed within his own ground, would effect the purpose the year after they were planted.

THE Art of Landscape Gardening is in no instance more intimately connected with that of painting than in whatever relates to perspective, or the difference between the real and apparent magnitude of the objects, arising from their relative situations ; for without some attention to perspective, both the dimensions and the distances of objects will be changed and confounded. Few instances having occurred to me where this can be more forcibly elucidated than in the ground at THE FORT, near Bristol ; I shall avail myself of the following observations to shew what can, and what can not be done, by a judicious application of the laws of perspective.

When I first visited THE FORT, I found it surrounded by vast chasms in the ground, and immense heaps of earth and broken rock : these had been made to form the cellars and foundations to certain additions to the city of Bristol, which were afterwards relinquished. The first idea that presented itself was to restore the ground to its original shape ; but a little reflection on the character and situation of the place, naturally led me to inquire whether *some* considerable advantage might not be derived from the mischief which had thus been already done.

Few situations command so varied, so rich, and so extensive a view as THE FORT; situated on the summit of a hill
which looks over the vast city of Bristol, it formerly surveyed
the river, and the beautiful country surrounding it, without
being incommoded by too much view of the city itself: but
the late prodigious increase of buildings had so injured the
prospect from this house, that its original advantages of
situation were almost destroyed, and there was some reason
to doubt whether it could ever be made desirable either as a
villa or as a country residence; because it was not only
exposed to the unsightly rows of houses in Park-street and
Berkeley-square, but it was liable to be overlooked by the
numerous crowds of people who claimed a right of footpath
through the park, immediately before the windows.   It was,
therefore, as public as any house in any square or street of
Bristol.   If the earth had been simply put back to the places
from whence it had been taken, the expense of its removal
would have been greater than the method which occurred to
me as more advisable; *viz.* to fill up the chasms partly, by
levelling the sides into them, and raising a bank with a wall
to exclude the footpath, as shewn in the annexed section
[fig. 38], where the dotted line shews the original shape of
the ground; the zig-zag line, holes from fifteen to twenty feet
deep; the shaded line, the shape of the ground as altered.

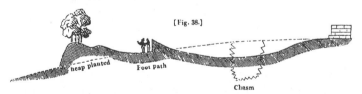

[Fig. 38.]

heap planted   Foot Path   Chasm

By this expedient we hide the objectionable part of the
view, and by planting the raised heap of earth we produce a
degree of privacy and seclusion in this newly created valley
within the pleasure-grounds, which was never before known
or expected in this open situation.   The pleasure-ground,
immediately near the house, is separated from the park by a
wall, against which the earth is everywhere laid as before
described, so as to carry the eye over the heads of persons
who may be walking in the adjoining footpath.   This wall

T

not only hides them from the house, but also prevents their overlooking the pleasure-ground. Yet, notwithstanding this great utility, this absolute necessity, the appearance of such a wall, from the park, gives an air of confinement, and the only expedient by which this might be well remedied, would be a total change in the character of the place, or, rather, by altering the house to make it what its name and situation denote: for if THE FORT were restored to its original character of a castle or fortress, this wall, instead of being objectionable, would then act as a terrace, and contribute to the general effect of extent, and the magnificence of the whole.*

The drawings [figs. 39 and 40] represent the view from the house, as it appeared before and after the improvement; upon the slide [fig. 39] are shewn five rods or poles, all of

[Fig. 39. View from the house at the Fort, near Bristol, as it appeared before it was improved.]

which are supposed to be ten feet high, and placed at different distances from the eye; these shew the difference in the apparent height of the same object in the different situations, and, of course, what may be expected from trees planted of any given size at each place: from hence, it is evident, that a young tree at No. 1, will hide nothing for many years except the park wall. A tree of the same size at No. 2, will do little more: this is confirmed, also, by the large trees already growing there; but at No. 3, where a heap of earth

* A drawing is inserted in the *Red Book* to shew the manner of thus altering the house; but the plate in this work is sufficient to explain the process used in ascertaining the possibility of so planting out the view of the neighbouring houses as to exclude what ought to be hid, without hiding what ought to be seen.

has been thrown up to a considerable height, a tree of twenty feet would hide most of the houses; and in like manner at No. 4 and No. 5, immediate effects may be produced, by judiciously planting, to shew the distant objects over or under the branches of trees in the foreground.

[Fig. 40. View from the Fort, as proposed to be improved.]

Although, from the nature of this work, it is difficult to preserve any connecting series of arrangement, yet it may not be improper, in this place, to mention a few remarkable instances of removing earth and altering the shape of the surface of ground, especially as there is no part of my profession attended with so much expense, or more frequently objected to, because so often mismanaged.

Where a ridge of ground very near the eye intercepts the view of a valley below, it is wonderful how great an effect may be produced by a very trifling removal of the ridge only; thus, at Moccas Court, a very small quantity of earth concealed from the house the view of that beautiful reach of the river Wye, which has since been opened. At Oldbury Court the view is opened into a romantic glen, by the same kind of operation. At Catchfrench the same thing is advised, to shew the opposite hills; and in this instance it may appear surprising, that the removal of a few yards of earth was sufficient to display a vast extent of distant prospect.

But this effect must depend on the natural shape of the surface near the eye; for example, if the shape be that of the upper line [fig. 41] A, the object at F cannot be seen without the removal of all the earth between the dotted line and the surface; but if the shape be that of в, the removal of the

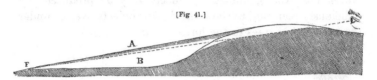

[Fig 41.]

part not shaded will not be sufficient to shew the valley; and it is not always desirable to see the whole surface; on the contrary, it is better that a part should be concealed than that the whole should be shewn foreshortened, which is always the case in looking down or up an inclined plane.*

The most arduous operations of removing ground are generally those where the geometric taste of gardening had distorted the natural surface, and where it would now be attended with much greater trouble and expense to restore the ground to its original shape, than had been formerly dedicated to make those slopes and regular forms, which are more like the works of a military engineer than of a painter or a gardener.

Few instances have occurred to me where great expense in moving ground was requisite to produce pleasing effects, and it is always with reluctance that I advise much alteration in the surface of ground, because, however great the labour, or expensive the process, it is a part of the art from which the professor can derive but little credit, since his greatest praise must be, that the ground looks, when finished, as if art had never interfered. " Ars est celare artem." [The excellence of art is in its concealment.]

When I was first consulted, at SUNDRIDGE PARK, by

* Having often seen great expense incurred by removing ground to shew the whole surface of a valley from the top of a hill, it may not be improper to explain that such an effort is seldom useful or desirable. To the painter it is impossible to represent ground thus foreshortened, and the first source of beauty, in the composition of a landscape, is the separation of distinct distances; the imagination delights in filling up those parts of the picture which the eye cannot see; and thus, in a landscape, while we do not see the bottom of a deep glen, we suppose it deeper than it really is; but when its whole shape is once laid open, the magic of fancied rocks and rattling torrents is reduced, perhaps, to the mortifying discovery of a dry valley or a swampy meadow.

Mr. Lind, the former possessor, the house, which has since been pulled down, stood on the south side of the valley; and those who knew the spot despaired of finding a situation for a house, on the opposite side of the valley, that the rooms might have a southern aspect, as the bank was too steep to admit of any building. My much respected friend, the present possessor, was aware of this circumstance, and by art we have produced a situation which nature denied. The earth was lowered thirty feet perpendicularly, at the spot on which the house was built, and so disposed at the foot of the hill that no trace of artificial management is now to be discovered.*

Among the greatest examples of removing ground, may be mentioned the work going on at BULSTRODE, under the direction of his Grace the Duke of Portland himself; whose good taste will not suffer any part of that beautiful park to be disguised by the misjudging taste of former times, and who, by opening the valleys and taking away a great depth of earth from the stems of the largest trees, which had been formerly buried, is, by degrees, restoring the surface of the ground to its original and natural shape.†

As connected with the subject of moving ground, I shall extract from my *Red Book* of WENTWORTH the following observations, concerning the great work at that place which had so long been carrying on under the direction of the late Marquis of Rockingham.

---

* The house, and the hill on which it stands, are exactly in due proportion to each other; and the former is so fitted to the situation and views which it commands, that I regret having shared with another the reputation of designing and adapting this very singular house to circumstances which cannot well be explained but upon the spot; having given a drawing and description of the scene to Mr. Angus, in justice to his work, I will not insert any view of this house; but its distance is so short from the capital, that, like many others, my best reference will be to the place itself.

In thus referring to places improved under my direction, it is not to be supposed that they are at all times accessible to *idle curiosity;* but the same good taste, and the same liberality of sentiment, which induce a proprietor to consult the professor of an art, will naturally operate in favour of *scientific inquiry.*

† In this great work are occasionally employed, among the more efficient labourers, a hundred children, from ten to fifteen years old, who are thus early trained to habits of wholesome industry, far different from the foul air and confinement of spinning in a cotton-mill; to the benevolent observer no object can be more delightful than park scenery thus animated.

Of the view from the portico at WENTWORTH HOUSE [figs. 42 and 43], my opinion is so contrary to that of many others who have advised a farther removal of the hill, that I hope it will not be improper to state very fully the reasons on which I ground this opinion,—*viz.* that so far from such an operation being equivalent to the trouble by which it must be executed, I would not advise its removal, if it could be much more easily effected, because—

1. The outline of the horizon beyond this hill is almost a straight line, and would be very offensive when shewn over another straight line parallel to it.

2. The view of the valley beyond, however rich in itself, is too motley to form a part of the proper landscape from such a palace as Wentworth House, although, from many situations in the park, it is a very interesting feature.

3. The vast plain, which has with so much difficulty been obtained in front of the house, is exactly proportionate to the extent of the edifice, and tends to impress the ideas of magnificence which so great a work of art is calculated to inspire. Such a plain forms an ample base for the noble structure which graces its extremity; the building and the plain are evidently made for each other, and, consequently, to increase the dimensions of either seems unnecessary.

The foregoing reasons relate to the hill as considered from the house only; I shall now consider it in other points of view.

[Fig. 42. View from Wentworth House, before it was improved, and while the improvements were going forward.]

Wentworth Park consists of parts, in themselves truly great and magnificent. The woods, the lawns, the water, and the buildings, are all separately striking; but, considered as a whole, there is a want of connexion and harmony in the composition; because parts in themselves large, if disjoined, lose their importance. This, I am convinced, is the effect of too great an expanse of unclothed lawn, but when the young trees shall have thrown a mantle over this extensive knoll, all the distant parts will assume one general harmony, and the scattered masses of this splendid scenery will be connected and brought together into one vast and magnificent whole.

The use of a plantation on this hill, in the approach from Rotherham, is evident, from the effect of a small clump which will form a part of this great mass, and which now hides the house, till, by the judicious bend round that angle, the whole building bursts at once upon the view.

It can readily be conceived, that before the old stables were removed there might appear some reason for not plant-ing this hill; not because it was too near the front, but because the view, thus bounded by a wood on one side, and the large pile of old stables on the other, would be too confined. That objection is removed with the stables, and now a wood on this hill will form a foreground, and lead the eye to each of those scenes, which are too wide apart ever to be considered as one landscape. In the adjoining sketch [fig. 43] I have endeavoured to shew the effect of planting

[Fig. 43. View from Wentworth House, shewing the effect intended to be produced by the proposed alterations.

this hill, leaving part of the rock to break out among the trees. In a line of such extent, and where the angle nearest the house will be rather acute, it may be necessary to hide part, and to soften off the corner of the plantation by a few scattered single trees, in the manner I have attempted to represent.

Among the future uses of the hill plantation, it may be mentioned, that the shape which the ground most naturally seems to direct, for the outline of this wood, is such as will,

hereafter, give opportunity to form the most interesting walk
that imagination can suggest; because, from a large crescent
of wood, on a knoll, the views must be continually varying;
while, by a judicious management of the small openings, and
the proper direction of the walks, the scenery in the park will
be shewn under different circumstances of foreground, with
increased beauty.

CHAPTER II.

Optics or Vision—At what Distance Objects appear largest—Axis of Vision
—Quantity or Field of Vision—Ground apparently altered by the situation of the Spectator—Reflections from the Surface of Water explained and applied—Different Effects of Light on different Objects—
Example.

LANDSCAPE GARDENING being connected with optics
or vision, or rather with the application of their rules to
practical improvement, it may not be improper to devote a
chapter to the following observations.

There is a certain point of distance from whence every object appears at its greatest magnitude.   This subject was originally discussed, in consequence of observing that a particular
rock at PORT ELIOT appeared higher or lower, at different
distances.   The inquiry into the cause of this difference led
me to propose a question to several ingenious friends.

Query, At what distance does any object appear at its
greatest height?

' The general optical distinction of the magnitude of objects is into real and apparent; the real being what its name
imports, and the apparent, not that which may ultimately result
to the mind, but that which is immediately impressed on the
eye.   This is measured by a plain and certain rule, namely, the
angle which is formed at the eye, by lines drawn from the
extremities of the object.   The apparent height of a man,
therefore, at a quarter of a mile distance, is not the conception which we form of his height, but the opening or angle of
the two lines above-mentioned, viz. of the two drawn from the
extremities of the object to our eye.   This apparent height,
therefore, of any object, will be measured always upon the
simplest principles; and will vary according to, first, the distance of the object; secondly, the inclination it makes with
the horizon; and, thirdly, our relative elevation or depression.
Any two of the above three things continuing the same, the
apparent magnitude will decrease with the third, though not
in exact proportion to it.

U

' Thus the object being perpendicular to the horizon, and our elevation remaining the same, its apparent height will decrease with the distance. Our elevation and the distance remaining the same, the apparent height of the object will decrease with its inclination to the horizon. The inclination and distance being the same, the angle, or apparent height, will decrease with our elevation or depression, supposing our height was, at first, the middle point of the object. This last being liable to some exceptions, the general rule is, that the distance from the object, measured by a perpendicular to it, being the same, the point at which its apparent height will be greatest, is, where the perpendicular from the eye falls upon the centre.

' The apparent height of a body, as upon the same principles any other of its dimensions, is a matter of easy consideration ; its inclination, its distance, and the relative position of the observer being known. The difficulty is to know what the conception is that we shall form of the height and magnitude of an object; according to different circumstances, its apparent height, as well as its real height, remaining the same. This, you will see, belongs to wholly different principles, and such as cannot be reduced to certain rules ; it appears, too, from hence, that the question has little or nothing to do with mathematical principles, at least beyond those simple ones which I have just stated. Of other principles, the consideration is more diversified : much may be ascribed to the habit, which we probably have, of estimating the height of objects, not by the angle, formed by lines to the summit and the base, when the base is below us, but by that formed between a line from the summit and a line parallel to the horizon ; in this way our conception of the magnitude may be less, while the apparent magnitude may be greater. A thousand other causes may likewise operate, amongst which will be some that belong to what is called aerial perspective, or those rules by which we judge of the distance or dimensions of objects, not by their outline on the retina, but by their colour and distinctness. The existence and operation of these can hardly be found, but by a careful examination and comparison of particular instances.'

The concluding paragraph in this letter, from one of the

most able men of the age, encouraged me to examine and
compare particular instances, as they fell under my own observation, and from a variety of these I am led to conclude, that,
among those numerous causes here said to operate, independent of mathematical principles, one may proceed from the
position of the eye itself; which is so placed as to view a certain portion of the hemisphere without any motion of the
head.   This portion has been differently stated by different
authors, varying from sixty to ninety degrees.

The question before us relates to the height, and not to
the general magnitude of the object, these being separate considerations; because the eye is capable of surveying more in
breadth than in height; but it is also capable of seeing much
farther below its axis than above it, as shewn by the following
profile [fig. 44].   From hence it appears, that the projection of the forehead and eyebrow causes great difference betwixt the angle A B and the angle A C, and that the line
parallel to the horizon A, which I shall call the *axis of vision*,
does not fall in the centre of the opening betwixt the extreme
rays B and C.

[Fig. 44.]

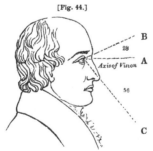

Doubtless these angles may vary in different individuals, from
various causes, such as the prominency of the eye, the habit
or usual position of the head, &c. yet the upper angle A B
will seldom be greater than one half of the lower angle A C;
and I have ascertained, with some precision, that I could not
distinguish objects more than twenty-eight degrees above my
axis of vision, although I can distinctly see them fifty-seven
degrees below it.   From hence I conclude, that the distance
at which an object appears at its greatest height, is, when the
axis of vision and the summit of the object form an angle of

about thirty degrees; because, under this angle, the eye perceives its full extent without moving the head, yet not without some effort of the eye itself to comprehend the whole of the object.

To this theory it may, perhaps, be objected, that, in the act of seeing, the motion of the head is too rapid to effect any material difference; but it will be found, on examining this subject attentively, that the object is seen in a new point of view, from the instant the head is moved, because the rays no longer meet at the same centre; and, therefore, the effect of such vision on the mind, is rather a renewal in succession of similar ideas, than the same single idea simultaneously excited: and this difference may be compared to that between seeing a landscape reflected in a mirror at rest, and the same landscape when the mirror has been removed from its original position.*

From frequent observation of the difference between seeing an object with and without moving the head, I am inclined to believe, that, by the latter, the mind grasps the whole idea at once; but, by the former, it is rather led to observe the parts separately: hence are derived many of those ideas of apparent magnitude or proportion which induce us to pronounce, at the first glance, whether objects are great or small. I should, therefore, answer the question, "At what distance does any object appear at its greatest height?" by saying, when the spectator is at such a distance, that the line drawn from his eye to the top of the object, forms an angle of not less than twenty-eight degrees with the axis of vision; and thus, supposing the eye to be five feet six inches from the ground, the distance will be according to the following diagram [fig. 45].

The scientific observer will always rejoice at discovering any law of Nature by which the judgment is unconsciously directed. At a certain distance from the front of any building, we admire the general proportions of the whole: but if the

* Perhaps this difference may be more familiarly explained by observing, that, when a lark ascends in the air, we have no difficulty in keeping the bird in sight so long as we continue our head in the first position; but from the moment the head is moved, we have to search for the object again, and often in vain, through the vast expanse of sky.

building can only be viewed within those angles of vision
already described, it is the several parts which first attract our

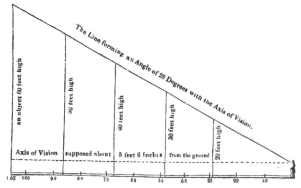

[Fig. 45. Scale of feet, shewing the distance of the spectator from the several objects.]

notice, and we generally pronounce that object large, the whole
of which the eye cannot at once comprehend.

Hence it is commonly observed by those who have seen
both St. Peter's, at Rome, and St. Paul's, at London, that
the latter appeared the largest at the first glance, till they
became aware of the relative proportion of the surrounding
space ; and I doubt whether the dignity of St. Paul's would
not suffer if the area round the building were increased, since
the great west portico is in exact proportion to the distance
from whence it can now be viewed, according to the preceding
table of heights and distances : but if the whole church could
be viewed at once, like St. Peter's, the dome would overpower
the portico, as it does in a geometrical view of the west
front.*

The field of vision, or the portion of landscape which the
eye will comprehend, is a circumstance frequently mistaken
in fixing the situation for a house ; since a view seen from
the windows of an apartment will materially differ from the
same view seen in the open air.   In one case, without moving
the head, we see from sixty to ninety degrees ; or, by a single

* I have sometimes thought that this same rule of optics may account for
the pleasure felt at first entering a room of just proportions, such as twenty
by thirty, and fifteen feet high ; or, twenty-four by thirty-six, and eighteen
feet high ; or, the double cube, when it exceeds twenty-four feet.

motion of the head, without moving the body, we may see every object within one hundred and eighty degrees of vision. In the other case the portion of landscape will be much less, and must depend on the size of the window, the thickness of the walls, and the distance of the spectator from the aperture. Hence it arises, that persons are frequently disappointed, after building a house, to find that those objects which they expected would form the leading features of their landscape are scarcely seen, except from such a situation in the room as may be inconvenient to the spectator; or, otherwise, the object is shewn in an oblique and unfavourable point of view. This will be more clearly explained by the following diagram [fig. 46].

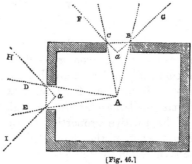

[Fig. 46.]

It is evident, that a spectator at A can only see, through an aperture of four feet, those objects which fall within the opening B C, in one direction, and D E in the other, neither comprehending more than twenty or thirty degrees. But if he removes to *a*, near the windows, he will then see all the objects, within the angle F G, in one direction, or H I in the other; yet it is obvious, that, even from these spots, that part of the landscape which lies betwixt the extreme lines of vision F and H, will be invisible, or at least seen with difficulty, by placing the eye much nearer to the window than is always convenient.

From hence it follows, that to obtain so much of a view as may be expected,* it is not sufficient to have a cross light, or

---

* Of this I observed a curious instance at HOOTON HOUSE, from whence a distant view of Liverpool, and its busy scenery of shipping, is not easily seen without opening the windows, while the difference of a few yards, in the original position of the house, would have obviated the defect, while it improved its general situation.

windows, in two sides of the room, at right angles with each other; but there must be one in an oblique direction, which can only be obtained by a bow window : and although there may be some advantage in making the different views from a house distinct landscapes, yet as the villa requires a more extensive prospect than a constant residence, so the bow window is peculiarly applicable to the villa.   I must acknowledge that its external appearance is not always ornamental, especially as it is often forced upon obscure buildings, where no view is presented near great towns, and oftener is placed like an uncouth excrescence upon the bleak and exposed lodging houses at a watering place ; but in the large projecting windows of old Gothic mansions, beauty and grandeur may be united to utility.

The apparent shape of the ground will be altered by the situation of the spectator.   This is a subject of much importance to the landscape gardener, although not generally studied.

[Fig. 47.]

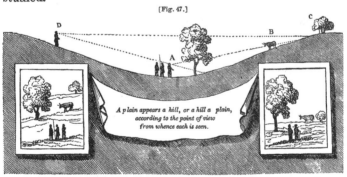

A plain appears a hill, or a hill a plain, according to the point of view from whence each is seen.

In hilly countries, where the banks are bold, a road in a valley is always pleasing, because it seems natural, and carries with it the idea of ease and safety ; but, in a country that is not hilly, we ought rather to shew the little * inequalities of ground to advantage.   The difference betwixt viewing ground

* That I may not be misunderstood, as recommending a road over hill and dale to shew the extent or beauty of a place, I must here observe, that nothing can justify a visible deviation from the shortest line, in an approach to a house, but such obstacles as evidently point out the reason for the deviation.

from the bottom of a valley, or the side of a hill, will be best explained by the following diagram [fig. 47], where the rules of perspective again assist the scientific improver.

The spectator at A, in looking up the hill towards c, will lose all the ground that is foreshortened; and every object which rises higher than five feet (*i. e.* the height of his eye), will present itself above his horizon, if the slope is exactly an inclined plane, or hanging level; but as the shape of ground here delineated more frequently occurs, he will actually see the sky, and, consequently, the utmost pitch of the hill, beneath the body of the animal placed at B, and part of the thorn at c, become invisible.

This accounts for the highest mountains losing their importance, when seen only from the base; while, on the contrary, a plain or level surface (for instance the sea) appears to rise considerably when viewed from an eminence. Let us suppose another spectator to be placed at D, it is evident that this person will see no ground foreshortened but that below him, while the opposite hill will appear to him far above the head of the man at A, and above the cow at B. In the section, the dotted lines are the respective horizons of the two spectators, and the sketches shew the landscape seen by each, in which the forked tree may serve as a scale to measure the height of each horizon.

The reflections of objects in water are no less dependant on the laws of perspective, or of vision, than the instances already enumerated.

If the water be raised to the level of the ground beyond it, we lose all advantage of reflection from the distant ground or trees: this is the case with pieces of water near the house, in many places, for all ponds, on high ground, present a constant glare of light from the sky; but the trees beyond can never be reflected on the surface, because the angle of incidence and the angle of reflection are always equal: and the surface of the water will always be a perfect horizontal plane. This I shall farther explain by the following lines [fig. 48].

The spectator at *a*, in looking on the upper water, will see only sky; because the angle of incidence *b*, and that of reflection c being equal, the latter passes over the top of the trees d, on lower ground: but the same spectator a, in look-ing on the lower water, will see the trees e reflected on its surface, because the line of reflection passes through them, and not over them, as in the first instance.

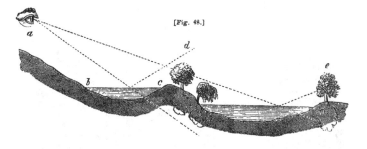

[Fig. 48.]

There are other circumstances belonging to reflection on the surface of water, which deserve attention, and of which the landscape gardener should avail himself in the exercise of his art. Water in motion, whether agitated by wind or by its natural current, produces little or no reflection; but in arti-ficial rivers, the quiet surface doubles every object on its shores, and, for this reason, I have frequently found that the surface could be increased in appearance by sloping its banks: not only that which actually concealed part of the water, but also the opposite bank; because it increased the quantity of sky reflected on the surface.

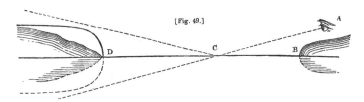

[Fig. 49.]

Example. The spectator at [fig. 49] A, sees the sky re-flected only from B to C, while the opposite bank is round; but if sloped to the shaded line, less of the bank will be reflected in the water, and the quantity of sky seen in the water, will be from B to D; and as the brilliancy of still water depends on the sky reflected on its surface, the quantity of water will be apparently increased.

As properly belonging to this chapter, may be mentioned a curious observation, which occurred in the view of the Thames, from PURLEY. In the morning [see fig. 50], when

[Fig. 50.   View of the Thames from Purley, in the morning.]

the sun was in the east, the landscape appeared to consist of wood, water, and distant country, with few artificial accompaniments; but in the evening, when the sun was in the west, objects presented themselves which were in the morning scarcely visible. In the first instance, the wood was in a solemn repose of shade, the water, reflecting a clear sky, was so brilliantly illuminated, that I could trace the whole course of the river; the dark trees were strongly contrasted by the vivid green of the meadows, and the outline of distant hills was distinctly marked by the brightness of the atmosphere. I could scarcely distinguish any other objects; but these formed a pleasing landscape, from the breadth or contrast of light and shade.

In the evening the scene was changed; dark clouds reflected in the water rendered it almost invisible, the opposite hanging wood presented one glare of rich foliage; not so beautiful in the painter's eye, as when the top of each tree was relieved by small catching lights: but the most prominent features were the buildings, the boat, the path, the pales, and even the distant town of Reading, now strongly gilded by the opposite sun. [See fig. 51.]

On comparing this effect with others, which I have frequently since observed, I draw this conclusion : that certain objects appear best with the sun behind them, and others with

[Fig. 51. View of the Thames from Purley, in the evening.]

the sun full upon them; and it is rather singular, that to the former belong all *natural* objects, such as woods, trees, lawn, water, and distant mountains; while to the latter belong all *artificial* objects, such as houses, bridges, roads, boats, arable fields, and distant towns or villages.

In the progress of this work I shall have occasion to call the reader's attention to the principles here assumed, and which, in certain situations, are of great importance, and require to be well considered.

CHAPTER III.

Water—it may be too naked or too much clothed—Example from WEST
  WYCOMBE—Digression concerning the Approach—Motion of Water—
  Example at ADLESTROP—Art must deceive to imitate Nature—Cascade
  at THORESBY—The Rivulet—Water at WENTWORTH described—A
  River easier to imitate than a Lake—A bubbling Spring may be imi-
  tated—a Ferry-boat at HOLKHAM—A rocky Channel at HAREWOOD.

THE observations in the preceding chapter concerning the
reflection of sky on the surface of water, will account for that
brilliant and cheerful effect produced by a small pool, frequently placed near a house, although in direct violation of
Nature : for since the ground ought to slope, and generally
does slope, from a house, the water very near it must be on the
side of a hill, and of course artificial. Although I have never
proposed a piece of water to be made in such a situation, I
have frequently advised that small pools so unnaturally placed
should be retained, in compliance with that general satisfaction which the eye derives from the glitter of water, however
absurd its situation.

It requires a degree of refinement in taste bordering on
fastidiousness, to remove what is cheerful and pleasing to the
eye, merely because it cannot be accounted for by the common
laws of Nature ; I was, however, not sorry to discover some
plea for my compliance, by considering, that although water
on a hill is generally deemed unnatural, yet all rivers derive
their sources from hills, and the highest mountains are known
to have lakes or pools of water near their summits.

We object, therefore, not so much to the actual situation,
as to the artificial management of such water. We long to
break down the mound of earth by which the water is confined ; although we might afterwards regret the loss of its
cheerful glitter ; and hence, perhaps, arises that baldness in
artificial pools, so disgusting to the painter, and yet so pleas-

ing to the less accurate observer. The latter delights in a
broad expanse of light on the smooth surface, reflecting a brilliant sky; the former expects to find that surface ruffled by
the winds, or the glare of light in parts obscured by the reflection of trees from the banks of the water; and thus, while the
painter requires a *picture*, the less scientific observer will be
satisfied with a *mirror*.

During great part of last century WEST WYCOMBE was
deemed a garden of such finished beauty, that to those who
formerly remembered the place, it will seem absurd to suggest
any improvement. But time will equally extend its changing
influence to the works of nature and to those of art, since the
PLANTER has to contend with a power—

> " A hidden power! at once his friend and foe!
> 'Tis VEGETATION! Gradual to his groves
> She gives their wished effects, and that displayed,
> O! that her power would pause; but, active still,
> She swells each stem, prolongs each vagrant bough,
> And darts, with unremitting vigour bold,
> From grace to wild luxuriance."
>
>   MASON.

Thus, at WEST WYCOMBE, those trees and shrubs which
were once its greatest ornament, have now so far outgrown
their situation, that the whole character of the place is altered; and instead of that gaiety and cheerfulness inspired by
flowering shrubs and young trees, gloom and melancholy seem
to have reared their standard in the branches of the tallest
elms, and to shed their influence on every surrounding object:
on the house, by lessening its importance; on the water,
by darkening its surface; and on the lawn, by lengthened
shadows.

The prodigious height of the trees near the house has not
merely affected the character, but also the very situation of
the house.   Instead of appearing to stand on a dry bank, considerably above the water (as it actually does), the house,
oppressed by the neighbouring trees, became damp, and appeared to have been placed in a gloomy bottom, while the
water was hardly visible, from the dark reflection of the trees
on its surface, and the views of the distant hills were totally
concealed from the house.

It is a fortunate circumstance for the possessor, where improvement can be made rather by cutting down than by planting trees. The effect is instantly produced, and as the change in the scenery at this place has actually been realized before I could make a sketch to explain its necessity, the following drawing serves to record my reason for so boldly advising the use of the axe. I am well aware that my advice may subject me to the criticism of some, who will regret the loss of old trees, which, like old acquaintances, excite a degree of veneration, even when their age and infirmity have rendered them useless, perhaps offensive, to all but their youthful associates. The tedious process of rearing and planting woods, and the dreadful havoc too often made by injudiciously felling large trees, ought certainly to inspire caution and diffidence; but there is in reality no more temerity in marking the trees to be taken down than those to be planted, and I trust there has not been a single tree displaced at WEST WYCOMBE, which has not tended to improve the healthfulness, the magnificence, and the beauty of the place.

Most of the principal rooms having a north aspect, the landscape requires peculiar management not generally understood.* Lawn, wood, and water, are always seen to the greatest advantage with the sun behind them, because the full glare of light between opposite trees destroys the contrast of wood and lawn; while water never looks so brilliant and cheerful when reflecting the northern, as the southern sky: a view, therefore, to the north would be dull and uninteresting without some artificial objects, such as boats or buildings, or distant corn-fields, to receive the opposite beams of the sun.

A sketch (in the *Red Book*)† shewed the effect of taking down trees to admit the distant woods, and by removing those on the island, and of course their reflection, the water becomes more conspicuous; in addition, the proposed road of approach, with carriages occasionally passing near the banks of the lake, will give animation to the view from the saloon.

---

* This subject has been explained in the preceding chapter.
† A view of the house across the water, not here inserted, being exactly the reverse of that which represents the view towards the house, which is inserted.

The views of WEST WYCOMBE, inserted in this work [figs. 52 and 53], being taken from the proposed approach, I shall here beg leave to make a short digression, explaining

[Fig. 52. View from the approach to Wycombe House before it was altered, by cutting down the trees in the island, &c.]

my reasons for that line, founded on some general principles respecting an approach, although it has no other reference to

[Fig 53. View of Wycombe House, as it will appear when the proposed alterations in the approach are carried into effect.]

the water, than as it justifies its course in passing the house to arrive at its object.

If the display of magnificent or of picturesque scenery in a park be made without ostentation, it can be no more at variance with good taste than the display of superior affluence in the houses, the equipage, the furniture, or the habiliments of wealthy individuals.   It will, therefore, I trust, sufficiently justify the line of approach here proposed, to say, that it passes through the most interesting parts of the grounds, and will display the scenery of the place to the greatest advantage, without making any violent or unnecessary circuit to include objects that do not naturally come within its reach.   This I deem to be a just and sufficient motive, and an allowable display of property without ostentation.

The former approach to the house [fig. 52] was on the south side of the valley, and objectionable for two reasons: 1st, it ascended the hill, and, after passing round the whole of the buildings, it descended to the house, making it appear to stand low: 2nd, by going along the side of the hill, little of the park was shewn, although the road actually passed through it; because, on an inclined plane,* the ground which either rises on one side or falls on the other, becomes foreshortened and little observed, while the eye is directed to the opposite side of the valley, which, in this instance, consisted of enclosures beyond the park.   On the contrary, the proposed new approach, being on the north side of the valley, will shew the park on the opposite bank to advantage [fig. 53], and by ascending to the house, it will appear in its true and desirable situation upon a sufficient eminence above the water: yet backed by still higher ground, richly clothed with wood, this view of the house will also serve to explain and, I hope, to justify the sacrifice of those large trees which have been † cut down upon the island, and whose dark shadows being reflected on the water, excluded all cheerfulness.   [See fig. 52.]

The water at WEST WYCOMBE, from the brilliancy of its colour, the varieties of its shores, the different courses of its

---

* This is explained in Chapter II.

† Mr Brown has been accused of cutting down large old trees, and afterwards planting small ones on the same spot; the annexed plate [our fig. 53] may serve to vindicate the propriety of his advice.

channel, and the number of its wooded islands, possessed a
degree of pleasing intricacy which I have rarely seen in artificial pools or rivers; there appears to be only one improvement necessary to give it all the variety of which it is capable.
The glassy surface of a still calm lake, however delightful, is
not more interesting than the lively brook rippling over a
rocky bed; but when the latter is compared with a narrow
stagnant creek, it must have a decided preference; and as this
advantage might easily be obtained in view of the house, I
think it ought not to be neglected.

It may perhaps be objected, that to introduce rock scenery
in this place would be unnatural; but if this artifice be properly executed, no eye can discover the illusion; and it is
only by such deceptions that art can imitate the most pleasing
works of nature. By the help of such illusion we may see the
interesting struggles of the babbling brook, which soon after

> ———— " spreads
> Into a liquid plain, then stands unmov'd,
> Pure as the expanse of heaven."

This idea has been realized in the scenery at ADLESTROP,
where a small pool, very near the house, was supplied by a
copious spring of clear water. The cheerful glitter of this
little mirror, although on the top of the hill, gave pleasure to
those who had never considered how much it lessened the
place, by attracting the eye and preventing its range over the
lawn and falling ground beyond. This pool has now been
removed; a lively stream of water has been led through a
flower-garden, where its progress down the hill is occasionally
obstructed by ledges of rocks, and after a variety of interesting circumstances it falls into a lake at a considerable distance,
but in full view both of the mansion and the parsonage, to
each of which it makes a delightful, because a natural, feature
in the landscape.

Few persons have seen the formal cascade at THORESBY in
front of the house, and heard its solemn roar, without wishing
to retain a feature which would be one of the most interesting
scenes in nature, if it could be divested of its disgusting and
artificial formality; but this can only be effected by an equally
violent, though less apparent, interference of art; because,

without absolutely copying any particular scene in Nature,
we must endeavour to imitate the causes by which she produces her effects, and the effects will be natural.

The general cause of a natural lake, or expanse of water,
is an obstruction to the current of a stream by some ledge or
stratum of rock which it cannot penetrate; but as soon as the
water has risen to the surface of this rock, it tumbles over
with great fury, wearing itself a channel among the craggy
fragments, and generally forming an ample basin at its foot.
Such is the scenery we must attempt to imitate at THORESBY.*

Having condemned the ill-judged interference of art in
the disposition of the ground and water at THORESBY, it may,
perhaps, be objected, that I now recommend an artificial
management not less extravagant, because I presume to introduce some appearance of rock scenery in a soil where no rock
naturally exists; but the same objection might be made with
equal propriety to the introduction of an artificial lake in a
scene where no lake before existed. When under the guidance of Le Nôtre and his disciples, the taste for geometric
gardening prevailed, nature was totally banished or concealed
by the works of art. Now, in defining the shape of land or
water, we take nature for our model; and the highest perfection of landscape gardening is, to imitate nature so judiciously, that the interference of art shall never be detected.

L'Arte che tutto fa nulla se scopre.

[The art which effects everything, discovers itself nowhere.]

A rapid stream, violently agitated, is one of the most
interesting objects in nature. Yet this can seldom be enjoyed
except in a rocky country; since the more impetuous the
stream, the sooner will it be buried within its banks, unless
they are of such materials as can resist its fury. To imitate
this natural effect, therefore, in a soil like that of THORESBY,

---

* No drawing is inserted of this cascade, because the whole has been so
well executed, that the best reference is to the spot itself, which will, I trust,
long continue to prove my art " above the pencil's power to imitate."

In forming this cascade, huge masses of rock were brought from the
crags of Creswell, one, in particular, of many tons weight, with a large tree
growing in its fissures; the water has been so conducted by concealed
leaden pipes, that in some places it appears to have forced its way through
the ledges of the rocks.

we must either force the stream above its level, and deprive it of natural motion, or introduce a foundation of stones disposed in such a manner as to appear the rocky channel of the mountain stream.   The former has been already done in forming the lake, and the latter has been attempted, according to the fashion of geometric gardening, in the regular cascade ; where a great body of water was led underground from the lake to move down stairs, into a scalloped basin, between two bridges immediately in front of the house.

The violence done to Nature by the introduction of rock scenery at THORESBY is the more allowable, since it is within a short distance of Derbyshire, the most romantic county in England; while, from the awful and picturesque scenery of Creswell Crags, such strata and ledges of stone, covered with their natural vegetation, may be transported thither, that no eye can discover the fraud.

It is scarcely possible for any admirer of nature to be more enthusiastically fond of her romantic scenery than myself; but her wildest features are seldom within the common range of man's habitation.   The rugged paths of alpine regions will not be *daily* trodden by the foot of affluence, nor will the thundering cataracts of Niagara seduce the votaries of pleasure *frequently* to visit their wonders; it is only by a pleasing illusion that we can avail ourselves of those means which Nature herself furnishes, even in tame scenery, to imitate her bolder effects; and to this illusion, if well conducted, the eye of genuine taste will not refuse its assent.

" La Nature fuit les lieux frequentés, c'est au sommet des montagnes, au fond des forèts, dans les isles desertes, qu'elle etale ses charmes les plus touchants, ceux qui l'aiment et ne peuvent l'aller chercher si loin, sont reduits à lui faire violence, et à la forcer en quelque sorte à venir habiter parmi eux, et tout cela ne peut se faire sans un peu d'illusion."—J. J. ROUSSEAU.

[Nature flies from frequented places; it is on the summit of mountains, in the depths of forests, and in desert islands, that she displays her most affecting charms; those who love her, and who cannot go so far in search of her, are reduced to the necessity of constraining her, and forcing her to take up her habitation among them; but this cannot be done without a certain degree of illusion.]

––––––

One of the views from the house at THORESBY looked towards

> –––––––––– " the long line
> Deep delv'd of flat canal, and all that toil,
> Misled by tasteless fashion, could achieve,
> To mar fair Nature's lineaments divine."
>
> MASON.

As, in this instance, I shall have occasion to propose a different idea to that suggested by Mr. Brown, I must beg leave to explain the reasons on which I ground my opinion.

Amidst the numerous proofs of taste and judgment which that celebrated landscape gardener has left for our admiration, he frequently mistook the character of running water; he was too apt to check its progress, by converting a lively river into a stagnant pool, nay, he even dared to check the progress of the furious Derwent at Chatsworth, and transform it into a tame and sleepy river unworthy the majesty of that palace of the mountains. Such was his intention with respect to the stream of water which flows through THORESBY PARK; but since the lake presents a magnificent expanse of water, the river below the cascade should be restored to its natural character: *a rivulet in motion.*

––––––

At WENTWORTH, although the quantity of water is very considerable, yet it is so disposed as to be little seen from the present approach [see fig. 54], and when it is crossed in the drive on the head between two pools, the artificial management destroys much of its effect: they appear to be several distinct ponds, and not the series of lakes which nature produces in a mountainous country. But the character of this water should rather imitate one large river than several small lakes; especially as it is much easier to produce the appearance of continuity than of such vast expanse as a lake requires. The following sketch [fig. 55] is a view of the scenery presenting itself under the branches of trees, which act as a frame to the landscape [and with certain trees, shewn in fig. 54, removed].

[Fig. 54. View in the park at Wentworth, before certain trees were removed, which exclude a view of the water from the approach.]

[Fig. 55. View from the approach in the grounds at Wentworth, on the supposition that certain trees and bushes, shewn in fig. 54, are removed.]

To preserve the idea of a river, nothing is so effectual as a bridge; instead of dividing the water on each side, it always tends to lengthen its continuity, by shewing the impossibility of crossing it by any other means, provided the ends are well concealed, which is fortunately the case with respect to this water. Although the upper side of the bridge would be very little seen, because the banks are everywhere planted; yet as the bridge would not be more than fifty yards long, it would be more in character with the greatness of the place to have such a bridge as would nowhere appear a deception, and in this case the different levels of the water (being only five feet) would never be discovered.

The *rippling motion* of water is a circumstance to which improvers have seldom paid sufficient attention. They generally aim at a broad expanse and depth, not considering that a narrow shallow brook in motion, over a gravelly bottom, is not less an object of beauty and worthy of imitation; the deep dell, betwixt the boat-house and the bridge, might be rendered very interesting, by bringing a lively brook along the valley; the embouchure of this brook should be laid with gravel, to induce cattle to form themselves in groups at the edge of the water, which is one of the most pleasing circumstances of natural landscape. It sometimes happens, near large rivers, that a clear spring bubbles from a fountain, and pours its waters rapidly into the neighbouring stream; this is always considered a delightful object in nature, yet I do not recollect it has ever been imitated by art; it would be very easy to produce it in this instance, by leading water in a channel from the upper pool, and after passing underground, by tubes, for a few yards, let it suddenly burst through a bed of sand and stones, and being thus *filtered by ascent*, it would ripple along the valley till it joined the great water. Milton was aware of this contrast between the river and the rill, where he mentions, amongst the scenery of his Allegro,

" Shallow brooks and rivers wide."

As applicable to the subject of this chapter, I shall insert the following extract from the *Red Book* of HOLKHAM:

" The opposite banks in the middle part of the lake being the most beautiful ground in HOLKHAM PARK, it is a desirable object to unite them, without the long circuit which must be made by land round either end of the lake.

·" A bridge, however elegant for the sake of magnificence, or however simple for the sake of convenience, would be improper, because it would destroy the effect of the lake, and give it the character of a river, which its round and abrupt terminations render improbable. I, therefore, propose to unite these opposite shores by a ferry-boat of a novel con-

struction, so contrived as to be navigated with the greatest
safety and ease, as explained by the following sketch [fig. 56].

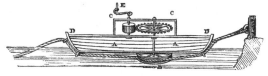

[Fig 56.]

" The ferry-boat to be a broad flat-bottomed punt, A, at
the bottom is a pulley-shaped wheel and axis, B, about a
yard in diameter, carrying a rope fastened to the two opposite
sides of the lake, which will sink to admit the passing of other
boats; this wheel is put in motion by the correspondent one
above it, which has five times as many teeth as the pinion c,
consequently, at every five turns of the winch E, the wheel
makes one revolution, and the boat advances three yards, or
three times the diameter of the wheel; at each end of the
boat the rope must pass through rings of brass, smoothly
polished, which will always guide it to one certain spot.   The
whole machinery, which is very simple, and not likely to be
out of order, may be covered by a box C C, to form a convenient seat in the centre of the ferry-boat, and the surface or
deck of this boat D may be covered with gravel and cement,
having a hand-rail on each side; thus it will in a manner
become a movable part of the gravel walk."

Where two pieces of water are at some distance from each
other, and of such different levels that they cannot easily be
made to unite in one sheet: if there be a sufficient supply to
furnish a continual stream, or only an occasional redundance
in winter, the most picturesque mode of uniting the two,
is by imitating a common process of Nature in mountainous
countries, where we often see the water, in its progress from
one lake to another, dashing among broken fragments, or
gently gliding over ledges of rock, which form the bottom of
the channel: this may be accomplished at HAREWOOD, where
the most beautiful stone is easily procured; but in disposing
the ledges of rock, they should not be laid horizontally, but

with the same slanting inclination that is observed, more or less, in the bed of the neighbouring river. A hint of such management is shewn under this bridge [see fig. 57], the design of which may serve as a specimen of architecture, neither too much nor too little ornamented for rock scenery, in the neighbourhood of a palace.

[Fig. 57.   Bridge in the grounds of Harewood Hall, shewing the artificial disposition of masses of stone in imitation of ledges of rock in the bed of the river.]

CHAPTER IV.

Of PLANTING for immediate and for future Effect—Clumps—Groups—
Masses—New Mode of planting Wastes and Commons—the browsing
Line described—Example MILTON ABBEY—Combination of Masses to
produce great Woods—Example COOMBE LODGE—Character and Shape
of Ground to be studied—Outline of New Plantations.

THE following observations on planting are not intended
to pursue the minute detail so copiously and scientifically
described in EVELYN's Sylva, and so frequently quoted, or
rather repeated from him, in modern publications; I shall
merely consider it as a relative subject : and being one of the
chief ornaments in landscape gardening when skilfully appropriated, I shall divide it into two distinct heads : the first including those single trees or groups which may be planted of
a larger size to produce *present effect;* the second comprehending those masses of plantations destined to become woods
or groves for *future generations.*

Since few of the practical followers of Mr. Brown possessed that force of genius which rendered him, according to
MASON,

——— " The living leader of thy powers,
Great Nature"———

it is no wonder that they should have occasionally copied
the means he used, without considering the effect which he
intended to produce. Thus Brown has been treated with
ridicule by the contemptuous observation, that all his improvements consisted in *belting, clumping,* and *dotting ;* but I
conceive the two latter ought rather to be considered as *cause*
and *effect,* than as two distinct ideas of improvement; for the
disagreeable and artificial appearance of young trees, when
protected by what is called a cradle fence, together with the
difficulty of making them grow thus exposed to the wind,
induced Mr. Brown to form small clumps fenced round, con-

z

taining a number of trees calculated to shelter each other, and to promote the growth of those few which might be ultimately destined to remain and form a group.

This I apprehend was the origin and intention of those clumps, and that they never were designed as ornaments in themselves, but as the most efficacious and least disgusting manner of producing single trees and groups to vary the surface of a lawn, and break its uniformity by light and shadow.

In some situations, where great masses of wood, and a large expanse of open lawn prevail, the contrast is too violent, and the mind becomes dissatisfied by the want of unity; we are never well pleased with a composition in natural landscape, unless the wood and lawn are so blended that the eye cannot trace the precise limits of either; yet it is necessary that each should preserve its original character in broad masses of light and shadow; for although a large wood may be occasionally relieved by clearing small openings to break the heaviness of the mass, or vary the formality of its outline, yet the general character of shade must not be destroyed.

In like manner the too great expanse of light on a lawn must be broken and diversified by occasional shadow, but if too many trees be introduced for this purpose, the effect becomes frittered, and the eye is offended by a deficiency of composition, or, as the painter would express it, of a *due breadth of light and shade.* Now it is obvious that, in newly formed places, such a redundance of trees will generally remain from former hedgerows, that there can seldom be occasion to increase the number of single trees, though it will often be advisable to combine them into proper groups.

It is a mistaken idea, scarcely worthy of notice, that the beauty of a group of trees consists in odd numbers, such as five, seven, or nine; a conceit which I have known to be seriously asserted. I should rather pronounce, that no group of trees can be natural in which the plants are studiously placed at equal distances, however irregular in their forms. Those pleasing combinations of trees which we admire in forest scenery, will often be found to consist of forked trees, or at least of trees placed so near each other that the branches intermix, and by a natural effort of vegetation the stems of the trees themselves are forced from that perpendicular direction,

which is always observable in trees planted at regular distances from each other. No groups will therefore appear
natural unless two or more trees are planted very near each

[Fig. 58. Artificial groups of trees, as in Park scenery, treated in the ordinary but imperfect manner.]

other,* whilst the perfection of a group consists in the combination of trees of different age, size, and character.

[Fig. 59. Natural groups of trees, as they appear in Forest scenery.]

The two sketches annexed [figs. 58 and 59] exemplify
this remark; the first represents a few young trees pro-

* To produce this effect two or more trees should sometimes be planted
in the same hole, cutting their roots so as to bring them nearer together;
and we sometimes observe great beauty in a tree and a bush thus growing
together, or even in trees of different characters, as the great oak and ash
at Welbeck, and the oak and beech in Windsor Forest. Yet it will generally be more consonant to nature if the groups be formed of the same
species of trees.

tected by cradles, and though some of them appear nearer together than others, it arises from their being seen in perspective, for I suppose them to be planted (as they usually are) at nearly equal distances. In the same landscape I have supposed the same trees grown to a considerable size, but from their equi-distance, the stems are all parallel to each other, and not like the group below, where, being planted much nearer, the trees naturally recede from each other. A few low bushes or thorns produce the kind of group in the lower sketch, consisting of trees and bushes of various growth. It may be observed that the single tree, and every part of the upper sketch, is evidently artificial, and that the lower one is natural, and like the groups in a forest.

Another source of variety may be produced by such opaque masses of spinous plants as protect themselves from cattle; thus stems of trees seen against lawn or water are comparatively dark, while those contrasted with a background of wood appear light. This difference is shewn in both these sketches: the stems of the trees *a a* appear light, and those at *b b* are dark, merely from the power of contrast, although both are exposed to the same degree of light.

Where a large tract of waste heath or common is near the boundary of a park, if it cannot be enclosed, it is usual to dot certain small patches of trees upon it, with an idea of improvement; a few clumps of miserable Scotch firs, surrounded by a mud wall, are scattered over a great plain, which the modern improver calls " clumping the common." It is thus that Hounslow Heath has been clumped; and even the vast range of country formerly the Forest of Sherwood, has submitted to this meagre kind of misnamed ornament.

It may appear unaccountable that these examples, which have not the least beauty either of nature or art to recommend them, should be so generally followed; but alteration is frequently mistaken for improvement, and two or three clumps of trees, however bad in themselves, will change the plain surface of a flat common. This I suppose has been the cause of planting some spruce firs on MAIDEN EARLY Common, which fortunately do not grow; for if they succeeded, the contrast is so violent between the wild surface of a heath, and the spruce appearance of firs, that they would be misplaced:

besides, the spiral firs are seldom beautiful, except when their lower branches sweep upon the ground, and this could never be the case with those exposed to cattle on a common.

A far better method of planting waste land, where enclosures are not permitted, has been adopted with great success in Norfolk, by my much valued friend the late Robert Marsham, Esq., of Stratton. Instead of firs surrounded by a mud bank, he placed deciduous trees of every kind, but especially birch, intermixed with thorns, crabs, and old hollies, cutting off their heads and all their branches about eight feet from the ground: these are planted in a *puddle* and the earth laid round their roots in small hillocks, which prevent the cattle from standing very near to rub them; and thus I have seen groups of trees which looked like bare poles the first year, in a very short time become beautiful ornaments to a dreary waste.

[Fig. 60. A clump, as usually seen on commons, compared with Mr. Marsham's mode of planting stumps of trees and thorns.]

This sketch [fig. 60] shews the difference between the sort of clump so often seen on a common, and that mode of plant-

[Fig 61. Stumps of trees and thorns planted in Mr. Marsham's manner, after a few years' growth.]

ing stumps of trees and thorns recommended in the foregoing page; the appearance at first is not very promising, but in a few years they will become such irregular groups and natural thickets as are here represented [fig. 61], while the formal clump of firs will for ever remain an artificial object.

Mr. Gilpin, in his Forest Scenery, has given some specimens of the outlines of a wood, one of which is not unlike that beautiful screen which bounds the park to the north of MILTON ABBEY, and which the first of the annexed sketches [figs. 62, 63, and 64,] more accurately represents. We have here a very pleasing and varied line formed by the tops of trees, but, from the distance at which they are viewed, they seem to stand on one straight base line, although many of the trees are separated from the others by a considerable distance: the upper outline of this screen is so happily varied, that the eye is not offended by the straight line at its base; but there is another line which is apt to create disgust in flat situations, and for this reason—all trees unprotected from cattle will be stripped of their foliage to a certain height, and where the surface of the ground is perfectly flat, and forms one straight line, the stems of trees thus brought to view by the browsing of cattle, will present another straight line parallel to the ground, at about six feet high, which I shall call

THE BROWSING LINE.*

Whether trees be planted near the eye or at a distance from it, and whether they be very young plants or of the greatest stature, this browsing line will always be parallel to the surface of the ground, and being just above the eye, if the heads of single trees do not rise above the outline of more distant woods, the stems will appear only like stakes of different sizes scattered about the plain; this is evidently the

* All trees exposed to cattle are liable to this browsing line, although thorns, crabs, and other prickly plants, will sometimes defend themselves: the alder, from the bitterness of its leaves, is also an exception; but where sheep only are admitted, the line will be so much below the eye, that it produces a different effect, of which great advantage may sometimes be taken, especially in flat situations.

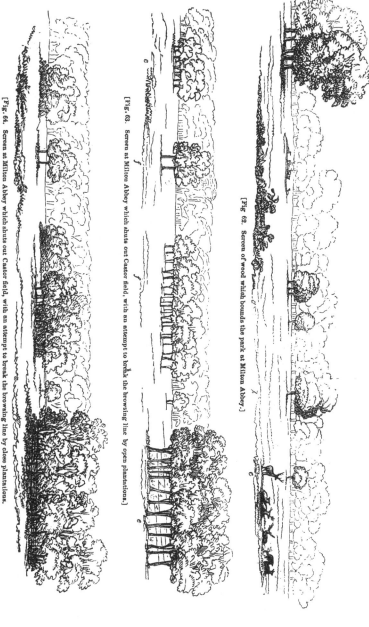

[Fig. 62. Screen of wood which bounds the park at Milton Abbey.]

[Fig. 63. Screen at Milton Abbey which abuts out Castor field, with an attempt to break the browsing line by open plantations.]

[Fig. 64. Screen at Milton Abbey which abuts out Castor field, with an attempt to break the browsing line by close plantations.]

effect of those single thorns or trees in the sketch [fig. 62] marked *a, b, c.*

In the sketches [figs. 63 and 64], I have represented a view of that long screen at MILTON ABBEY, which shuts out Castor field, and which is certainly not a pleasing feature, from its presenting not only a straight line at the bottom, but the trees being all of the same age; the top outline is also straight. This screen forms the background of a view taken from the approach, and [figs. 63 and 64] represents the difference between an attempt to break the uniformity of the plain by open or by close plantations.

The trees of this screen are of such a height, that we can hardly expect, in the life of man, to break the upper outline by any young trees, except they are planted very near the eye, as at *e* [in fig. 63]; because those planted at *f* or *g* [in the same figure] will, by the laws of perspective, sink beneath the outline of the screen; it is therefore not in our power to vary the upper line, and if the plantations be open, the browsing line will make a disagreeable parallel with the even surface of the ground; this can only be remedied by preventing cattle from browsing the underwood, which should always be encouraged in such situations; thus, although we cannot vary the upper line of this screen, we may give such variety to its base as will, in some measure, counteract the flatness of its appearance.

[Fig. 65. Trees of different ages, characters, and heights, but each having the browsing line formed by cattle, of the same height.]

The browsing line being always at nearly the same distance of about six feet from the ground, it acts as a *scale,* by which the eye measures the comparative height of trees at any distance; for this reason the importance of a large tree may be injured by cutting the lower branches above this usual standard. It is obvious that the foregoing trees [see fig. 65] are of different ages, characters, and heights, yet the browsing line is the same in all, and furnishes a natural scale by which we at once decide on their relative heights at various distances.

Let us suppose the same trees pruned or trimmed by man, [as in fig. 66], and not by cattle, and this scale will be destroyed: thus, a full grown oak may be made to look like an

[Fig. 66. Trees of different ages, characters, and heights, pruned by man, in which the scale formed by the browsing line is destroyed.]

orchard-tree, or by encouraging the under branches to grow lower than the usual standard, a thorn or a crab-tree may be mistaken for an oak, at a distance.

The last tree in the foregoing example [fig. 66], is supposed to be one of those tall elms which, in particular counties, so much disfigure the landscape ; it is here introduced for the sake of the following remark. I am sorry to have observed, that when trees have long been used to this unsightly mode of pruning, it is difficult, or indeed impossible, to restore their natural shapes, because if the lower branches be suffered to grow, the tops will also decay; and therefore they must either be continued tall by occasionally cutting off the lateral branches, or they must be converted into pollards by cutting off their tops.

Single trees, or open groups, are objects of great beauty when scattered on the side of a steep hill, because they may be made to mark the degree of its declivity, and the shadows of the trees are very conspicuous ; but on a plain the shadows are little seen, and therefore single trees are of less use.

———

I am now to speak of plantations for future, rather than for immediate effect, and instead of mentioning large tracts of land which have been planted under my directions, where a naked, or a barren country, has been clothed without difficulty or contrivance, I shall rather instance a subject requiring peculiar management, especially as, from its vicinity to a high road, I cannot perhaps produce a better example than the following extract furnishes:—

"COOMBE LODGE, seen from the turnpike-road, does not at present give a favourable impression; for though the view from the house, consisting of the opposite banks of Basildon, is richly wooded, the place itself is naked; and it is difficult to remove this objection without sacrificing more land to the purposes of beauty than would be advisable, or even justifiable.

Both the situation and the outline of the house at COOMBE LODGE have been determined with judgment: the situation derives great advantage from its southern aspect, and from the views which it commands; and the house derives importance from its extended front.   Both these circumstances, however, contribute to the bad opinion conceived of the place when viewed from the road, which is the point from whence its defects are most apparent.

The front towards the road faces the south, and is therefore lighted by the sun during the greatest part of the day; but being backed by lawn and arable land, and not relieved by wood, the effect of sunshine is equally strong on the back ground as on the house, because there is not a sufficient opposition of colour to separate these different objects; but if, on the contrary, the house be opposed to wood, it will then appear light and conspicuous, the attention being principally directed to the mansion, while the other parts of the scene will be duly subordinate.

It is also proper that the grounds should accord with the size and style of the place, and that the mansion be surrounded by its appropriate appendages.   At present the character of the house, and that of the place, are at variance: the latter is that of a farm, but the character of the house is that of a gentleman's residence, which should be surrounded by pleasure-grounds, wood, and lawn; and although great credit is due to those gentlemen who patronise farming by their example, as well as by their influence, it would be a reflection on the good taste of the country to suppose that the habitation of the gentleman ought not to be distinguished from that of the farmer, as well in the character of the place as by the size of the house.

I shall not on this occasion enter into a discussion of the difference between a scene in nature, and a landscape on the painter's canvas; nor consider the very different means by which the painter and the landscape gardener produce the same effect: I shall merely endeavour to shew how far the same principles would direct the professors of either art in the improvement of COOMBE LODGE, and more particularly in the form and character of the wood to the north of the house.

Breadth, which is one of the first principles of painting, would prompt the necessity of planting the whole of the hill behind the house; but the improver, who embellishes the scene for the purposes of general utility and real life, must adopt what is convenient as well as beautiful. The painter, when he studies the perfection of his art, forms a correct picture, and takes beauty for his guide. The improver consults the genius of the scene, and connects beauty with those useful supporters, economy and convenience; and as COOMBE LODGE would not be relieved by one large wood without a great sacrifice of land, the effect must be produced by planting a part only, whilst the judgment must be influenced by two principles belonging to the sister art, *breadth* and *intricacy.*

Breadth directs the necessity of large masses, or continued lines of plantation, whilst intricacy suggests the shape and direction of the glades of lawn, and teaches how to place loose groups of trees, and separate masses of brushwood, where the outline might otherwise appear hard; and, by occasional interruptions to the flowing lines of grass, with suitable recesses and projections of wood, intricacy contrives to 'lead the eye a wanton chase,' producing variety without fritter, and continuity without sameness.

There is another principle to guide the improver in planting the hill in question, which may be derived from the art of painting, and belongs to perspective. It is evident, that if the whole bank were planted, its effect would be good from every point of view: it is no less evident, that where it is necessary to regard economy in planting, and, as in the present instance, to produce the effect of clothing by several lines of wood, instead of one great mass; that effect from some points of sight may be good, from some indifferent, and from others bad; it is therefore necessary to consider how those lines of plantation, which produce a good effect from the house, will appear in perspective from different heights and from different situations, and this question has been determined by various circumstances of the place itself.

This subject was elucidated by as many drawings as there were stations described; but as most of them were taken from the public road between Reading and Wallingford, the effect of these plantations will be seen from thence; and I have availed myself, as much as possible, of those examples which, from their proximity to a public road, are most likely to be generally observed."

If the more common appearances in nature were objects of our imitation, we should certainly plant the valleys and not the hills, since nature generally adopts this rule in her spontaneous plantations; but it is " la belle nature," or those occasional effects of extraordinary beauty, which nature furnishes as models to the landscape gardener. And although a wood on the summit of a bleak hill may not be so profitable, or grow so fast, as one in the sheltered valley, yet its advantages will be strongly felt on the surrounding soil. The verdure will be improved when defended from winds, and fertilized by the successive fall of leaves, whilst the cattle will more readily frequent the hills when they are sheltered and protected by sufficient screens of plantation. *

In recommending that the hills should be planted, I do not mean that the summits should be covered by a patch or clump; the woods of the valleys should, on the contrary, seem to climb the hills by such connecting lines as may neither appear meagre nor artificial, but, following the natural shapes of the ground, produce an apparent continuity of wood falling down the hills in various directions.

---

* This remark is verified at ASTON, where it is found that more cattle are fed in the park from the improved *quality* of the pasture, since the *quantity* has been reduced by the ample plantations made within the last ten years.

——— " Rich the robe,
And ample let it flow, that Nature wears
On her thron'd eminence! where'er she takes
Her horizontal march, pursue her step
With sweeping train of forest; hill to hill
Unite, with prodigality of shade."

MASON.

During the first few years of large plantations in a naked country, the outline, however graceful, will appear hard and artificial; but when the trees begin to require thinning, a few single trees or groups may be brought forward.   The precise period at which this may be advisable must depend on the nature of the soil: but so rich is the ground in which plantations were made at ASTON, about ten years since, that this management has already been adopted with effect.   Although it will again be repeated in the chapter treating of fences, I must observe in this place, that, instead of protecting large plantations with hedges and ditches, I have generally recommended a temporary fence of posts and rails, or hurdles on the outside, and either advise a hedge of thorns to be planted at eight or ten yards distance from the outline, or rather that the whole plantation be so filled with thorns and spinous plants, that the cattle may not penetrate far when the temporary fences shall be removed, and thus may be formed that beautiful and irregular outline so much admired in the woods and thickets of a forest.

CHAPTER V.

Woods.—Whateley's Remarks exemplified at SHARDELOES.—Intricacy—
Variety—A Drive at BULSTRODE traced, with Reasons for its Course—
Further example from HEATHFIELD PARK—A Belt—On thinning Woods
—Leaving Groups—Opening a Lawn in great Woods—Example
CASHIOBURY.

" OBSERVATIONS on Modern Gardening," by the late Mr.
Whateley, contain some remarks peculiarly applicable to the
improvement of woods, and so clearly expressive of my own
sentiments, that I beg to introduce the ample quotation
inserted in the note, * especially as the annexed drawings
[figs. 67 and 68] convey specimens of these rules, which require but little further elucidation.

* " The *outline* of a wood may sometimes be great, and always be beautiful ; the first requisite is irregularity.   That a mixture of trees and underwood should form a long straight line, can never be natural, and a succession
of easy sweeps and gentle rounds, each a portion of a greater or less circle,
composing altogether a line literally serpentine, is, if possible, worse : it is
but a number of regularities put together in a disorderly manner, and equally
distant from the beautiful, both of art and of nature.   The true beauty of an
outline consists more in breaks, than in sweeps ; rather in angles, than
rounds ; in variety, not in succession.
  " The outline of a wood is a continued line, and small variations do not
save it from the insipidity of sameness ; one deep recess, one bold prominence,
has more effect than twenty little irregularities : that one divides the line
into parts, but no breach is thereby made in its unity ; a continuation of
wood always remains, the form of it only is altered, and the extent is
increased : the eye, which hurries to the extremity of whatever is uniform,
delights to trace a varied line through all its intricacies, to pause from stage
to stage, and to lengthen the progess.
  " The parts must not, however, on that account, be multiplied till they
are too minute to be interesting, and so numerous as to create confusion.
A few large parts should be strongly distinguished in their forms, their
directions, and their situations ; each of these may afterwards be decorated
with subordinate varieties, and the mere growth of the plants will occasion
some irregularity ; on many occasions more will not be required.
  " Every variety in the outline of a wood must be a *prominence* or a
recess ; breadth in either is not so important as length to the one, and depth
to the other ; if the former ends in an angle, or the latter diminishes to a
point, they have more force than a shallow dent or a dwarf excrescence, how
wide soever : they are greater deviations from the continued line which they
are intended to break, and their effect is to enlarge the wood itself.
  " An inlet into a wood seems to have been cut, if the opposite points of
the entrance tally, and that shew of art depreciates its merit : but a differ-

The beech-woods in Buckinghamshire derive more beauty from the unequal and varied surface of the ground, on which they are planted, than from the surface of the woods themselves; because they have generally more the appearance of copses, than of woods : and as few of the trees are suffered to arrive to great size, there is a deficiency of that venerable dignity which a grove always ought to possess.

These woods are evidently considered rather as objects of profit, than of picturesque beauty; and it is a circumstance to be regretted, that pecuniary advantage and ornament are seldom strictly compatible with each other. The underwood cannot be protected from cattle, without fences, and if the fence be a live hedge, the trees lose half their beauty, while they appear confined within the unsightly boundary. To remedy this defect, the quick-fence at SHARDELOES has, in many places, been removed, and a rail placed at a little distance within the wood; but the distance is so small, that the original outline is nearly as distinct as if the fence were still visible, and the regular undulations of those lines give an artificial appearance to the whole scenery.

A painter's landscape depends upon his management of *light* and *shade :* if these be too smoothly blended with each other, the picture *wants force ;* if too violently contrasted, it is called *hard.* The light and shade of natural landscape require no less to be studied than that of painting. The shade of a landscape gardener is wood, and his lights proceed either from a lawn, from water, or from buildings. If, on the lawn, too many single trees be scattered, the effect becomes *frittered,* broken, and diffuse ; on the contrary, if the general surface of the lawn be too naked, and the outline of the woods form an

ence only in the situation of those points, by bringing one more forward than the other, prevents the appearance, though their forms be similar.

"Other points which distinguish the great parts, should, in general, be strongly marked ; a short turn has more spirit in it than a tedious circuity ; and a line, broken by angles, has a precision and firmness which in an undulated line are wanting : the angles should, indeed, be a little softened ; the rotundity of the plant, which forms them, is sometimes sufficient for that purpose ; but if they are mellowed down too much they lose all meaning.

" Every variety of outline, hitherto mentioned, may be traced by the *underwood* alone ; but, frequently, the same effects may be produced with more ease, and much more beauty, by a *few trees* standing out from the thicket, and belonging, or seeming to belong, to the wood, so as to make a part of its figure."

uniform heavy boundary, between the lawn and the horizon,
the eye of taste will discover an unpleasing harshness in the
composition, which no degree of beauty, either in the shape
of the ground, or in the outline of the woods, can entirely
counteract.   In this state, the natural landscape, like an unfinished picture, will appear to want the last touches of the
master : this would be remedied on the canvas, in proportion
as the picture became more highly finished ; but, on the
ground, it can only be effected by taking away many trees
in the front of the wood, leaving some few individually and
more distinctly separated from the rest : this will give the
finishing touches to the outline, where no other defect is
apparent.

The eye, or rather the mind, is never long delighted with
that which it surveys without effort, at a single glance, and
therefore sees without exciting curiosity or interest.   It is not
the vast extent of lawn, the great expanse of water, or the long
range of wood, that yields satisfaction ; for these, if shapeless,
or, which is the same thing, if their exact shape, however
large, be too apparent, only attract our notice by the space
they occupy, " to fill that space with objects of beauty, to
" delight the eye after it has been struck, to fix the attention
" where it has been caught, to prolong astonishment into
" admiration, are purposes not unworthy of the greatest
" designs."

This can only be effected by *intricacy*, the due medium
between uniformity on the one hand, and confusion on the
other ; which is produced by throwing obstacles in the way to
amuse the eye, and to retard that celerity of vision, so natural,
where no impediments occur to break the uniformity of objects.   Yet while the hasty progress of the eye is checked, it
ought not to be arrested too abruptly.   The mind requires a
continuity, though not a sameness; and, while it is pleased
with succession and variety, it is offended by sudden contrast,
which destroys the unity of composition.

There is a small clump at *b* [fig. 67], which is of great use
in breaking the outline of the wood beyond it ; and there is
a dell or scar in the ground at *c*, that may also be planted for
the like purpose.   It is a very common expedient to mend an
outline, by adding new plantation in the front of an old one :

but, although the improver may plant large woods, with a view to future ages, yet something appears due to the present day. If by cutting down a few trees, in the front of a large wood,

[Fig. 67. View from the house at Shardeloes, shewing the beech woods before they were broken and varied]

the shape of its outline may immediately be improved in a better manner than can be expected from a solitary clump a century hence, it is surely a more rational system of improvement, than so long to endure a patch, surrounded by an unsightly fence, in the distant hope of effects which the life of man is too short to realize.

There is a part of the wood at d so narrow as to admit the light between the stems of the trees; this naturally suggests the idea of adding new plantation. But the horizon is already uniformly bounded by wood, and the mind is apt to affix the idea of such boundary being the limit of the park, as strongly as if the pale itself were visible; on the contrary, the ground falling beyond this part, and a range of wood sweeping over the brow of the hill, it is better to clear away some of the trees, to increase the apparent extent of lawn. Instead of destroying the continuity of wood, this will increase its quantity; because the tops of the trees being partly seen over the opening, the imagination will extend the lawn beyond its actual boundary, and represent it as surrounded by the same chain of woods.

I have often heard it asserted, as a general maxim in gardening, that hills should be planted, and valleys cleared of wood. This idea, perhaps, originated, and ought only to be implicitly followed in a flat or tame country, where the hills are so low as to require greater height by planting, and the valleys so shallow, that trees would hide the neighbouring

2 B

hills: but, whenever the hills are sufficiently bold to admit of ground being seen, between large trees in the valley, and those on the brow of the hill, it marks so decided a degree of elevation, that it ought sedulously to be preserved. Instead, therefore, of removing the trees in the valley, at *e*, I should prefer shewing more of the lawn above them, by clearing away some of the wood on the knoll at *f*, which I have distinguished by the pavilion shewn in [fig. 68]: such a building would have many

[Fig. 68. View from the house at Shardeloes, on the supposition that certain proposed improvements have been carried into execution.]

uses, besides acting as an ornament to the scenery, which seems to require some *artificial* objects to appropriate the woods to the magnificence of place; because wood and lawn may be considered as the *natural* features of Buckinghamshire.

The *Red Book* of SHARDELOES contains a minute description of the rides made in the woods, with the reasons for every part of their course; but as this subject is more amply treated in my remarks on BULSTRODE, the following extract is accompanied with a map, on which the course of an extensive drive is minutely described. This park must be acknowledged one of the most beautiful in England, yet I doubt whether Claude himself could find, in its whole extent, a single station from whence a picture could be formed. I mention this as a proof of the little affinity between pictures and scenes in nature.

It is not uncommon to conduct a drive either round a park, or into the adjoining woods, without any other consideration than its length; and I have frequently been carried through a belt of plantation, surrounding a place, without one remarkable object to call the attention from the trees, which are every where mixed in the same unvaried manner.

Although the verdure, the smoothness of the surface, and
nature of the soil at Bulstrode, are such as to make every part
of the park pleasant to drive over; yet there is a propriety in
marking certain lines of communication, which may lead from
one interesting spot to another; and though a road of approach
to a house ought not to be circuitous, the drive is necessarily
so; yet this should be under some restraint. By the assistance of the map [fig. 69], I shall describe the course of the
drive at Bulstrode; and, however devious it may appear on
paper, it will, I trust, be found to possess such a variety as
few drives can boast; and that no part of it is suggested without sufficient reasons for its course.

I would not here be understood to infer, that every park
can boast those advantages which Bulstrode possesses, or
that every place offers sufficient extent and variety for such a
drive appropriated to pleasure only; but this is introduced as
an archetype or example, from whence certain principles are
reduced to practice. Some of my observations, in the course
of this description, may appear to have been anticipated by
Mr. Whateley, and if I may occasionally deliver them as my
own sentiments, I hope the coincidence in opinion with so
respectable a theorist, will not subject me to the imputation
of plagiarism.

COURSE OF THE DRIVE AT BULSTRODE.

Taking the departure from the house, along the valley, towards the north, it passes the situation proposed for a cottage
at [fig. 69] No 1, from thence ascends to the summit of the
chalk cliff, that overhangs the dell at No. 2, and making a
sharp turn at No. 3, to descend with ease, it crosses the head
of the valley, and enters the rough broken ground, which is
curious for the variety of plants at No. 4.

From the several points, Nos. 1, 2, and 3, the view
along the great valley is nearly the same, but seen under
various circumstances of foreground: at No. 4 it crosses the
approach from London, and passes through an open grove,
No. 5.

The drive now sweeps round on the knoll at No. 6, along
a natural terrace, from which the opposite hill and the house
appear to great advantage. From hence, crossing the valley,

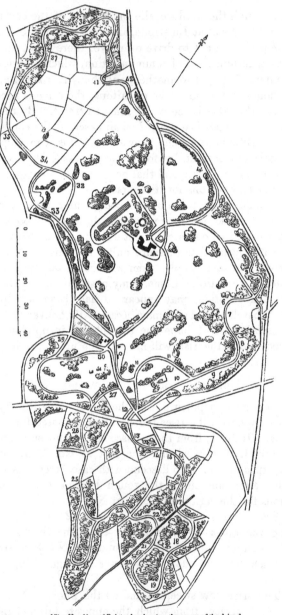

[Fig. 69. Map of Bulstrode, shewing the course of the drive.]

A. The old Entrance.
B. The Court in which the
    Entrance is to be.
C. The American Garden.

D. The Flower Garden.
E. The Nursery for Flowers,
F. The Great West Terrace
G. The South Terrace and Pheasantry.

No. 7, among the finest trees in the park, it passes a deep romantic dell at No. 8, which might be enlivened by water, as a
drinking pool for the deer, and then, as it will pass at No. 9,
near the side of the Roman camp, I think the drive * should be
made on one of the banks of the Vallum; because it is a circumstance of antiquity worthy to be drawn into notice; and,
by being elevated above the plain, we shall not only see into
the intrenchment, but remark the venerable trees which enrich
its banks; these trees are the growth of many centuries, yet
they lead the mind back to the far more ancient date of this
encampment, when the ground must have been a naked surface.   Another advantage will also be derived from carrying
the drive above the level of the plain.   *The eye being raised
above the browsing* † *line, the park wall will be better hid by
the lower branches of intermediate trees.*   At No. 10 the drive
is less interesting, because the surface is flat; *but such occasional tameness gives repose,* ‡ and serves to heighten the
interest of subsequent scenery; yet at this place, if the drive
be made to branch along the Vallum, it will pass over the most
beautiful part of the park, on a natural terrace, at No. 11,
and this will join the inner drive, returning down the valley
towards the kitchen-garden.

I am now to speak of the great woods called Fentum's,
Piper's, Column's, Walk Wood, and Shipman's, in which a
serpentine drive has been formerly cut, which no one would
desire to pass a second time, from its length, added to the
total absence of interest or variety of objects; but following
the taste which supposes "Nature to abhor a straight line,"
this drive meanders in uniform curves of equal lengths, and the
defect is increased by there being only one connexion with the
park, while the other end of the drive finishes at a great distance across Fulmer Common.   The first object, therefore, of
improvement will be to form such a line of connexion with the
park as may make it seem a part of the same domain, and this
would be more easily done if the hollow wayroad under the

* This great work being in a progressive state, the reader will observe
that some parts of this drive are mentioned as not yet completed.
† The browsing line is explained in Chapter IV. page 174.
‡ The excess of variety may become painful, and therefore, in a long
drive, some parts should be less interesting, or, if possible, should excite no
interest, and be indifferent without exciting disgust.

park wall could be removed; because otherwise the drive
must cross the road twice at No. 12, as I suppose it to enter
a field at No. 13, which might be planted to connect it with
the Broomfield copse, No. 14, from whence, after crossing
several interesting small enclosures, with forest-like borders,
it enters and sweeps through the wood, Little Fentums,
No. 16, to join the old drive, or at least such parts of it
as can be made subservient to a more interesting line.   After
crossing a valley and streamlet at No. 17, and another
at No. 18, it should ascend the hill of Piper's Wood,
in which there are at present no drives, and at No. 19, a
branch may lead on to the common, as a green way to London.
The drive sweeping round to No. 20, opens on a view of the
village and valley of Fulmer, with a series of small ponds,
which, in this point of view, appear to be one large and beautiful piece of water: this scene may be considered the most
pleasing subject for a picture, during the whole course of the
drive.   This would be a proper place for a covered seat, with
a shed behind it for horses or open carriages; * but it should
be set so far back as to command the view under the branches
of trees, which are very happily situated for the purpose at
No. 20.

From hence the drive descends the hill, in one bold line,
No. 21, with a view towards the opposite wood across the
valley.   Having again ascended the hill, in wood, there are
some parts of the present drive which might be made interesting by various expedients.   At No. 22, one side of the drive
might be opened to shew the opposite hanging wood in glades
along the course of the drive.   At No. 23, a shorter branch
might be made to avoid the too great detour, though there is
a view into the valley of Fulmer, at No. 24, worthy to be preserved. †   In some parts *the width of the drive might be varied,*

---

* In long drives such attention to convenience is advisable; a thatched
hovel of Doric proportions, may not only be made an ornament to the
scenery, but it will often serve for a shelter from sudden storms in our uncertain climate; for this reason it should be large enough to contain several
open carriages.

† I have distinguished, by Italics, some peculiar circumstances of
variety, from having observed great sameness in the usual mode of conduct-
ing a drive through a belt of young plantation, where trees of every species
are mixed together.   There is actually more variety in passing from a grove

and some of the violent curvatures corrected; in others, *the best trees might be singled out,* and *little openings made, to be fed* by sheep occasionally; and another mode of producing variety would be *to take away certain trees,* and *leave others, where any particular species abound :* thus, in some places, *the birches only might be left,* and all the oaks and beech and other plants removed, to make, in time, a specimen of Birkland forest, while there are some places where the *holly and hawthorn might be encouraged,* and all taller growth give place to these low shrubs, with irregular shapes of grass flowing among them. This would create a degree of variety that it is needless to enlarge upon.

The course of the drive through Shipman's Wood, No. 26, may be brought lower down the hill, to keep the two lines as far distant from each other as possible, and also to make the line easier round the knoll at No. 28, though an intermediate or shorter branch may also diverge, at No. 27, towards the valley. There is some difficulty in joining this drive with the park without going round the gardener's house ; but as the kitchen-garden must be seen from this part of the drive, and as it forms a leading feature in the establishment of Bulstrode, it will sometimes become part of the circuit to walk through it, and the carriages may enter the drive again at No. 31 ; I have, therefore, described two ways, No. 29, and No. 30, as I suppose the bottom of this valley to be an orchard, through which the drive may pass, or make the shorter line along the garden wall, to No. 31.

The course, along the valley, is extremely interesting; and, as some consider the farm-yard and premises a part of the beauty, as well as the comfort, of a residence in the country, I have supposed one branch of the drive, No. 32, to pass near a large tree, and the other to go on the bank at No. 33, and cross the corner of Hedgerley green, which I suppose might be planted round the gravel pit ; but *when the drive enters the farm enclosures, it ought, if possible, to follow the course of the hedges, and not to cross a field diagonally.* From No. 34 to No. 35, is perfectly flat, and follows the line of the hedges to

of oaks to a grove of firs, or a scene of brushwood, than in passing through a wood composed of a hundred different species of trees as they are usually mixed together.

the corner at No. 35, where a new scene presents itself, *viz.* a view towards the village of Hedgerley, in a valley, surrounded with woody banks. The drive now skirts along the hedge, and passes, at No. 36, a farm-house, which might be opened to the field, and then enters Wapsey's Wood, in which the first bold feature will present itself at No. 37, where the drive may come so near the edge as to shew the view along the valley, and the amphitheatre of wood surrounding these small enclosures: it then passes through the wood to a very large oak, at No. 38, which may be brought into notice by letting the drive go on each side of it, and afterwards, following the shape of the ground, it sweeps round the knoll at No. 39, with a rich view of the opposite bank, across the high road, seen under large trees; it then ascends the hill by the side of a deep dell at No. 40, and makes a double at No. 41, to cross the valley, that it may skirt round the knoll on the furze hill at No. 42, from whence it descends into the valley at No. 43, and either returns to the house, by the approach from Oxford, or is continued under the double line of elms at No. 44, to ascend by the valley from whence the drive began.

To some persons this description may appear tedious; to others it will, perhaps, furnish amusement, to trace the course of such a line on the map; but I have purposely distinguished, by Italics, some observations, containing principles which have not before been reduced to practical improvement.

———

HEATHFIELD PARK is one of those subjects from whence my art can derive little credit: the world is too apt to mistake *alteration* for *improvement*, and to applaud every *change*, although no higher beauty is produced. The character of this park is strictly in harmony with its situation; both are splendid and magnificent; yet a degree of elegance and beauty prevails, which is rarely to be found where greatness of character and loftiness of situation are predominant: because magnificence is not always united with convenience, nor extent of prospects with interesting and beautiful scenery. The power of art can have but little influence in increasing the natural advantages of Heathfield Park. It is the duty of the improver to avail himself of those beauties

which nature has profusely scattered, and by leading the
stranger to the most pleasing stations, to call his attention to
those objects which, from their variety, novelty, contrast, or
combination, are most likely to interest and delight the mind.
On this foundation ought to be built the future improvement
of Heathfield Park; not by doing violence to its native genius,
but by sedulously studying its true character and situation:
certain roads, walks, or drives, may collect the scattered
beauties of the place, and connect them with each other in
lines, easy, natural, and graceful.

A common error, by which modern improvers are apt to
be misled, arises from the mistake so often made by adopting
extent for *beauty.* Thus the longest circuit is frequently preferred to that which is most interesting; not indeed by the
visitors, but by the fancied improver of a place. This, I
apprehend, was the origin, and is always the tedious effect, of
what is called a *Belt;* through which the stranger is conducted,
that he may enjoy the drive, not by any striking points of
view or variety of scenery, but by the number of miles over
which he has traced its course, and instead of leading to those
objects which are most worthy our attention, it is too common to find the drive a mere track round the utmost verge of
the park; and if any pleasing features excite our notice, they
arise rather from chance than design.

To avoid this popular error, therefore, I shall endeavour to
avail myself of natural beauties in this drive, without any unnecessary circuit calculated to surprise by its extent. I shall
rather select those points of view which are best contrasted
with each other, or which discover new features, or the same
under different circumstances of foreground; beguiling the
length of the way by a succession of new and pleasing objects.

If the circuitous drive round a place becomes tedious by
its *monotony,* we must equally avoid too great sameness or
confinement in any road which is to be made a path of pleasure: a short branch from the principal drive, although it
meets it again at a little distance, relieves the mind by its
variety, and stimulates by a choice between two different objects; but we must cautiously avoid confusion, lest we cut a
wood into a labyrinth. The principal road at Heathfield
leads towards the tower, the other is no less interesting where

2 c

it bursts out on one of those magnificent landscapes so pleasing in nature, yet so difficult to be represented in painting;
because quantity and variety are apt to destroy that unity of
composition which is expected in an artificial landscape: for
it is hardly possible to convey an adequate and distinct idea
of those numerous objects so wonderfully combined in this
extensive view; the house, the church, the lawns, the woods,
the bold promontory of Beachy Head, and the distant plains
bounded by the sea, are all collected in one splendid picture,
without being crowded into confusion.

This view is a perfect *landscape*, while that from the tower
is rather a *prospect;* it is of such a nature as not to be well
represented by painting; because its excellence depends upon
a state of the atmosphere which is very hostile to the painter's
art. An extensive prospect is most admired when the distant
objects are most clear and distinct; but the painter can represent his distances only by a certain haziness and indistinctness,
which is termed aërial perspective. I cannot dismiss this subject without expressing the pleasure which was excited in my
mind, on finding a lofty tower erected by the present possessor, and consecrated as a tribute of respect and gratitude
to that gallant Commander, for his public services, who
derived his title of HEATHFIELD from this domain, and his
military glory from the rock of Gibraltar. Over the door is
inscribed in large letters, made from the metal of the gun-
boats destroyed,

CALPES DEFENSORI.

[TO THE DEFENDER OF GIBRALTAR.]

In the woodland counties, such as Hertfordshire, Herefordshire, Hampshire, &c. it often happens that the most
beautiful places may rather be formed by *falling*, than by
planting trees; but the effect will be very different, whether
the axe be committed to the hand of genius or the power of
avarice. The land-steward, or the timber-merchant, would
mark those trees which have acquired their full growth and
are fit for immediate use, or separate those which he deems
to stand too near together, but the man of science and of taste

will search with scrutinizing care for groups and combinations, such as his memory recalls in the pictures of the best masters ; these groups he will studiously leave in such places as will best display their varied or combined forms : he will also dis-cover beauties in a tree which the others would condemn for its decay; he will rejoice when he finds two trees whose stems have long grown so near each other that their branches are become interwoven; he will examine the outline formed by the combined foliage of many trees thus collected in groups, and removing others near them, he will give ample space for their picturesque effect : sometimes he will discover an aged thorn or maple at the foot of a venerable oak, these he will respect, not only for their antiquity, being perhaps coeval with the father of the forest, but knowing that the importance of the oak is comparatively increased by the neighbouring situ-ation of these subordinate objects; this will sometimes hap-pen when young trees grow near old ones, as when a light airy ash appears to rise from the same root with an oak or an elm.   These are all circumstances dependent on the sportive accidents of nature, but even where art has interfered, where the long and formal line of a majestic avenue shall be sub-mitted to his decision, the man of taste will pause, and not always break their venerable ranks, for his hand is not guided by the levelling principles or sudden innovations of modern fashion; he will reverence the glory of former ages, while he cherishes and admires the ornament of the present, nor will he neglect to foster and protect the tender sapling, which promises, with improving beauty, to spread a grateful shade for future " tenants of the soil."

To give, however, such general rules for thinning woods as might be understood by those who have never attentively and scientifically considered the subject, would be like attempting to direct a man who had never used a pencil, to imitate the groups of a Claude or a Poussin.*

On this head I have frequently found my instructions

* It is in the act of removing trees and thinning woods that the land-scape gardener must shew his intimate knowledge of pleasing combinations, his genius for painting, and his acute preception of the principles of an art which transfers the imitative, though permanent beauties of a picture, to the purposes of elegant and comfortable habitation, the ever-varying effects of light and shade, and the inimitable circumstances of a natural landscape.

opposed, and my reasons unintelligible to those who look at a
wood, as an object of gain; and for this reason I am not sorry
to have discovered some arguments in favour of my system, of
more weight, perhaps, than those which relate to mere taste
and beauty: these I shall beg leave to mention, not as the
foundation on which my opinion is built, but as collateral
props to satisfy those who require such support.

1st. When two or more trees have long grown very near
each other, the branches form themselves into one mass, or
head; and if any part be removed, the remaining trees will
be more exposed to the power of the wind, by being heavier
on one side, having lost their balance. 2nd. If trees have long
grown very near together, it will be impossible to take up the
roots of one without injuring those of another: and, lastly,
although trees at equal distances may grow more erect, and
furnish planks for the use of the navy, yet not less valuable to
the ship-builder are those naturally crooked branches, or knees,
which support the decks, or form the ribs, and which are
always most likely to be produced, from the outside trees of
woods, or the fantastic forms which arise from two or more
trees having grown very near each other in the same wood, or
in hedgerows.

It is therefore not inconsistent with the considerations of
profit, as well as picturesque effect, to plant or to leave trees
very near each other, and not to thin them in the usual manner without caution.

In some places belonging to ancient noble families, it is not
uncommon to see woods of vast extent intersected by vistas
and glades in many directions; this is particularly the case at
BURLEY and at CASHIOBURY. It is the property of a straight
glade or vista to lead the eye to the extremity of a wood, without attracting the attention to its depth.

I have occasionally been required to fell great quantities of
timber, from other motives than merely to improve the landscape; and in some instances this work of necessity has produced the most fortunate improvements. I do not hesitate to
say, that some woods might be increased five-fold in apparent
quantity, by taking away a prodigious number of trees, which

are really lost to view; but unless such necessity existed, there is more difficulty and temerity in suggesting improvement by cutting down, however profitable, and however suddenly the effect is produced, than by planting, though the latter be tedious and expensive.

I have seldom found great opposition to my hints for planting, but to cutting down trees innumerable obstacles present themselves; as if, unmindful of their value, and heedless of their slow growth, I should advise a *military abbatis*, or one general sweep, denuding the face of a whole country. What I should advise both at BURLEY and at CASHIOBURY,* would be to open some large areas within the woods, to produce a spacious internal lawn of intricate shape and irregular surface, preserving a sufficient number of detached trees or groups, to continue the general effect of one great mass of wood.

* This advice has been followed at Cashiobury since the above pages were written, and the effect is all that I had promised to myself.

CHAPTER VI.

Of Fences—The Boundary—The Separation—Example from SHEFFIELD
PLACE—Fence to Plantations only temporary—The double Gate—
LINES of Fences—of Roads—of Walks—of Rivers—all different.

THAT the boundary-fence of a place should be concealed
from the house, is among the few general principles admitted
in modern gardening; but even in this instance, want of precision has led to error; the necessary distinction is seldom
made between the fence which encloses a park, and those
fences which are adapted to separate and protect the subdivisions within such enclosure. For the concealment of the
boundary various methods have been adopted, on which I
shall make some observations.

1. A plantation is certainly the best expedient for hiding
the pales; but in some cases it will also hide more than is required. And in all cases, if a plantation surround a place in
the manner commonly practised under the name of a *belt*, it
becomes a boundary scarce less offensive than the pale itself.
The mind feels a certain disgust under a sense of confinement
in any situation, however beautiful; as Dr. Johnson has forcibly illustrated, in describing the feeling of Rasselas, in the
happy valley of Abyssinia.

2. A second method of concealing a fence is, by making it
of such light materials as to render it nearly invisible; such
are fences made of slender iron and wire painted green.

3. A third method is, sinking the fence below the surface
of the ground, by which means the view is not impeded, and
the continuity of lawn is well preserved. Where this sunk
fence or fosse is adopted, the deception ought to be complete;
but this cannot be where grass and corn-lands are divided by
such a fence: if it is used betwixt one lawn and another, the
mind acquiesces in the fraud even after it is discovered, so
long as the fence itself does not obtrude on the sight. We

must therefore so dispose a fosse, or ha! ha! that we may look across it and not along it.   For this reason a sunk fence must be straight and not curving, and it should be short, else the *imaginary* freedom is dearly bought by the *actual* confinement, since nothing is so difficult to pass as a deep sunk fence.

4. A fourth expedient I have occasionally adopted, and which (if I may use the expression) is a more bold deception than a sunk fence, *viz.*, a light hurdle instead of paling ; the one we are always used to consider as a fixed and immovable fence at the boundary of a park or lawn ; the other only as an occasional division of one part from the other : it is a temporary inconvenience, and not a permanent confinement.

It is often necessary to adopt all these expedients in the boundaries and subdivisions of parks ; but the disgust excited at seeing a fence may be indulged too far, if in all cases we are to endeavour at concealment ; and therefore the various situations and purposes of different sorts of fences deserve consideration.

However we may admire natural beauties, we ought always to recollect, that, without some degree of art and management, it is impossible to prevent the injury which vegetation itself will occasion : the smooth bowling-green may be covered by weeds in a month, while the pastured ground preserves its neatness throughout the year.   There is no medium between the *keeping* of *art* and of *nature*, it must be either one or the other, art or nature, that is, either mowed, or fed by cattle ; and this practical part of the management of a place forms one of the most difficult points of the professors of art, because the line of fence, which separates the dressed ground from the pasture, is too often objectionable ; yet there is not less impropriety in admitting cattle to feed in a flower-garden, than in excluding them from such a tract of land as might be fed with advantage.

At SHEFFIELD PLACE, the beautiful and long meadow in Arno's Vale is a striking example of what I have mentioned ; because, if it were possible, or on the principle of economy advisable, to keep all this ground as neatly rolled and mowed as the lawn near the house, by which it would always appear as it does the first week after the hay is carried off ; yet I contend, that the want of animals and animation deprives it of half its real charms; and although many beauties must be relinquished by curtailing the number of walks, yet others may be obtained, and the whole will be more easily kept with proper neatness by judicious lines of demarcation, which shall separate the grounds to be fed, from the grounds to be mown; or rather by such fences as shall, on the one hand, protect the woods from the encroachments of cattle, and, on the other, let the cattle protect the grass-land from the encroachment of woods; for such is the power of vegetation at SHEFFIELD PLACE, that every berry soon becomes a bush, and every bush a tree.

From this luxuriant vegetation the natural shape of the vale is obliterated, the gently-sloping banks are covered with wood, and the narrow glade in the bottom is choked with spreading larches. It is impossible to describe by words, and without a map, how this line of demarcation should be effected; but I am sure many acres might be given to cattle, and the scenery be improved, not only by such moving objects, but also by their use in cropping those vagrant branches which no art could watch with sufficient care and attention. It is to such accidental browsing of cattle that we are indebted for those magical effects of light and shade in forest scenery, which art in vain endeavours to imitate in pleasure-grounds.

Perhaps the brook might be made the natural boundary of Arno's Vale, where a deep channel immediately at the foot of the hill, with or without posts and rails, would make an effectual fence. It will perhaps be objected, that a walk by the side of such a fence would be intolerable; yet surely this watercourse, occasionally filled with a lively stream, is far preferable to a dry channel; and yet the only walk from the house at present is by the side of what may be so called: and, far from considering this a defect, I know it derives much of its interest from this very circumstance. A gravel-walk is an artificial convenience, and that it should be protected is one of its first requisites: therefore, so long as good taste and good sense shall coincide, the eye will be pleased where the mind is satisfied. Indeed, in the rage for destroying all that appeared artificial in the ancient style of gardening, I have frequently regretted the destruction of those majestic terraces which marked the precise line betwixt nature and art.

To describe the various sorts of fences suitable to various purposes, would exceed the limits and intentions of this work: every county has its peculiar mode of fencing, both in the construction of hedges and ditches, which belong rather to the farmer than the landscape gardener; and in the different forms and materials of pales, rails, hurdles, gates, &c., my object is rather to describe such application of common expedients as may have some degree of use or novelty.

Among these I shall first mention, that, instead of surrounding a young plantation with a hedge and ditch, with live quick or thorns, I generally recommend as many, or even more, thorns than trees, to be intermixed in the plantation, and the whole to be fenced with posts and rails, more or less neat, according to the situation; but, except near the house, I never suppose this rail to continue after the trees (with the aid of such intermixed thorns) are able to protect themselves against cattle; and thus, instead of a hard marked outline, the woods will acquire those irregularities which we observe in forest scenery, where in some few instances the trees are choked by the thorns, though in many they are nursed and reared by their protection.

It often happens that a walk in a plantation or shrubbery is crossed by a road or driftway; this has been ingeniously obviated (I believe originally by Mr. Brown), by making one pass over the other, and where the situation requires such expense, a subterraneous passage may either be made under the carriage-road, as I have done at WELBECK, at GAYHURST, and at other places, or a foot-bridge may be carried over the road, as I have frequently advised: but a more simple expedient will often answer the purpose, which I shall describe with the help of the annexed sketch [fig. 70], representing the ground plan of the intersected roads.

Two light gates, like the rail-fence to the plantation, are so hung to the posts A and B, that they will swing either to the posts D or C, and thus they will either close the spaces D B and A C, leaving open the walk, or they must be shut so as to close the spaces A D and B C, leaving open the road or driftway; for this purpose the posts A and B, to which the gates are hung, should be round, and the hinge turn on a pivot at

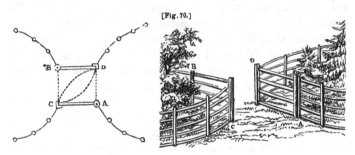

[Fig. 70.]

the top; the other two posts may be square, or with a rebate to receive the gate.

In the course of this work, I may have frequent occasion to mention the necessity of providing a fence near the house, to separate the dressed lawn from the park, or feeding ground: various ingenious devices have been contrived to reconcile, with neatness and comfort, the practice introduced by Mr. Brown's followers, of setting a house in a grass-field.

The sunk fence, or ha! ha! in some places, answers the purpose; in others a light fence of iron or wire, or even a wooden rail, has been used with good effect, if not too high; but generally near all fences the cattle make a dirty path, which, immediately in view of the windows, is unsightly; and where the fence is higher than the eye, as it must be against deer, the landscape seen through its bars becomes intolerable. After various attempts to remedy these defects by any expedient that might appear natural, I have at length boldly had recourse to artificial management, by raising the ground near the house about three feet, and by supporting it with a wall of the same materials as the house. In addition to this, an iron rail on the top, only three feet high, becomes a sufficient fence, and forms a sort of terrace in front of the house, making an avowed separation between grass kept by the scythe, and the park fed by deer or other cattle, while at a little distance it forms a base line or deep plinth, which gives height and consequence to the house.

This will, I know, be objected to by those who fancy that everything without the walls of a house should be natural; but a house is an artificial object, and, to a certain distance around the house, art may be avowed: the only difference of opinion will be, where shall this line of utility, separating art from nature, commence ? Mr. Brown said, at the threshold of the door : yet he contradicted himself when he made, as he always did, another invisible line beyond it. On the contrary, I advise that it be near the house, though not quite so near : and that the line should be artificially and visibly marked.*

When Mr. Brown marked the outline of a great wood sweeping across hill and valley, he might indulge his partiality for a serpentine or graceful curve, which had been then newly introduced by Hogarth's idea respecting the line of beauty : but it may be observed, that a perfectly straight line, drawn across a valley diagonally, appears to the eye the same as this line of fancied beauty, and therefore, in many cases, the line should be straight. I have already hinted in this Chapter, that the fence of a wood or plantation should be considered as merely temporary, that is, till the thorns planted among the trees can supersede its use. Wherefore, it is of little conse- quence in what manner a hurdle, or rough posts and rails, without any hedge or ditch, may be placed : a straight line is ever the shortest, and I have often preferred it, especially as I know that a few trees or bushes at each end of such a line will prevent the eye from looking along its course.

Sometimes it happens, from the intermixture of property, or other causes, that the fence is obliged to make a very acute angle ; this may occasionally be remedied by another line of fence fitting to its greatest projection; and as this same prin- ciple may be extended to roads, walks, or rivers, I shall ex- plain it.

The sharp elbow or projection of the fence [fig. 71] A, ceases to be offensive if another fence can be joined to it, as at

* Examples of this may be seen at BULSTRODE, at MICHEL GROVE, at BRENTREE HILL, &c.

B, and the same with the line of road or walk; the branch obviates the defect.

[Fig. 71.]

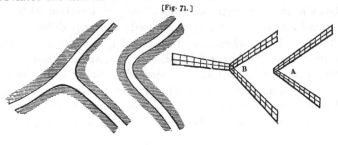

It has been observed by the adversaries of the art, that exactly the same line will serve either for a road or a river, as it may be filled with gravel or with water. This ridicule may perhaps be deserved by those engineers who are in the habit of making navigable canals only, but the nice observer will see this material difference,—

The banks of a natural river are never equidistant, the water in some places will spread to more than twice the breadth it does in others; this pleasing irregularity depends on the shape of the ground through which it flows: a river seldom proceeds far along the middle of a valley, but generally keeps on one side, or boldly stretches across to the other, as the high ground resists, or the low ground invites its course: these circumstances in natural rivers should be carefully imitated in those of art, and not only the effects, but even the causes, if possible, should be counterfeited, especially in the form of the shores: thus, the convex side of the river at A [in fig. 72], should have its shores convex, or steep; and the

[Fig. 72.]

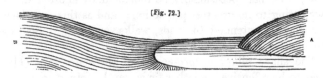

concave side of the river at B, should have its shores concave, or flat; because, by this means, the course of the river is accounted for.

There is another circumstance, with respect to lines, deserving attention. The course of a river may frequently shew two or more different bends, which do not so intersect each other as to impede the view along it; and these may be increased in proportion to the breadth of the river: but in a road, or a walk, especially if it passes through a wood or plantation, a second bend should never be visible.

The degree of curve in a walk, or road, will therefore depend on its width; thus looking along the narrow line of walk, you will not see the second bend: but in the same curve, if the road be broader, we should naturally wish to make the curve bolder by breaking from it, according to the dotted line from A to B in the diagram [fig. 73].

[Fig. 73.]

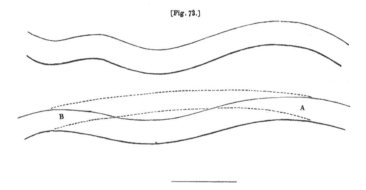

When two walks separate from each other, it is always desirable to have them diverge in different directions, as at *a* [in fig. 74], rather than give the idea of reuniting, as at *b*.

[Fig. 74.]

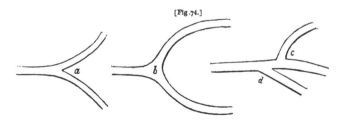

Where two walks join each other, it is generally better that they should meet at right angles, as at *c*, than to leave the sharp point, as in the acute angle at *d*.

The most natural course for a road, or walk, is along the banks of a lake, or river; yet I have occasionally observed great beauty in the separation of these two lines; as where the water sweeps to the left, and the road to the right, or *vice versâ:* the true effect of this circumstance I have often attempted to represent on paper, but it is one of the many instances in which the reality and the picture excite different sensations.

This Chapter might have included every necessary remark relative to fences, whether attached to parks or farms; but as I wish to enlarge upon the distinction between the improvements designed for ornament, and those for profit, or gain, I shall endeavour to explain these different objects, as they appear to me opposite in their views, and distinct in their characteristics. Both are, indeed, subjects of cultivation; but the cultivation in the one is husbandry, and in the other decoration.

CHAPTER VII.

Ferme ornée, a Contradiction—Farm and Park distinct Objects—Experimental, or useful Farm—Beauty and Profit seldom compatible.

THE French term *Ferme ornée*, was, I believe, invented by Mr. Shenstone, who was conscious that the English word Farm would not convey the idea which he attempted to realize in the scenery of the Leasowes. That much celebrated spot, in his time, consisted of many beautiful small fields, connected with each other by walks and gates, but bearing no resemblance to a farm as a subject of profit. I have never walked through these grounds without lamenting, not only the misapplication of good taste, but that constant disappointment which the benevolent Shenstone must have experienced in attempting to unite two objects so incompatible as ornament and profit. Instead of surrounding his house with such a quantity of ornamental lawn or park only, as might be consistent with the size of the mansion, or the extent of the property, his taste, rather than his ambition, led him to ornament the whole of his estate; vainly hoping that he might retain all the advantages of a farm, blended with the scenery of a park. Thus he lived under the continual mortification of disappointed hope, and, with a mind exquisitely sensible, he felt equally the sneer of the great man, at the magnificence of his attempt, and the ridicule of the farmer, at the misapplication of his paternal acres.

Since the removal of court yards and lofty garden walls from the front of a house, the true substitute for the ancient magnificence destroyed is the more cheerful landscape of modern park scenery; and, although its boundary ought in no case to be conspicuous, yet its actual dimensions should bear some proportion to the command of property by which the mansion is supported. If the yeoman destroys his farm by making what is called a *Ferme ornée*, he will absurdly sacrifice his income to his pleasure: but the country gentleman can only ornament his place by separating the features of farm and park; they are so totally incongruous as not to admit of any union but at the expense either of beauty or profit. The following comparative view will tend to confirm this assertion.

The chief beauty of a *park* consists in uniform verdure; *undulating** lines contrasting with each other in variety of forms; trees so grouped as to produce light and shade to display the varied surface of the ground; and an undivided range of pasture. The animals fed in such a park appear free from confinement, at liberty to collect their food from the rich herbage of the valley, and to range uncontrolled to the drier soil of the hills.

The *farm*, on the contrary, is for ever changing the colour of its surface in motley and discordant hues; it is subdivided by straight lines of fences. The trees can only be ranged in formal rows along the hedges; and these the farmer claims a right to cut, prune, and disfigure. Instead of cattle enlivening the scene by their peaceful attitudes, or sportive gambols, animals are bending beneath the yoke, or closely confined to fatten within narrow enclosures, objects of profit, not of beauty [see fig. 75].

[Fig. 75. View of Farm Lands, shewing the bad effect of hedgerows and ridges with reference to Park scenery.]

This reasoning may be further exemplified by an extract from the *Red Book* of ANTONY.

The shape of the ground at ANTONY is naturally beautiful, but attention to the farmer's interest has† almost obliterated

---

* I am aware that the word *undulating* is seldom applied to solid bodies, but I know no other word so expressive of that peculiar shape of ground consisting of alternate concave and convex lines flowing into each other.

† In this, as in many other cases, I transcribe from the *Red Book*, as if my plans were not yet executed.

all traces of its original form; since the line of fence, which the farmer deems necessary to divide *arable* from *pasture land*, is unfortunately that which, of all others, tends to destroy the union of hill and valley. It is generally placed exactly at the point where the undulating surface changes from convex to concave, and, of course, is the most offensive of all intersecting lines; for it will be found that a line of fence, following the

[Fig. 76. The Farm lands shewn in fig. 75, with the hedgerows and ridges removed and planted in the style of Park scenery.]

shape of the ground, or falling in any direction from the hill to the valley, although it may offend the eye as a boundary, yet it does not injure, and, in some instances, may even improve the beautiful form of the surface. No great improvement, therefore, can be expected at Antony, until almost all the present fences be removed, although others may be placed in more suitable directions [see figs. 76 and 77].

I am aware that, in the prevailing rage for agriculture, it is unpopular to assert, that a farm and a park may not be united; but, after various efforts to blend the two, without

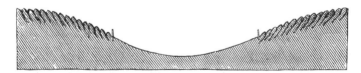

[Fig. 77. Section of fig. 75, in which the hills are shewn in aration, and the valley between them in pasture.

violation of good taste, I am convinced that they are, and must be distinct objects, and ought never to be brought together in the same point of view.

To guard against misrepresentation, let me be allowed to say, each may fill its appropriate station in a gentleman's

2 E

estate : we do not wish to banish the nectarine from our des-
serts, although we plant out the wall which protects it; nor
would I expunge the common farm from the pleasures of the
country, though I cannot encourage its motley hues and
domestic occupations to disturb the repose of park scenery.
It is the *union*, not the *existence*, of beauty and profit, of
laborious exertion and pleasurable recreation, against which
I would interpose the influence of my art; nor let the fastidious objector condemn the effort, till he can convince the
judgment that, without violation of good taste, he could introduce the dairy and the pig-sty (those useful appendages of
rural economy) into the recesses of the drawing-room, or the
area of the saloon.

The difficulty of uniting a park and a farm arises from this
material circumstance, that the one is an object of beauty, the
other of profit. The scenery of both consists of ground,
trees, water, and cattle; but these are very differently arranged. And since a park is less profitable than arable land,
the more we can diminish the quantity of the former, provided
it still be in character with the style of the mansion, the less
we shall regret the sacrifice of profit to beauty.

The shape and colour of corn-fields, and the straight lines
of fences, are so totally at variance with all ideas of picturesque beauty, that I shall not venture to suggest any hints on
the subject of a farm, as an ornament; yet I think there
might be a distinction made between the farm of a tenant,
who must derive benefit from every part of his land, and that
occupied by a gentleman for the purposes of amusement or
experiment.

It is usual in Hampshire, and, indeed, in the neighbourhood
of many forests, to divide the enclosures of a farm by rows of
copse-wood and timber, from ten to twenty yards wide; at a
little distance these rows appear united, and become one rich
mass of foliage. This kind of subdivision I should wish to be
generally adopted on experimental farms. The advantages of
such plantations will be,—

Shady and pleasant walks through the farm—to afford
shelter to corn, and protect the cattle which are grazed on
the farm—to give the whole, at a distance, the appearance of
one mass of wood—to make an admirable cover for game;

and, lastly, if it should ever hereafter be thought advisable to extend the lawn, such plantations will furnish ample choice of handsome trees to remain single or in groups, as taste or judgment shall direct.

In some counties the farms consist chiefly of grass-land, but even a dairy-farm must be subdivided into small enclosures: and although it is not necessary that a lawn near a mansion should be fed by deer, yet it is absolutely necessary that it should have the appearance of a park, and not that of a farm; because, in this consists the only difference betwixt the residence of a landlord and his tenant, the gentleman and the farmer: one considers how to make the greatest immediate advantage of his land; the other must, in some cases, give up the idea of profit, for the sake of that beauty which is derived from an air of liberty, totally inconsistent with those lines of confinement and subdivision which are characteristic of husbandry.

Since the beauty of pleasure-ground, and the profit of a farm, are incompatible, it is the business of taste and prudence so to disguise the latter and to limit the former, that park scenery may be obtained without much waste or extravagance; but I disclaim all idea of making that which is most beautiful also most profitable: a ploughed field, and a field of grass, are as distinct objects as a flower-garden and a potato-ground. The difference between a farm and a park consists not only in the number of fences and subdivisions, but also in the management of the lines in which the fences of each should be conducted. The farmer, without any attention to the shape of the ground, puts his fences where they will divide the uplands from the meadows; and, in subdividing the ground, he aims only at square fields, and consequently straight lines, avoiding all angles or corners. This is the origin of planting those triangular recesses in a field surrounded by wood, which the farmer deems useless; but which, to the eye of taste, produce effects of light and shade.

There is no mistake so common as that of filling up a recess in a venerable wood with a miserable patch of young plantation. The outline of a wood can never be too boldly indented, or too irregular; to make it otherwise, by cutting off the projections, or filling up the hollows, shews a want of taste, and is as incongruous as it would be to smooth the furrowed bark of an aged oak.

In a park, the fences cannot be too few, the trees too majestic, or the views too unconfined. In a farm, small enclosures are often necessary; the mutilated pollard or the yielding willow, in the farmer's eye, are often preferable to the lofty elm or spreading oak; whilst a full crop of grain, or a copious swath of clover, is a more gladdening prospect than all the splendid scenery of wood and lawn from the windows of a palace. Small detached farms, adapted to useful and laborious life, unmixed with the splendours of opulence, but supporters of national wealth, are indeed objects of interest in every point of view; they want not the adventitious aid of picturesque effect to attract peculiar notice; to a benevolent mind they are more than objects of beauty; they are blessings to society; nor is it incompatible with the pursuit of pleasure, sometimes to leave the boundaries of the park, and watch the exertions of laudable industry, or visit the cottages

" Where cheerful tenants bless their yearly toil."

The monopolist only can contemplate with delight his hundred acres of wheat in a single enclosure; such expanded avarice may *enrich the man*, but will impoverish and distress, and (I had almost added) will ultimately *starve mankind*.

CHAPTER VIII.

Of Pleasure-Grounds — Flower-Gardens, Example, Bulstrode — Valley Field — Nuneham — Greenhouse and Conservatory belong to a Flower-Garden — Various Modes of attaching them to a House — Difficulty — Objection — Attempt to make them Gothic.

In the execution of my profession, I have often experienced great difficulty and opposition in attempting to correct the false and mistaken taste for placing a large house in a naked grass-field, without any apparent line of separation between the ground exposed to cattle and the ground annexed to the house, which I consider as peculiarly under the management of art.

This line of separation being admitted, advantage may be easily taken to ornament the lawn with flowers and shrubs, and to attach to the mansion that scene of "embellished neatness," usually called a pleasure-ground.

The quantity of this *dressed ground* was formerly very considerable. The royal gardens of Versailles, or those of Kensington Palace, when filled with company, want no other animation; but a large extent of ground without moving objects, however neatly kept, is but a melancholy scene. If solitude delight, we seek it rather in the covert of a wood, or the sequestered alcove of a flower-garden, than in the open lawn of an extensive pleasure-ground.

I have therefore frequently been the means of restoring acres of useless garden to the deer or sheep, to which they more properly belong.

This is now carrying on with admirable effect at Bulstrode, where the gardens of every kind are on a great scale, and where, from the choice and variety of the plants, the direction of the walks, the enrichment of art, and the attention to every circumstance of elegance and magnificence, the pleasure-ground is perfect as a whole, while its several parts may furnish models of the following different characters of taste in gardening: the *ancient garden*, the *American garden,*

the *modern terrace-walks*, and *the flower-garden :* the latter
is, perhaps, one of the most varied and extensive of its kind,
and therefore too large to be otherwise artificial, than in the
choice of its flowers, and the embellishments of art in its
ornaments.

Flower-gardens on a small scale may, with propriety, be
formal and artificial; but in all cases they require neatness
and attention.   On this subject I shall transcribe the following passage from the *Red Book* of VALLEY FIELD.*

To common observers, the most obvious difference between
Mr. Brown's style and that of ancient gardens, was the change
from straight to waving or serpentine lines.   Hence, many of
his followers had supposed good taste in gardening to consist
in avoiding all lines that are straight or parallel, and in adopting forms which they deem more consonant to nature, without considering what objects were natural and what were
artificial.

[Fig. 78. View of the flower-garden at Valley Field]

This explanation is necessary to justify the plan which I
recommended for the canal in this flower-garden [see fig. 78];

* " Although I have never seen VALLEY FIELD, myself, yet it flatters
me to learn, that, under the direction of my two sons, by taking advantage
of the deep romantic glen and wooded banks of the river which flows
through the grounds, and falls into the Frith of Forth at a short distance

for, while I should condemn a long straight line of water in an open park, where everything else is natural, I should equally object to a meandering canal or walk, by the side of a long straight wall, where everything else is artificial.

A flower-garden should be an object detached and distinct from the general scenery of the place ; and, whether large or small, whether varied or formal, it ought to be protected from hares and smaller animals by an inner fence : within this enclosure rare plants of every description should be encouraged, and a provision made of soil and aspect for every different class. Beds of bog-earth should be prepared for the American plants: the aquatic plants, some of which are peculiarly beautiful, should grow on the surface or near the edges of water. The numerous class of rock-plants should have beds of rugged stone provided for their reception, without the affectation of such stones being the natural production of the soil; but, above all, there should be poles or hoops for those kind of creeping plants which spontaneously form themselves into graceful festoons, when encouraged and supported by art.

Yet, with all these circumstances, the flower-garden, except where it is annexed to the house, should not be visible from the roads or general walks about the place.   It may therefore be of a character totally different from the rest of the scenery, and its decorations should be as much those of art as of nature.

The flower-garden at NUNEHAM,* without being formal, is highly enriched, but not too much crowded with seats, temples, statues, vases, or other ornaments, which, being works of art, beautifully harmonize with that profusion of flowers and curious plants which distinguish the flower-garden from natural landscape, although the walks are not in straight lines.

But at VALLEY FIELD, where the flower-garden is in front

from the house, an approach has been made, which, for variety, interest, and picturesque scenery, may vie with anything of the kind in England ; while it remains a specimen of the powers of landscape gardening, in that part of Scotland where the art had been introduced only by those imitators of Mr. Brown's manner, who had travelled into the north.   His own improvements were confined to England."

* Earl Harcourt, although possessing great good taste, gives the whole merit of this garden to Mason the poet, as he does of his pleasure-grounds to Brown.   Thus, superior to that narrow jealousy which would deny the just tribute of praise to the professor, his lordship is satisfied with having been the liberal friend and patron of merit.

of a long wall, the attempt to make the scene natural would
be affected; and, therefore, as two great sources of interest in
a place are *variety* and *contrast*, the only means by which
these can be introduced are in this flower-garden, which, as a
separate object, becomes a sort of episode to the general and
magnificent scenery.

The river being everywhere else a lively stream, rattling
and foaming over a shallow bed of rock or gravel, a greater
contrast will arise from a smooth expanse of water in the
flower-garden: to produce this must be a work of art, and,
therefore, instead of leading an open channel from the river to
supply it, or making it appear a natural branch of that river,
I recommend that the water should pass underground, with
regulating sluices or shuttles to keep it always at the same
height. Thus the canal will be totally detached from the
river, and become a distinct object, forming the leading feature of the scene to which it belongs; a scene purely artificial,
where a serpentine canal would be as incongruous as a serpentine garden wall, or a serpentine bridge; and, strange as it
may appear, I have seen such absurdities introduced, to avoid
nature's supposed abhorrence of a straight line.

The banks of this canal, or fish-pond, may be enriched with
borders of curious flowers, and a light fence of green laths will
serve to train such as require support, while it gives to the
whole an air of neatness and careful attention.

But, as the ends of this water should also be marked by
some building, or covered seat, I have supposed the entrance
to the flower-garden to be under a covered passage of hoops,
on which may be trained various sorts of creeping plants; and
the farther end may be decorated by an architectural building, which I suppose to consist of a covered seat between two
aviaries.

It will perhaps be objected, that a long straight walk can
have little variety; but the greatest source of variety in a
flower-garden is derived from the selection and diversity of its
shrubs and flowers.

There is no ornament of a flower-garden more appropriate
than a conservatory, or a green-house, where the flower-garden

is not too far from the house; but, amongst the refinements of modern luxury may be reckoned that of attaching a green-house to some room in the mansion, a fashion with which I have so often been required to comply, that it may not be improper, in this work, to make ample mention of the various methods by which it has been effected in different places.

At Bowood, at Wimpole, at Bulstrode, at Attingham, at Dyrham Park, at Caenwood, at Thoresby, and some other large houses of the last century, green-houses were added, to conceal offices behind them, and they either became a wing of the house, or were in the same style of architecture: but these were all built at a period when only orange-trees and myrtles, or a very few other green-house plants were introduced, and no light was required in the roof of such buildings. In many of them, indeed, the piers between each window are as large as the windows.

Since that period, the numerous tribe of geraniums, ericas, and other exotic plants, requiring more light, have caused a very material alteration in the construction of the green-house; and, perhaps, the more it resembles the shape of a nurseryman's stove, the better it will be adapted to the purposes of a modern green-house.

Yet such an appendage, however it may increase its interior comfort, will never add to the external ornament of a house of regular architecture: it is therefore generally more advisable to make the green-house in the flower-garden, as near as possible to, without forming a part of, the mansion; and in these situations great advantage may be taken of treillage ornaments to admit light, whilst it disguises the ugly shape of a slanting roof of glass.

There is one very material objection to a green-house immediately attached to a room constantly inhabited, viz. that the smell and damp from a large body of earth in the beds, or pots, is often more powerful than the fragrance of the plants; therefore the conservatory should always be separated from the house, by a lobby, or small anti-room. But the greatest objection arises from its want of conformity to the neighbouring mansion, since it is difficult to make the glass roof of a conservatory architectural, whether Grecian or Gothic.

An arcade is ill adapted to the purpose, because, by the

form of an arch, the light is excluded at the top, where it is most essential in a green-house; for this reason, the flat Gothic arch of Henry the Eighth is less objectionable, yet in such buildings we must suppose the roof to have been taken away to make room for glass; of this kind is the conservatory in front of RENDLESHAM HOUSE.

In the adaptation of ancient forms to modern uses and inventions, we are often under the necessity of deviating from the rules of true Gothic. Under such circumstances it is perhaps better to apply old expedients to new uses, than to invent a new and absurd style of Gothic or Grecian architecture. At PLAS-NEWYD, where the house partakes of a Gothic character, I suggested the addition of a green-house, terminating a magnificent enfilade through a long line of principal apartments. The hint for this model is taken from the chapter-rooms to some of our cathedrals, where an octagon roof is supported by a slender pillar in the middle, and if this were made of cast-iron, supporting the ribs of a roof of the same material, there would be no great impropriety in filling the interstices with glass, while the side window-frames might be

[Fig. 79. Pavilion and green-house suggested for the Gothic mansion at Plas Newyd, as seen by moonlight.]

removed entirely in summer, making a beautiful pavilion at that season, when, the plants being removed, a green-house is generally a deserted and unsightly object.   The effect of this building by moonlight is shewn in the foregoing sketch [fig. 79]; and there are many summer evenings when such a pavilion would add new interest to the magnificent scenery of water and mountains with which PLAS-NEWYD everywhere abounds.*

* In a conversation I had the satisfaction to enjoy with the late Earl of Orford, at Strawberry Hill, he shewed me the gradual progress of his knowledge in Gothic architecture, by various specimens in that house, in which he had copied the forms of mouldings without always attending to the scale or *comparative proportion ;* and his lordship's candour pointed out to me the errors he had at first committed. This error, in the imitators of Gothic, often arises from their not considering the difference of the materials with which they work : if, in the mullions of a window, or the ribs of a ceiling, they copy, in wood or plaster, ornaments originally of stone, they must preserve the same massive proportions that were necessary in that material, or they must paint it like wood, and not like stone : but if the architects of former times had known the use we now make of cast-iron, we should have seen many beautiful effects of lightness in their works ; and surely in ours, we may be allowed to introduce this new material for buildings, in the same manner that we may fairly suppose they would have done, had the invention been known in their time : *but wherever cast-iron is used in the construction, it ought to be acknowledged* as a support, either by gilding, or bronze, or any expedient that may shew it to be *metal,* and not *wood* or *stone,* otherwise it will appear unequal to its office.

CHAPTER IX.

Defence of the Art—Difference between Landscape Gardening and Painting
—Further Answer to Messrs. PRICE and KNIGHT—Cursory Remarks on
Views from Rooms, Appropriation, Extent, &c.—Example from ATTING-
HAM—Pictures may imitate Nature, but Nature is not to copy Pictures.

AT the time my former publication was in the press, the
Art of Landscape Gardening was attacked by two gentlemen,
Mr. Knight,* of Herefordshire, and Mr. Price,† of Shrop-
shire; and I retarded its publication till I could take some
notice of the opinions of these formidable, because ingenious,
opponents.

Having since been consulted on subjects of importance in
those two counties, I willingly availed myself of opportunities
to deliver my sentiments, as particular circumstances occurred,
and therefore, with permission of the respective proprietors, I
insert the following observations from the *Red Books* of SUF-
TON COURT, in Herefordshire, and ATTINGHAM, in Shrop-
shire:—

"My opinion concerning the improvement of SUFTON
COURT involving many principles in the art of landscape
gardening, I take this opportunity of justifying my practice,
in opposition to the wild theory which has lately appeared;
and shall therefore occasionally allude to this new system when
it bears any relation to our objects at SUFTON COURT.

Having already published a volume on the subject of
landscape gardening, it will be unnecessary to explain the

* Mr. Knight has endeavoured to ridicule all display of extent of property, which I consider one of the leading principles of the art.   I contend,
that it is impossible to annex the same degree of importance to a modern
house, however large, by the side of a high road, that may be justly given
to one surrounded by an extensive park.   To this principle of improvement
I have given the name of *appropriation*.

† Mr. Price builds a theory of improvement on the study of the best
pictures, without considering how little affinity there is between the confined
landscape exhibited on canvas, and the extensive range which the eye at
once comprehends; and argues, that the best works of the painter should be
models for the improver.

motives which induced me to adopt this name for a profession, as distinct from the art of landscape painting, as it is from the art of planting cabbages, or pruning fruit-trees.* The slight, and often gaudy sketches by which I have found it necessary to elucidate my opinions, are the strongest proofs that I do not profess to be a landscape painter; but to represent the scenes of Nature in her various hues of blue sky, purple mountains, green trees, &c., which are often disgusting to the eye of a connoisseur in painting.

The best painters in landscape have studied in Italy, or France, where the verdure of England is unknown: hence arises the habit acquired by the connoisseur, of admiring brown tints and arid foregrounds in the pictures of Claude and Poussin, and from this cause he prefers the bistre sketches to the green paintings of Gainsborough. One of our best landscape painters studied in Ireland, where the soil is not so yellow as in England; and his pictures, however beautiful in design and composition, are always cold and chalky.

Autumn is the favourite season of study for landscape painters, when all nature verges towards decay, when the foliage changes its vivid green to brown and orange, and the lawns put on their russet hue. But the tints and verdant colouring of spring and summer will have superior charms to those who delight in the perfection of nature, without, perhaps, ever considering whether they are adapted to the painter's landscape.

It is not from the colouring only, but the general composition of landscapes, that the painter and landscape gardener will feel the difference in their respective arts; and although each may occasionally assist the other, yet I should no more advise the latter, in laying out the scenery of a place, to copy

---

* " In the art of gardening, the great *materials* of the scene are pro-
" vided by Nature herself, and the artist must satisfy himself with that
" degree of expression which she has bestowed.
" In a landscape, on the contrary, the painter has the choice of the
" circumstances he is to represent, and can give whatever force or extent he
" pleases to the expression he wishes to convey. In gardening, the materials of
" the scene are few, and those few unwieldy, and the artist must often content
" himself with the reflection, that he has given the best disposition in his
" power to the scanty and intractable materials of Nature. In a landscape,
" on the contrary, the whole range of scenery is before the eye of the
" painter."—*Allison.*

the confined field of vision, or affect the careless graces of
Claude or Poussin, than I should recommend, as a subject
proper for a landscape painter, the formal rows or quincunx
position of trees in geometric gardening.   It has been wittily
observed, that ' the works of nature are well executed, but
in a bad taste;' this, I suppose, has arisen from the propensity
of good taste, to display the works of nature to advantage ;
but it does not hence follow that art is to be the standard for
nature's imitation ; neither does it disgrace painting, to assert
that nature may be rendered more pleasing than the finest
picture; since the perfection of painting seldom aims at exact
or individual representation of nature.   A panorama gives a
more natural idea of ships at sea than the best picture of
Vandervelde ; but it has little merit as a painting, because it
too nearly resembles the original, to please as an effort of
imitative art.   My sketches, if they were more highly finished,
would be a sort of panorama, or *fac-simile*, of the scenes they
represent, in which little effect is attempted on the principle
of composition in painting; but, like a profile shadow or
sillouette, they may please as portraits, while they offend the
connoisseur as paintings.   The art I profess is of a higher
nature than that of painting, and is thus very aptly described
by a French author.

' —il est, à la poèsie et à la peinture, ce que la realité est
à la description, et l'original à la copie.'

The house at Sufton Court having been built long
before I had the honour of being consulted, its aspects,
situation, and general arrangement, do not properly come
under my consideration.   Yet, as I shall suggest a hint for
altering the windows in the drawing-room, I must consider
the different landscapes in each direction.

The views towards the south and west are extensive, and,
under certain circumstances of light and weather, often wonderfully beautiful; but, as distant prospects depend so much
on the state of the atmosphere, I have frequently asserted,
that the views from a house, and particularly those from the
drawing-room, ought rather to consist of objects which evidently belong to the place.   To express this idea, I have used
the word *appropriation*, by which I mean, such a portion of
wood and lawn as may be supposed to belong to the proprietor

of the mansion, occupied by himself, not so much for the purposes of gain, as of pleasure, and convenience: this, of course, should be grass, whether fed by deer, by sheep, or by other cattle, and its subdivisions, if there be any, ought not to be permanent. I am ready to allow that this part of modern gardening has often been egregiously mistaken and absurdly practised; I find no error so difficult to counteract as the general propensity for extent, without sufficient attention to the size, style, or character of the house, or of the surrounding estate.

Extent and beauty have ever appeared to me distinct objects; and a small place, in which the boundary is not obtrusive, may be more interesting, and more consonant to elegance and convenience, than a large tract of land, which has no other merit than that it consists of many hundred acres, or is encompassed by a pale of many miles in circuit, while, perhaps, within this area, half the land is ploughed in succession.

The drawing-room, at present, looks towards the south, but there appear to be several reasons for altering its aspect; 1st. because the hall and dining-room command the same prospect, but more advantageously; 2nd. because the windows being near the hall door, a carriage-road, which must occasionally be dirty, becomes a bad foreground; and, lastly, the view towards the east will not only be different from the others, but is of such a nature as to appear wholly appropriate to the place, and, therefore, in strict harmony with the quiet home scene of a country residence: it consists of a beautiful lawn or valley, having its opposite bank richly clothed with wood, which requires very little assistance to give it an irregular

[Fig. 80.   View from the drawing-room at Sufton Court, before a row of trees was removed, and some masses of wood partially broken.]

and pleasing outline; and is one of the many subjects, more capable of delighting the eye in nature, than in a picture.

The sketch [fig. 80] shews, with accuracy, the situation of the
several trees which ought to be removed.

It has been laid down, by a recent author before named,
as a general rule for improvement, to plant largely and cut
down sparingly: this is the cautious advice of timidity and
inexperience; for, in some situations, improvement may be
effected by the axe rather than by the spade, of which this
sketch furnishes an instance: the trees in a straight line, at
the bottom of the hill, have in vain been encumbered by young
trees, planted with a view of breaking their formal row; while
in reality they produce the contrary effect. I rather advise
boldly taking away all the young trees, and part of the old
ones, but particularly an oak, which not only hides the

[Fig. 81. View from the drawing-room at Sufton Court, as it will appear when certain trees shewn in Fig. 80,
are removed.]

forked stem of a tree behind, but from its situation depresses
the other trees, and lessens the magnitude and importance
both of the hill and of the grove, by which its brow is
covered" [see fig. 81].

———————

" The situation of ATTINGHAM is at variance with its character; since it is impossible to annex ideas of grandeur and
magnificence to a mansion, with little apparent domain. The
flat lawn between the high road and the house, although very
extensive, yet, possessing no variety in the size of the trees,
and but little in the shape of ground; the eye is deceived in
its real distance.

By the laws of perspective, the nearer any object is to the
eye, the larger it will appear; also, the larger any object is,
the nearer it will appear to the eye: consequently, the magnitude of the house makes it appear nearer than it really is,

there being no intervening objects to divert the attention, or
to act as a scale, and assist the eye in judging of the distance.
For this reason, every stranger who sees this house from the
turnpike road, would describe it as a large house with very
little ground between it and the road. The first idea of
improvement would be, either to remove the house or the
road; but as neither of these expedients is practicable, we
must have recourse to art to do away with this false impression. This I shall consider as forming the basis of the alteration
proposed at ATTINGHAM.

In ancient Gothic structures, where lofty walls and various
courts intervened between the palace and the neighbouring
village, there was sufficient *dignity* or seclusion, without that
apparent extent of domain which a modern mansion requires;
but, since the restraint of ancient grandeur has given place to
modern elegance, which supposes greater ease and freedom,
the situation of a house in the country is more or less defective,
in proportion as it is more or less bounded or incommoded by
alien property. Thus a high road, a ploughed field, a barn, or
a cottage adjoining a large house, has a tendency to lessen its
importance; and hence originates the idea of extending park,
lawn, or pleasure-grounds, in every direction from the house:
hence, also, arises the disgust we feel at seeing the park pales,
and grounds beyond, when they are so near, or so conspicuous,
as to impress the mind with an idea of not belonging to the
place.

Perhaps the love of unity may contribute to the pleasure
we feel in viewing a park where the boundary is well concealed.
This desire of hiding the boundary introduced the modern
practice of surrounding almost every park with a narrow plantation or belt; which, if consisting of trees planted at the same
time, becomes little better than a mere hedgerow, and is
deservedly rejected by every man of taste; yet there are many
situations where a plantation becomes the natural boundary of
a park: such is the screen of wood on the highest ground to
the east of ATTINGHAM, where it forms a pleasing outline to
the landscape, without exciting a wish to know whether it is
the termination of the property.

In consequence of the apparent want of extent in the park
or lawn at ATTINGHAM, it was suggested to add many hundred

acres of land to the east, by removing the hedges of the adjoining fields. This would have increased the real, without extending the apparent magnitude of the park : but I contend, that oftentimes it is the *appearance*, and not the *reality* of extent, which is necessary to satisfy the mind ; for the size of the park has little reference to that of the estate of the proprietor. The land attached to a villa, near a city, may with propriety be surrounded by pales, or a wall, for the sake of privacy and seclusion ; but it is absurd to enclose more of a distant domain than is necessary for the beauty of the place : besides, if this park or lawn had been extended a mile farther to the east, the confinement to the south, which is in the front of the house, would not have been done away, and, consequently, to the traveller passing the road, the apparent extent would not have been increased ; and without some striking or beautiful feature, extent alone is seldom interesting.

If large trees, river scenery, or bold inequality of ground, can be included, by enlarging a park, they are sufficient motives ; but views of distant mountains, which may be seen as well from the high road, are not features that justify extensive lawn over a flat surface. *

To do away the impression of confinement at ATTINGHAM, the park should be extended across the road, and thus the stranger will be induced to believe he passes through, and not at the extremity of the park. Secondly, some striking and interesting features should be brought into notice, such as the junction of the Severn and the Terne, which may be actually effected within the limits of the park ; and particularly the great arch across the Terne, of which no adequate advantage is at present taken. There are, also, some large trees, and many interesting points of view, which well deserve attention, in a plan professing to increase the number of beautiful circumstances, rather than the number of acres in the park.

In opposition to Mr. Price's idea, that all improvement of scenery should be derived from the works of great painters, I shall observe, that there are, at present, very near the house,

* One great error, in Mr. Brown's followers, has been the unnecessary extent of parks. It is my opinion, that, provided the boundary can be properly disguised, the largest parks need not exceed two or three hundred acres, else they are apt to become farms within a pale, or they are forests rather than parks.

some fragments of an old mill, and brick arches [see fig. 82], that make a charming study for a painter; the composition is not unlike a beautiful picture of Ruisdale's, at ATTINGHAM, which every man of taste must admire: of this scene, as it now exists, I have endeavoured to give a faint idea. Among the trees is seen part of the colonnade that joins the east wing to the body of the house: from the general character of this scenery, we cannot but suppose this to be a fragment of some ruined Grecian temple, and no part of a modern inhabited palace. Hence it is evident, that the mind cannot associate the ideas of elegance with neglect, or perfect repair and neatness with ruin and decay : such objects, therefore, however picturesque in themselves, are incongruous and misplaced, if near such a palace as ATTINGHAM.

[Fig. 82. Scene in the grounds at Attingham.]

Another mistake of the admirers of painters' landscape is, the difference in the quantity of a natural and an artificial composition : the finest pictures of Claude (and here again I may refer to a picture at ATTINGHAM) seldom consist of more than one-fifth of that field of vision which the eye can with ease behold, *without any motion of the head, viz.* about 20 degrees out of 90 ; and we may farther add, that, *without moving the body,* our field of vision is extended to 180 degrees.

Now it is obvious, that the picture of Claude, already mentioned, which is between four and five feet long, if it had been extended to 20 or 30 feet, would not have been so pleasing a composition; because, instead of a picture, it would have resembled a panorama. This I may further instance, in the view from the breakfast-room, consisting of a distant range of

mountains, by far too long for any picture. Yet a small part
of this view might furnish a subject for the painter, by supposing a tree to form the foreground of the landscape. Are we
then to plant such a tree, or a succession of such trees, to
divide the whole field of vision into separate landscapes ? and
would not such an attempt at improvement be like placing five
or six pictures of Claude in one long frame ? The absurdity
of the idea proves the futility of making pictures our models for
natural improvements: however I may respect the works of
the great masters in painting, and delight to look at nature
with a painter's eye, yet I shall never be induced to believe
that " the best landscape painter would be the best landscape
gardener."*

The river Terne, being liable to floods from every heavy
shower of rain which falls upon the neighbouring hills, has
formed a number of different channels and islands: some of
these channels are dry when the water is low, and some of the

* Since I began these remarks on Attingham, Mr. Price has published
a second volume of " Essays on the Picturesque," the whole of which is
founded on his enthusiasm for pictures ; and he very justly observes (page
269), " Enthusiasm always leads to the verge of ridicule, and seldom keeps
totally within it." Thus, not content with making the works of great
painters the standard for laying out grounds, they are also to furnish plans
and elevations for all our buildings, from the palace to the cottage : and
since we cannot be quite reconciled to their being in a state of ruin, which
would certainly be most picturesque, we must build them in such irregular
forms, that trees may be introduced in various hollows and recesses, to be
left for this purpose : these will, indeed, very soon contribute to produce
those weather stains, and harmonious tints, which are more grateful to the
painter's eye than polished marble ; as the green rust on copper coins is
more interesting to the antiquarian than the bright surface of gold or silver.
Mr. Price confesses, that two small difficulties occur, in putting these projects fully in practice, viz. that " he sees no examples of chimneys, and very
few of slanting roofs," but where fine pictures can be transferred from the
canvas to the real residence of man. How void of taste must that man be,
who could desire a chimney, or roof to his country-house, when we are told
that Poussin, and Paul Veronese, built whole cities without a single chimney, and with only one or two slanting roofs ! This idea of deriving all our
instruction from the works of great painters, is so ingenious and useful, that
it ought not to be confined to gardening and building. In our markets, for
instance, instead of that formal trim custom of displaying poultry, fish, and
fruit, for sale on different stalls, why should we not rather copy the
picturesque jumble of Schnyders and Rubens? Our kitchens may be furnished after the designs of Teniers and Ostade, our stables after Woover-
mans, and we may learn to dance from Watteau or Zuccarelli ; in short,
there is no individual, from the emperor to the cobbler, who may not find a
model for his imitation in the works of painters, if he will but consult the
whole series from Guido to Teniers.

islands are covered when the water is high. These irriguous appearances have charms in the eye of a landscape painter, who, from some detached parts, might select a study for a foreground, at a happy moment when the water is neither too high nor too low; but the landscape gardener has a different object to effect, he must secure a constant and permanent display of water, which may be seen at a distance, and which shall add brilliancy and grandeur to the character of the scenery: it is not an occasionally meandering brook that such a palace or such a bridge requires, but it is an ample river, majestically flowing through the park, and spreading cheerfulness on all around it.

Mr. Price has written an Essay to describe the *practical* manner of finishing the banks of artificial water: but I confess, after reading it with much attention, I despair of making any practitioner comprehend his meaning; indeed, he confesses that no workman can be trusted to execute his plans. It is very true, that large pieces of water may be made too trim and neat about the edges, and that often, in Mr. Brown's works, the plantations are not brought near enough to the water; but if the banks are finished smoothly at first, the treading of cattle will soon give them all the irregularity they require: and, with respect to plantations, we must always recollect, that no young trees can be planted without fences, and every fence near the water is doubled by reflexion; consequently, all rules for creating bushes to enrich the banks are nugatory, except where cattle are excluded.

The difficulty of clothing the banks of artificial water has been a source of complaint made against Mr. Brown, for having left them bare and bald: but the river at ATTINGHAM will be sufficiently enriched by the few trees already growing on its margin, and by the plantations proposed on the island, &c.

There is a part of the river Terne, above the house, where both its banks are richly clothed with alders, and every person of discernment must admire the beauty of this scene; but if the same were continued quite to the bridge, the river would be invisible from the house and from every part of the park: how, then, is it possible that the banks of water should everywhere be covered with wood? I contend, that a broad ample channel, in proportion to the bridge, will be far more in character with the style of the house and the bridge, than the more intricate, which, on paper, is perhaps more picturesque.

If it be ridiculous to imitate nature badly in a picture, how much more ridiculous will it appear to imitate a picture badly in nature; an imitation which, after all, must be left for half a century, to be finished by the slow process of "neglect and accident."

The water at ATTINGHAM having been completed, and a new channel made to connect the river Terne with the Severn, the improvement is obvious to every person who travels the great road to Shrewsbury: it is therefore needless to elucidate these observations by any views of the place, especially as painting can give but an imperfect idea of the situation commanding that extensive range of hills which separates England from Wales.

CHAPTER X.

Of ancient and modern Gardening—Authors—Change of Style—WIMPOLE
—Terraces—at the HASELLS—at COBHAM—Art and Nature considered
—Example, BURLEY ON THE HILL.

IT is not my intention to enter into a minute history of
gardening, or, pursuing the course of some other writers, to
trace back the gradual progress of the art from Brown to
Kent, from Kent to Le Nôtre, from him to the Italians, the
Romans, the Grecians, and, ultimately, to Adam, who was "the
first gardener;" but I shall confine myself to a few observations on the change in the *fashion* of gardens, to shew how
much of each different style may be preserved or rejected
with advantage ; and, lest it should appear to some readers
that my allusions are too frequent to the late theoretical
writers on landscape gardening, it is necessary to observe,
that many of the MSS. whence I now transcribe, were written
long before Mr. Knight's and Mr. Price's works appeared ; of
course the allusions relate to other authors on the subject,
whose sentiments these gentlemen seem to have taken up,
without acknowledging that they had ever read them.

It may not be uninteresting here to mention a few of the
authors who have written on gardening, especially as the
works of some are become scarce, and are not generally
known.

I scarcely need mention the late *Horace Walpole*, who, in
his lively and ingenious manner, has given both the history
and the rules of the art better than any other theorist.

The History of Gardening is very learnedly discussed, in
a brief inquiry into the knowledge the ancients possessed of
the art, by *Dr. Faulkner;* and the same subject is more
lightly, but not less correctly or elegantly, treated by my late
ingenious friend, *Daniel Malthus, Esq.,* in a preface to his

translation of " *D'Ermenonville* de la Composition des Pay-
sages." *

Every person the least interested in this study, must have
read the beautiful " Poems of *Mason*," and " *De Lisle*," the
" Oriental Gardening of *Sir William Chambers*," and the
" Observations on Modern Gardening, by *Mr. Whately ;*"
but, perhaps, few have seen that elaborate performance, in
five volumes quarto, published in German, and also in French,
under the title of " Theorie de l'Art des Jardins, par *M.
Hirschfeld*," a work in which are collected extracts from
almost every book, in every European language, that has any
reference to the scenery of nature, or to the art of landscape
gardening. †

_____

When gardening was conducted by the geometric principles of the school of Le Nôtre, the perfection of planting

* From this gentleman I received a letter in 1795, written in so playful
a style, and so much connected with the subject of this volume, that I will
venture to insert it, even though I should incur the imputation of vanity.

" DEAR SIR,
    " I have been lately very much pleased with a letter of yours to Mr.
Price, which is so easy, friendly, and gentleman-like, that it defeats at once
the pertness of your antagonists, before you enter into the question; at the
same time, I think it as perfect an answer as if it were more laboured, and
that you have put your finger on the very pith and marrow of the question.
Even in the little snatch of acquaintance we have had together, you may
have perceived that I am rather *too much* inclined to the Price and Knight
party, and yet I own to you, that I have been often so much disgusted by
the affected and technical language of connoisseurship, that I have been sick
of pictures for a month, and almost of Nature, when the same jargon was
applied to her. I know the abilities of the two gentlemen, and am sorry
they have made themselves such pupils of the Warburtonian school, as to
appear more like Luther and Calvin than a couple of west country gentlemen, talking of gravel walks and syringas. To be sure, one would imagine
they would have broiled poor Brown, but I hope not. I suppose you know
Mr. Knight's place, his elegant house, and the enchanting valley which lies
under it : no man wants to *dot* himself about with firs who has such woods
as those. He has done nothing to spoil it, and everything that he could
have done chastely to adorn it. He has three bridges that are admirable in
their way. I was diverted with one of the reviewers, who took him for a
poor Grub-street poet, who had never seen any more gardening than the
pot of mint at his windows."
    † If I were to enumerate all those who have occasionally mentioned
gardening as a relative subject of taste, I should hardly omit the name of
any author, either ancient or modern. Some of the most ingenious hints,
and even some just principles in the art, are to be found in the works of
*Theocritus, Homer, Virgil, Petrarch, Rousseau, Voltaire, Temple, Bacon,
Addison, Home, Gilpin, Allison,* &c.

was deemed to consist in straight lines of trees, or regular corresponding forms of plantation; and, as the effect of this style of gardening greatly depended on a level surface of ground, we often find that prodigious labour was employed to remove those inequalities which nature opposed to this ill-judging taste.

At WIMPOLE, the natural shape of the surface seemed to invite this fashion for geometric forms; the ground was covered, in every direction, with trees in straight lines, circles, squares, triangles, and in almost every mathematical figure. These had acquired the growth of a century, when the taste of gardening changed; and, as every absurd fashion is apt to run from one extreme to another, the world was then told, that " *Nature abhorred a straight line;*" that perfection in gardening consisted in waving lines, and that it was necessary to obliterate every trace of artificial interference. And now many a lofty tree, the pride and glory of our ancient palaces, was rooted up, because it stood on the same line with its fellows and contemporaries: and because these ranks of sturdy veterans could not,* like a regiment of soldiers, be marched into new shapes, according to the new system of tactics, they were unmercifully cut down; not to display beautiful scenery behind them, but merely to break .their ranks: while a few were spared which could be formed into *platoons*, this was called *clumping an avenue.*

The position of all the large trees on the plain near the house, at WIMPOLE, shews the influence of fashion in these different styles; the original lines may be easily traced by the trees which remain, and the later formed clumps are scattered about, like the ghosts of former avenues, or monstrous shapes which could not be subdued.

One great advantage of WIMPOLE arises from its comparative beauty, or the contrast between the place and its

---

* That this simile may not appear ludicrous, I should observe, that the ancient gardens were often made with a reference to military dispositions; or trees were sometimes planted in conformity to the order of certain. battles; thus, at Blenheim, the square, clumps planted before Brown saw the place, were in imitation of the famous battle from whence the place was named.   And in an old map of a place in Suffolk, which, I believe, was planned by Le Nôtre, the names of regiments were given to square clumps, or platoons, of trees, which on paper resembled the positions of an army.

environs. The counties of Cambridge and Huntingdon consist generally of flat ground, while the hills are open corn
fields thinly intersected by hedges. But WIMPOLE abounds in
beautiful shapes of ground, and is richly clothed with wood;
it is, therefore, like a flower in the desert, beautiful in itself,
but more beautiful by its situation. Yet no idea of this beauty
can be formed from the approach to the house; because the
plain is everywhere covered with lofty trees, which hide, not
only the inequalities of the ground, but also the depth of
wood in every direction; and although the original straight
lines of the trees have been partially broken, the intervals
shew none of the varied scenery beyond. I do not, therefore,
hesitate to say, that, by judiciously removing some hundred
trees, the place would be made to appear more wooded : for it
frequently happens, that a branch near the eye may hide a
group of twenty trees, or a single tree conceal a whole grove.

In thus recommending the liberal use of the axe, I hope I
shall not be deemed an advocate for that bare and bald system
of gardening which has been so justly ridiculed. I do not
profess to follow either Le Nôtre or Brown, but, selecting
beauties from the style of each, to adopt so much of the
grandeur of the former as may accord with a palace, and so
much of the grace of the latter as may call forth the charms
of natural landscape. Each has its proper situation; and
good taste will make fashion subservient to good sense.

" The modern rage for natural landscape has frequently
carried its admirers beyond the true limits of improvement,
the first object of which ought to be *convenience*, and the
next *picturesque beauty*.

My taste may, perhaps, be arraigned for asserting that the
straight terrace at the HASELLS * ought not to be disturbed:
although it is a remnant of geometric gardening of the last
century, yet it is an object of such comfort and convenience,
that it would be unpardonable to destroy it, for no other
reason than because a straight walk is out of fashion; this

---

* The *Red Books* of the HASELLS and COBHAM, from whence these observations were transcribed, were written in the year 1790, before Mr. Price
published his Essays.

would be acknowledging (what I protest against) that the art of landscape gardening ought to be under the dominion of fashion.

If this terrace were constantly an object of view, or very materially offensive to the general scenery of the place, its linear direction might cut the composition, and destroy its effect as a natural landscape : in its present situation it is merely a foreground, or frame, to a pleasing picture, and the view from hence is so fine, so varied, and so interesting, that the spectator must be fastidious indeed, who could turn away disgusted, because it is seen over a clipt hedge, or with a broad flat walk in its foreground. A beautiful scene will always be beautiful, whether we view it from an alcove, a window, or a formal terrace ; and the latter, in the height of summer, may sometimes answer the purpose of an additional room or gallery, when there is much company, who delight to saunter on such an esplanade ; while the intricacies of a winding path are better calculated for a solitary walk."

" The ancient dignity of character in the house at COBHAM would be violated by the too near intrusion of that *gay prettiness* which generally accompanies a garden walk ; yet convenience and comfort require such a walk at no great distance from the house.* I shall, perhaps, astonish some of the improvers in modern serpentine gardening, by declaring that, as an appendage to this ancient mansion, I would prefer the broad and stately mall along a straight line of terrace, to their too frequently repeated waving line of beauty.

This sort of walk may, I think, be still farther encouraged, where it already in some degree exists, to the north of the kitchen-garden, which, falling from the eye, might easily be concealed from the park by a shrubbery kept low ; not to intercept the view towards the opposite bank in the park, while it would give an imaginary increase of depth to the vale beneath. And, to remove the objection of returning by the same walk, a second terrace might be carried still higher on the bank, and, by the style and accompaniment of its plan-

---

* Twelve years ago, when I first delivered these opinions, they were deemed so contrary to modern practice, that I was cautious in defending them. I have since more boldly supported my original opinion, and rejoice that the good sense of the country admits their propriety.

tation, all sameness would easily be obviated, perhaps, by making one of them a *winter walk*, planted chiefly with evergreens and shrubs.

To justify my opinion, it is necessary to guard against a misconstruction of what I have advanced, lest I may be accused of reviving the old taste of gardening.

I do not recommend the terrace as an object of beauty in all cases, but of convenience; for the same reason that I advise the proximity of a kitchen-garden, provided the principal apartments do not look upon either.

Our ancestors were so apt to be guided by utility, that they at length imagined it was in all cases a substitute for beauty; and thus we frequently see ancient houses surrounded not only by terraces, avenues, and fish-ponds, but even stables, and the meanest offices, formed a part of the view from the windows of their principal rooms. I am far from recommending a return to these absurdities; yet, in the rage for picturesque beauty, let us remember that the landscape holds an inferior rank to the historical picture; one represents nature, the other relates to man in a state of society; if we banish winter comforts from the country seats of our nobility, we shall also banish their inhabitants, who generally reside there more in winter than in summer; and there is surely no object of greater comfort and utility belonging to a garden and a country mansion, than a dry, spacious walk for winter, sheltered by such trees as preserve their clothing, while all other plants are destitute of foliage.

" Vernantesque comas tristis ademit hyems." *

[Dreary winter has stripped off the green leaves.]

I will add the opinion of a very able commentator, who, mentioning "this self-evident proposition, that a rural scene

* " In the summer season the whole country blooms, and is a kind of garden, for which reason we are not so sensible of those beauties, that at this time may be everywhere met with; but when nature is in her desolation, and presents us with nothing but bleak and barren prospects, there is something unspeakably cheerful in a spot of ground which is covered with trees, that smile amidst all the rigours of winter, and give us a view of the most gay season, in the midst of that which is the most dead and melancholy."—*Spectator*, No. 477.

And the great Lord Bacon says, " In the royal ordering of gardens, there ought to be gardens for every month in the year."

in *reality*, and a rural scene on *canvas*, are not precisely one and the same thing," says, " that point in which they differ here, is not itself without a guiding principle : UTILITY sets up her claim, and declares that, however concurrent the genuine beauty of nature and picture may be, the garden scene is hers, and must be rendered conformable to the purposes of human life ; if to these every consonant charm of painting be added, she is pleased; but by no means satisfied, if that which is convertible to use be given absolutely to wildness."—*Elements of Criticism.*

The natural situation of BURLEY differs from that of every other large place which has fallen under my consideration. To say that the house stands on a lofty hill, would be giving a very imperfect idea of its situation ; on the contrary, it ought rather to be described as a magnificent palace, built on the extremity of a vast plain, or, what is called by geographers, a table mountain, from the brow of which it boldly commands an assemblage of wood, water, lawn, and distant country, spread magnificently at its base.

The view from the principal suite of apartments, however rich and varied in itself, becomes much more interesting by the power of contrast, because the great plain to the north affords no promise of such views, and, therefore, the surprise occasioned by this unexpected scenery, is a subject worthy the attention of the improver : the effects of surprise are seldom to be produced by *art*, and those who attempt to excite it by novelty, or contrast, are in danger of falling into puerile conceits.* But where, as in the present instance, much of the natural sublime exists, this effect should be increased by every means which does not betray the insignificance of art, when compared with the works of nature.

For this reason, if the approach were brought along the straight line of avenue, gradually ascending, the situation of BURLEY would lose much of its sublimity by anticipation.

* Like those described by Sir William Chambers, in his " Chinese Gardening."

The prevalence of fashion, in all subjects of taste, will at times have its influence, but as fashion is more the effect of whim and caprice, than of reason and argument, it has been my great object to rescue landscape gardening from its fascinating power; and, while accommodating myself to the wishes of those who consult me, to the customs of the times, or to the peculiarity of various situations and characters, I hope never to lose sight of the great and essential object of my profession, the elegance, the magnificence, and the convenience of rural scenes, appropriated to the uses of a * *gentleman's habitation.*

This may be equally effected, whether we revert to the formal fashion of straight walled gardening, or adopt the serpentine lines of modern improvers, under the pretended notion of imitating nature. But there is a certain dignity of style in BURLEY, which, like the cumbrous robes of our nobility, neither can nor ought to be sacrificed to the innovation of fashion or the affectation of ease and simplicity.

Mr. Burke justly observes, that " a true artist should put " a generous deceit on the spectators, and effect the noblest " designs by easy methods. Designs that are vast, only by " their dimensions, are always the sign of a common and low " imagination. No work of art can be great but as it de- " ceives;† to be otherwise is the prerogative of nature only." This precept seems to have been overlooked in the attempts to modernise BURLEY: the spacious court, surrounded by a colonnade, has been frequently quoted as a wonderful effort of art: and when the distant country was excluded by a wall, by the village, and by trees beyond it, this ample area was undoubtedly one of the most striking appendages of a palace.‡ But the moment one side of the quadrangle is opened to the adjacent country, it shrinks from the comparison, and the

---

* By this term I mean to express scenery, less rude and neglected than the forest haunts of wild animals, and less artificial than the farmer's field, laid out for gain, and not for appearance : or, in the words of a celebrated author, "to create a scenery more pure, more harmonious, and more expressive, than any that is to be found in nature itself."

[† See our note on this subject, p. 77.—J.C.L.]

‡ Lest this should look like an implied censure on the person by whose advice the wall was removed, I must acknowledge that, till I had seen the effect, I might have adopted the same error, in compliance with the prevailing fashion of opening lawns.

long fronts of opposite offices seem extended into the vast
expanse, without any line of connexion. This comparative
insignificancy of art is nowhere more strongly exemplified
than in the large wet docks of Liverpool and Hull : while
the margins of the river are left dry by the ebbing tides, we
look with astonishment at the capacious basins, filled with a
vast body of water; but when the tide flows to the same
level, and the floodgates are thrown open, the extent and
importance of the river convert these artificial basins into
creeks or mere pools. It is, therefore, only by avoiding a
comparison with the works of nature, that we can produce the
effect of greatness in artificial objects; and a large court surrounded by buildings, can have no pretensions to be deemed
a natural object.

After removing the wall, which formed the front of the
court, a doubt arose whether the present gate and porter's
lodge should or should not remain, and how to approach the
house to the greatest advantage.

There is a certain point* of distance from whence every
object appears at its greatest magnitude : but in cases where
symmetry prevails, the distance may be rather greater, because
exact correspondence of parts assists the mind in forming an
idea of the whole. I should therefore conceive, that the effect
of surprise, of magnificence, and of the sublime, in this effort
of art, is greatly injured by seeing the interior of this ample
court, before we arrive at the entrance gate ; because that is
nearly the spot where the eye is completely filled and gratified
by the surrounding objects. But as this view should not be
momentary, I suppose the road to continue from the gate in a
straight line, till it falls into a circle with the colonnade; and
here the broad road may be intercepted with posts and chains,
to direct carriages into that course which displays the whole
area to the greatest advantage, passing nearer to the side colonnade ; shewing that in perspective, and presenting the house
at the angle to shew its depth. The manner in which this is
effected by sweeping round the court, is not to be described
by painting; because every step varies the position of the
several parts, as they advance or recede perspectively.

* This subject has been discussed in Chapter II.

Hitherto I have spoken of the north, or entrance front, and court-yard of Burley, the whole of which I would treat only as a work of art, and, if possible, exclude all view of the country. But to the south, the prospect, or natural landscape, is the leading feature for our consideration.

The steep descent from the house has been cut into a number of terraces, each supported by a red brick wall [see fig. 83]; and if these several walls had been of stone, or

[Fig. 83. View of Burley with its ancient brick terrace walls]

architecturally finished like the old costly hanging gardens of France and Italy, they might, perhaps, have added more magnificence to the house than any improvement which modern gardening could suggest; but they are mean in their forms, diminutive in their height, and out of harmony in their colour. Yet the style of the house and the steepness of the declivity will not admit of their being all taken away to slope the ground, in the manner too often practised by modern improvers.

I therefore make a compromise between ancient and modern gardening, between art and nature, and by increasing the height, or rather the depth, from the upper terrace to the lower level of the ground, I make *that* the line of demarcation between the dressed ground and the park, in the manner

explained by the view of BURLEY [fig. 84]; and happy would
it be for the magnificence of English scenery, if many such
stately terraces near a palace had been thus preserved.

[Fig. 84. View of Burley, with the low brick terrace walls removed, and a stone terrace wall substituted.]

CHAPTER XI.

Miscellaneous—Endless Variety of Situation and Character—First Impressions—Roads—Example, STOKE PARK—Scenery in Wales—Example, RÜG—Ornaments—Entrances—HAREWOOD—BLAIZE CASTLE—Adaptation of ornamental Buildings — Ornaments—Decorations—Colours—Metals.

I HAVE occasionally been asked, when visiting a beautiful spot, " Which, of all the places I had seen, was the most beautiful?"   It is impossible to define those circumstances which, on different persons, make different impressions at first sight; perfection is no more to be found in the works of nature than in those of art.   Such is the equal providence of the great Author of nature, that every place has its beauties and its deformities, and, whether situated among the mountains of Wales, or on the margin of Clapham Common, it will *not only* be endeared to its proprietor, but to the discerning stranger, by some peculiar features of beauty.

The materials of natural landscape are ground, wood, and water, to which man adds buildings, and adapts them to the scene.   It is therefore from the artificial considerations of utility, convenience, and propriety, that a place derives its real value in the eyes of a man of taste: he will discover graces and defects in every situation; he will be as much delighted with a bed of flowers as with a forest thicket, and he will be as much disgusted by the fanciful affectation of rude nature in tame scenery, as by the trimness of spruce art in that which is wild : the thatched hovel in a flower-garden, or the *treillis bocage* [grove trellis, trellis-work arched over head] in a forest, are equally misplaced.

General principles, or general designs, which may be applicable to all situations, would be alike impossible.   The painter copies, in their respective places, the eyes, the nose, and mouth, of the individual, but, without adding character, his picture will not be interesting. The landscape gardener finds ground, wood, and water, but with little more power than the painter, of changing their relative position; he adds character, by the point of view in which he displays them, or by the ornaments of art with which they are embellished. To describe by words the various characters and situations of all the places in which I have been consulted, would be tedious, and to give views of each would alter the design of this work: I shall, therefore, dedicate this chapter to a miscellaneous assemblage of extracts from different *Red Books*, without aiming at connexion or arrangement. These may furnish examples of variety in the treatment of various subjects; while the reasons on which their treatment is founded will, I hope, be deemed so far conclusive, that some general principles may be drawn from them, tending to prove that *there are rules for good taste.*

There is no principle of the art so necessary to be studied, as the effects produced on the mind by the first view of certain objects, or, rather, that general disposition of the human mind, by which it is capable of strongly receiving *first impressions.* We frequently decide on the character of places, as well as of persons, with no other knowledge of either, than what is acquired by the first glance of their most striking features; and it is with difficulty, or with surprise, that the mind is afterwards constrained to adopt a contrary opinion.

Thus, if the approach to a house be over a flat plain, we shall pronounce the situation to be flat also, although the ground immediately near the house be varied and uneven; whilst, on the contrary, if the road winds its course over gentle hills and dales, and at length ascends a steep bank to the house, we shall always consider it as standing on an eminence, although the views from the house may be perfectly flat.

I have, therefore, watched, with nice attention, the first ideas which have occurred to me in visiting any new subject; and if a more intimate knowledge of it induces me afterwards to alter my opinion, I then inquire into the causes which influenced my former false judgment, that I may by this means increase or diminish them accordingly.*

One of the first objects of improvement should be, to adapt the character of the grounds to that of the house; and both should bear some proportion to the extent of property by which they are surrounded.

" At STOKE, in Herefordshire, the house and park are as perfectly separated from each other by a turnpike road, as if they were the property of different persons; and both are seen from that road in the most unfavourable points of view. Of the house little is visible except the roof and chimneys; and, with respect to the park, which naturally abounds with the most pleasing shapes of ground, richly clothed with wood, the road passes so immediately at the foot of the declivity, that the whole appears foreshortened, and all its beauties are entirely lost. To divert the course of this road, therefore, becomes the first object of improvement." †

I have, on several occasions, ventured to condemn as false taste, that fatal rage for destroying villages, or depopulating a country, under the idea of its being necessary to the import-ance of a mansion: from the same *Red Book* the following extract is taken :—

" As a number of labourers constitutes one of the requisites of grandeur, comfortable habitations for its poor dependants ought to be provided. It is no more necessary that these habitations should be seen immediately near the palace, than that their inhabitants should dine at the same table; but if their humble dwellings can be made a subordinate part of the

* The situation of the HASELLS, of BURLEY, and of STONEASTON, on the extremity of table-land, may serve as examples.

† This has been done, and the improvement to the place is equally felt by the proprietor, and conspicuous to every stranger who travels from Led-bury to Hereford. It seldom happens that both the public and the indivi-dual are benefited, by altering the course of a high road; but their mutual advantage ought to be studied. It often happens, that the basis of all im-provement depends on removing a public road, of which examples occurred in the following places : ABINGTON HALL, ADLESTROP, BAYHAM, KENWOOD, PANSHANGER, GARNONS, HASELLS: these I mention in preference to many others, because the improvement is obvious to the public.

general scenery, they will, so far from disgracing it, add to the dignity that wealth can derive from the exercise of benevolence. Under such impressions, and with such sentiments, I am peculiarly happy in being called upon to mark a spot for new cottages, instead of those which it is necessary to remove, not absolutely because they are too near the house, for that is hardly the case with those cottages in the dell, but because, the turnpike-road being removed, there will be no access for the inhabitants but through a part of the park, which cannot then be private. I must advise, however, that some one or more of the houses in this dell be left, and inhabited either as a keeper's house, a dairy, or a menagerie, that the occasional smoke from the chimneys may animate the scene. The picturesque and pleasing effect of smoke ascending, when relieved by a dark hanging wood in the deep recess of a beautiful glen like this, is a circumstance by no means to be neglected."

As an example of a place in a mountainous country, the following extract from the *Red Book* of Rüg, in North Wales, is subjoined: " At a period when the ancient family honours of a neighbouring country are rooted out with savage barbarity, I rejoice in an opportunity of contributing my assistance to preserve in this, every vestige of ancient or hereditary dignity ; and I should feel it a kind of sacrilege in taste to destroy an atom of that old, ruinous, and almost uninhabitable mansion at Rüg, if it were to be replaced by one of those gaudy scarlet houses, which we see spring up, like mushrooms, in the neighbourhood of large manufacturing towns. I am, however, restrained from indulging, to its full extent, my veneration for antiquity, by reflecting that modern comfort and convenience are the first objects to be consulted in the improvement of a modern residence ; and, therefore, I trust I shall neither incur the censure of those who know and feel the comforts of the age we live in, nor offend the genius of the place, by ' calling from the vasty deep the angry spirits' of Owen Glendwr of Burgontum, who formerly inhabited this domain.

" In a country like that of North Wales, abounding in

magnificent scenery, the views from the house should rather aim at comfort and appropriation of landscape, than extensive prospect; because the latter may be had from every field or public-road on the mountains; and the attempt to make a large park or domain would be fruitless, where a lawn of a thousand acres would appear but a small spot, compared with the wide expanse of country seen from the neighbouring hills. I should therefore advise the lawn to be confined within the compass of forty or fifty acres; yet, from the variety of its surface, and the diversity of objects it contains, there will be more real beauty, and even magnificence, within this small enclosure, than in other parks of many hundred acres.

However partial we may be to grand and extensive prospects, they are never advisable for the situation of a house, in which convenience and comfort should doubtless take the lead of every other consideration. The frequent rains, and violent storms of wind, to which all mountainous countries are exposed, have taught the inhabitants not only to choose warm valleys for their houses, but have also introduced a style of architecture peculiarly suited to those situations: the small towns of Llangollen and Corwen, as well as those in the mountains of Switzerland, have all low sheds, or penthouses, under which the inhabitants may take shelter from occasional driving storms. The arcade of Gothic architecture is infinitely more applicable to such situations than the lofty portico of Greece, which is rather calculated for those warm regions where man wants protection from the vertical beams of a burning sun. I hope, therefore, that both the character and situation of Rüg, will justify a* design for a new house, which may possess a degree of grandeur and magnificence not incompatible with modern convenience."

There is no circumstance in which bad taste is so conspicuous, as in the misuse of ornaments and decorations; an ob-

---

* This *Red Book* having been written in 1793, it was before I had the advantage of my son's architectural assistance; and the design here mentioned was that of my ingenious friend Mr. Wilkins, who built one of the best houses in England for Earl Moira, at Donnington, in a correct Gothic style, and under whom my son was at that time studying: for reasons, which I had no right to inquire into, the plan for the house was not adopted; in every other respect, however, my plans have there been followed in the most gratifying manner.

servation equally applicable to all the polite arts, and not less true with respect to eloquence, poetry, music, and painting, than to architecture and gardening.

Thus, for instance, a rural scene may be delightful without any building or work of art, yet, if judiciously embellished by artificial objects in character with the scene, the landscape will be more perfect; on the contrary, if encumbered by buildings in a bad taste, or crowded by such as are too large, too small, or in any respect inapplicable, however correct they may be as works of art, the scene will be injured, and thus a thatched hovel may be deemed an ornament, where a Corinthian temple would be misplaced, or *vice versâ*.

In this miscellaneous chapter may properly be inserted some specimens of various buildings, to elucidate the truth of an observation, which hardly seems to require enforcing; yet the frequent introduction of ornamental buildings, copied from books, without reference to the character and situation of the scenery, is not less fatal to the good taste of the country, than it would be to the life of individuals, to use medical prescriptions without inquiring into the nature and cause of diseases.

The facility with which a country carpenter can erect small buildings intended for ornament, may, perhaps, account for their frequency; but I am not ashamed to confess, that I have often experienced more difficulty in determining the form and size of a hovel, or a park entrance, than in arranging the several apartments of a large mansion; indeed, there is no subject on which I have so seldom satisfied my own judgment, as in that of an entrance to a park.

The custom of placing a gate between two square boxes, or, as it is called, a " pair of lodges," has always appeared to me absurd, because it is an attempt to give consequence to that which in itself is mean; the habitation of a single labourer, or, perhaps, of a solitary old woman, to open the gate, is split into two houses for the sake of childish symmetry;* and very often the most squalid misery is found in the person

* As this absurd fashion of a pair of lodges deserves to be treated with ridicule, I cannot help mentioning the witty comment of a celebrated lady, who, because they looked like tea-caddies, wrote on two such lodges in large letters, GREEN and BOHEA.

thus banished from society, who inhabits a dirty room of a
few feet square. *

It is the gate, and not the dwelling of the person who
opens it, that ought to partake of the character of the house,
where architectural display is necessary; and this principle
seems to point out the true mode of marking the entrance to
a place. Instead of depopulating villages, and destroying
hamlets in the neighbourhood of a palace, I should rather wish
to mark the importance of the mansion, and the wealth of its

* [The existence of so many lodges, containing accommodation of this
description throughout the country, by the sides of the public-roads; and of
equally miserable houses for gardeners, in the back sheds of hothouses in
kitchen-gardens, almost everywhere; shews how very little sympathy there
exists between the rich and the poor in England. The cause of this, we
believe to be, in most cases, want of reflection, and ignorance of the moral
fact, that the more extended our sympathy is for our fellow-creatures, the
greater will be our enjoyments. Another cause of the miserable accommodation in the lodges at gentlemen's gates, and also in gardeners' houses,
may be traced to the want of sympathy with those whom they consider
beneath them, on the part of architects, landscape gardeners, and builders.
The greater number of these persons being sprung from the people, necessarily
have more or less the character of *parvenus*, when introduced into the society of the higher classes. Observing in this class the contempt and disdain with which they look on the mass of the people, they naturally avoid
everything which may remind either themselves, or the society into which
they have been introduced, of their low origin. Hence they fear, that, to
advocate the cause of the class from which they sprang, to be thought to
care about their comfort, or to suggest improvements in their dwellings,
would remind the employer of their origin, and be thought derogatory to
their newly acquired station. An architect, or a landscape gardener, therefore, who has sprung from the people, is rarely found with the moral
courage necessary to propose, to the rich who employ him, ameliorations of
any kind for the poor. In the course of thirty years' observation we have
found this to hold good, both in Scotland and England, and in the former
country more particularly. How many improved plans of kitchen-gardens,
and new ranges of hothouses, have there not been carried into execution in
Scotland, since the commencement of the present century, and yet how few
improved gardeners' houses have been built within the same period. Mr.
Repton, having been born a gentleman, was under no such dread as that to
which we have alluded, and we find him continually advocating the improvement of cottages. It is clearly both the duty and interest of the higher
classes, to raise, by every means, the standard of enjoyment among all
that are under them. Humanity dictates this line of action, as well as prudence; for it would be easy to shew, that, if improvement did not pervade
every part of society, the breach between the extreme parts would soon become so great as to end in open rupture. The more the comforts, enjoyments,
and even luxuries, of every servant, from the highest to the lowest, are increased, the more will they be useful, assiduous, and attached to their masters. Every servant feels this, and by every master it either is or will be
felt.—*Gard. Mag. vol.* iv. *p.* 46. *See also the same work, vol.* viii. *p.* 257 *to*
266. J. C. L.]

domain, by the appearance of proper provision for its poor dependants ; the frequent instances I have witnessed, where the industrious labourer had many miles to walk from his daily task, have strongly enforced the necessity, not to say the humanity, of providing comfortable and convenient residences for those who may have employment about the grounds. It is thus that the real importance of a place might be distinguished by the number of cottages, or, rather, substantial houses, appropriated to the residence of those belonging to the place ; this would truly enrich the scenery of a country, by creating a village at the entrance of every park ; it is not by their number only, but by the attention to the neatness, comfort, and simple ornament of such buildings, that we should then judge of the style of the neighbouring palace ; and whether the houses were of clay and thatched, or embellished with the ornaments of architecture, there would be equal opportunity for the display of good taste.

The entrance to HAREWOOD PARK, from a large town of the same name, may serve as a magnificent specimen of this kind of importance ; and although, in this instance, the character and peculiar circumstances of this splendid palace are properly supported, by the regularity and substantial manner in which the town is built and ornamented, yet, in more humble situations, the same attention to the repair and neatness of the adjoining cottages, would confer adequate propriety to this mode of entrance. Various specimens of this attention may be seen in the roads near the following places :—BAB-WORTH, BETCHWORTH, BUCKMINSTER, CATTON, LIVERMERE,

[Fig. 85. Elevation of the proposed entrance to Harewood Hall.]

2 K

PANSHANGER, PRESTWOOD, STOKE PARK, SUTTONS, SCARIS-
BRIC, TENDRING, &c.

If the entrance to a park be made from a town, or village,
the gate may, with great propriety, be distinguished by an arch,
as in that of HAREWOOD [fig. 85], where the approach from
Weatherby, after passing along a straight road intended to be
planted on each side, is terminated by a town regularly built

[Fig. 86.   View of the town of Harewood, shewing the gateway to Harewood Hall at the farther end.]

of the most beautiful stone, at the end of which an arched
gateway forms the entrance to one of the finest palaces in
England [see fig. 86].

[Fig. 87.   View of the entrance to Blaize Castle before the lodge was built.]

In determining the sort of entrance proper for BLAIZE
CASTLE, the name of the place caused some difficulty; the

house to which the castle belongs, neither does nor ought to partake of any Gothic character, yet there appeared some incongruity in making the entrance in the Grecian style of architecture to accord with the house, which is nowhere seen from the road [see fig. 87], while the castle is a conspicuous feature, and gives a name to the place; I, therefore, recommended the design [fig. 88], as a proper object to attract

[Fig. 88. Entrance lodge to Blaize Castle.]

notice in the approach, which is one of the most interesting and romantic.*

An arched gateway at the entrance of a place is never used with so much apparent propriety as when it forms a part of a town or village, at least, it should be so flanked by lofty walls as to mark the separation between the public and the park, and increase the contrast; but when seen in contact with a low park-pale, or even an iron palisade, it appears to want connexion; it looks too ostentatious for its utility, and

* After passing through a wood, the road arrives at a cottage on the side of a hill, from whence the house appears, across a deep wooded glen which was deemed impassable. However, by cutting away the face of the rock in some places, and building lofty walls in others, to support the road, and by taking advantage of the natural projections and recesses to make the necessary curvatures, carriages now pass this tremendous chasm with perfect ease and safety.

Where man resides, Nature must be conquered by art: it is only the ostentation of her triumph, and not her victory, that ought to offend the eye of taste.

I doubt whether it would not lessen the pleasure we derive from viewing the magnificent Grecian arches at Burlington House and at Blenheim, if the side-walls were lower.*

In recommending the use of an arch, I must guard against being misunderstood, by mentioning several circumstances which I deem objectionable.

1st. The arch should not be a mere aperture in a single wall, but it should have depth in proportion to its breadth.

2nd. It should have some visible and marked connexion either with a wall, or with the town to which it belongs, and not appear insulated.

3rd. It should not be placed in so low a situation, that we may rather see over it than through it.

4th. Its architecture should correspond with that of the house, in style, if not in order; that is, the Grecian and Gothic should be kept separate, although the design may not be copied from the house. And,

Lastly. Neither the house should be visible from the entrance, nor the entrance from the house, if there be sufficient distance between them to make the approach through a park, and not immediately into a court-yard; the two last general rules are equally applicable to every sort of entrance, as well as that through an arch; yet there are certain situations where the latter cannot be avoided; of this, an instance occurred in STOKE PARK, Herefordshire, where the gate and the cottage near it were disguised by the portico, represented in the following sketch [fig. 89, in p. 254]; which forms a pavilion, or covered seat, adjoining to the walk in the shrubbery.

In various situations various expedients have been adopted; thus, at ANTONY, I recommended, near the gate, a cottage, over which is a room, to command the fine view of the harbour, &c. At St. JOHN's, in the Isle of Wight, two cottages covered with flowering creepers, attract the notice of all who visit the island; and while one is a comfortable residence for a family, the other consists of a room near the road side, from whence the mind derives peculiar satisfaction

---

* This remark is less applicable to a Gothic entrance, because, if it is correct, it may be supposed a fragment of some more extensive building; but a Grecian arch, in this country, must be modern, and cannot properly be a ruin, except by design.

in seeing the constant succession of visitors who leave their homes in search of happiness. In some places the cottage is more conspicuous, by dividing the road to the house from the public road, as at MILTON; but, in most cases, I have endeavoured to conceal the cottage, when it is quite solitary, among the trees, only shewing the gate of entrance.

Concerning gates, it may not be improper to mention my opinion, with reasons for it.

1st. As an entrance near a town, I prefer close wooden gates, for the sake of privacy, except where the view is only into a wood, and not into the open lawn.

2nd. The gates should be of iron, or close boards, if hanging to piers of stone or brick-work; otherwise an open or common field-gate of wood appears mean, or as if only a temporary expedient.

3rd. If the gates are of iron, the posts or piers ought to be conspicuous, because an iron gate hanging to an iron pier of the same colour, is almost invisible; and the principal entrance to a park should be so marked that no one may mistake it.

4th. If the entrance-gate be wood, it should, for the same reason, be painted white, and its form should rather tend to shew its construction, than aim at fanciful ornament of Chinese, or Gothic,* for reasons to be explained, in speaking of decorations.

It is not sufficient that a building should be in just proportions with itself; it should bear some relative proportion to the objects near it. The example here given [fig. 89] is the Doric portico at STOKE PARK, in Herefordshire, where the size of the building was regulated by a large oak and a young plantation near it: had this building been more lofty, it would have overpowered the young trees, by which it is

* That I may not appear too severe in my comments upon those fanciful forms *called Gothic*, I am not ashamed to acknowledge, that, when I first retired into the country, I began the improvements to my own residence in Norfolk, by putting a sharp-pointed window in a cottage seen from my house; and in my former work a design was inserted for a wooden-gate, which I then deemed applicable to the Gothic character, before I became better acquainted with subjects of antiquity.

surrounded, and a smaller building would have appeared diminutive so near to the neighbouring large oak; I there-

[Fig. 89. View of the Doric portico in Stoke Park, in which the size of the building is in harmony with the size of the adjoining trees.]

fore judged, that the best rule for the dimensions of the columns was rather less than the diameter of the oak, and this,

[Fig. 90. Gothic Cottage.]

of course, determined the whole proportion of the Doric portico.

So prevalent is the taste for what is called Gothic, in the neighbourhood of great cities, that we see buildings of every description, from the villa to the pigsty, with little pointed arches, or battlements, to look like Gothic; and a Gothic dairy is now become as common an appendage to a place, as were formerly the hermitage, the grotto, or the Chinese pavilion. Why the dairy should be Gothic, when the house is not so, I cannot understand, unless it arises from that great source of bad taste, to introduce what is called a pretty thing, without any reference to its character, situation, or uses. Even in old Gothic cottages we never see the sharp-pointed arch, but often the flat arch of Henry VIII., and perhaps there is no form more picturesque for a cottage than buildings of that date, especially as their lofty perforated chimneys not only contribute to the beauty of the outline, but tend to remedy the curse of the poor man's fire-side, a smoky house [see fig. 90].

[Fig. 91. Rustic thatched hovel, on the summit of a naked brow, commanding views in every direction.]

There are few situations in which any building, whether of rude materials or highly-finished architecture, can be properly introduced without some trees near it. Yet the summit of a naked brow, commanding views in every direction, may require a covered seat or pavilion; for such a situation, where an architectural building is proper, a circular temple with a dome, such as the temple of the Sybils, or that of Tivoli, is best calculated; but in rude scenery, as on a knoll or pro-

montory in a forest, the same idea may be preserved in a
thatched hovel supported by rude trunks of trees; yet, as the
beauty of such an object will greatly depend on the vegetation, it should be planted with ivy, or vines; and other
creeping plants should be encouraged to spread their foliage
over the thatch [see fig. 91].

[Fig. 92. Principal view from the house at Blaize Castle, which is considered too sombre for the character
of a villa.]

The principal view from the house at BLAIZE CASTLE, is
along that rich glen of wood through which the approach has
been made, as already described: in this view, the castle,
although perfectly in harmony with the solemn dignity of the
surrounding woods, increases, rather than relieves, that apparent solitude which is too *sombre* for the character of a villa
[see fig. 92].

Some object was wanting to enliven the scenery: a temple,
or a pavilion, in this situation, would have reflected light, and
formed a contrast with the dark woods; but such a building
would not have appeared to be inhabited; this cottage [fig. 93],
therefore, derives its chief beauty from that which cannot
easily be expressed by painting—the ideas of motion, animation, and inhabitancy, contrasted with those of stillness and
solitude.   Its form is meant to be humble, without meanness;
it is, and appears, the habitation of a labourer who has the
care of the neighbouring woods; its simplicity is the effect of

art, not of neglect or accident; it seems to belong to the mansion, and to the more conspicuous tower, without affecting to imitate the character of either.

[Fig. 93. View from the house at Blaize Castle, enlivened by a cottage in the distance.]

The propensity for imitation, especially where no great trouble or expense is incurred, has made treillage ornaments so common, that some observations concerning them may be expected in this work, especially as I believe I may have contributed originally to their introduction;* but I little thought how far this flimsy ornament might be misapplied.

The treillages of Versailles and Fontainbleau were of substantial carpentry, preserving architectural proportions, in which plants were confined and clipped to form a sort of vegetable and architectural *berceau*, or *cabinet de verdure;* these being made of strong wood, and painted, were more costly and more durable; and, as they only formed a frame for the plants, they might perish, without injuring the forms of these leafy buildings; but the English treillage is made of such slight materials, and so slightly put together, that they can hardly outlive the season for which they are erected; this, however, is no objection where they are used in flower-gardens, or where they are merely to be considered as garden

* To conceal a house near the entrance of a flower-garden at TAPLOW, I covered the whole with treillage many years ago.

2 L

sticks supporting plants; but, when added to architectural houses, and made the supporters of a heavy roof, or even a canvas awning, it looks as if the taste of the country were verging to its decline; since shade might be obtained by the same awning supported by iron, if architectural forms and projections are to be despised,* or discarded.

I should therefore suppose that no treillage ought to be introduced, except in situations where creeping plants may be fastened to the framing, which should be stout in proportion to its height, or its intentions [see fig. 94]: it is a com-

[Fig. 94. Greenhouse, with the piers covered externally with trellis-work.]

mon mistake to suppose a thing will look light by being slender; if it be not equal to its office by its apparent substance, it will look *weak*, not *light*; but the lattice-work is supposed to support nothing, and may therefore be of any dimensions, and, being always painted, it will be invisible at a distance.

---

* This observation is the result of having lately seen some houses containing rooms of admirable· proportion, and well connected together, but which externally appear to be built of lath and paper, or canvas; perhaps the late frequency of living in camps, or at watering places, may have introduced this unsubstantial mode of building, which looks as if it were only intended for the present generation, or, rather, for the present year.

Architectural Ornaments and Decorations.

I could wish, in speaking of architecture, if the use of language would admit of such distinction, to make a difference between the words *ornament* and *decoration.* The former should include every enrichment bearing *the semblance of utility;* the latter is supposed to have no relation whatever to the uses or construction of the building; thus, for instance, a house may answer all the purposes of habitation without a column, a pilaster, an entablature, a pediment, a dome, an arcade, or a balustrade, which I call the external *ornaments* of Grecian architecture.*

I include under the word *decorations*—statues, vases, basso-relievos, sculpture, &c., which have no use, but as additional enrichments to the ornaments of architecture; on the contrary, where these *decorations* are applied to plain buildings without *ornaments,* they are marks of bad taste.†

The ornaments of architecture must be correct in design, since no degree of costliness in their materials or their workmanship can compensate for any defect in proportion, order, or disposition. The eye of good taste will be equally offended

---

* That these ornaments, although not absolutely necessary, should appear to be *useful,* is evident, from the disgust we feel at seeing them improperly applied; as in a column without an entablature, or an arch supporting nothing, or a pediment without a roof; but I do not consider columns, or pilasters, as ornaments, when used, as we often see them, to the doors of houses; they may then more properly be called decorations in a bad taste. A column is the most sumptuous ornament of Grecian architecture, and should never be subordinate to any other part of the edifice; it should either belong to the entablature and cornice of the building, or it should be wholly omitted.

If the door requires a projecting covering, it is far better to support it by consoles, or cautlivres, or even small cast-iron pillars, without architectural pretensions, than by two diminutive columns, which bear no proportion to the buildings against which they are attached.

This observation, however, does not include those porticos to churches or public buildings, which form a colonnade on so extended a scale, that they become, in a manner, detached and principal; of this kind are the magnificent and useful colonnade at STOKE POGIES, and that added by the same architect to the garden front of FROGMORE.

† Instances of this often occur in the neighbourhood of large cities and towns, where the taste of a carpenter, and not of an architect, puts balustrades to houses without any entablatures, or, perhaps, places them in a garret window, while the plain parapet wall is loaded with Mercuries, vases, pine-apples, eagles, acorns, and round balls.

with columns too large or too small, too near or too far apart;
in short, with every deviation from the established rules of
the respective orders, whether such column be composed of
marble, of stone, or of plastered brick-work; the costliness of
the material makes no difference in the design; but this is
not the case with decorations.   The cheapness and facility
with which good designs may be multiplied in *papier mâché*,
or putty composition, have encouraged bad taste in the lavish
profusion of tawdry embellishment.

This consideration leads me to assert, that every species
of enrichment or decoration ought to be costly, either in its
materials or in its workmanship: and if we attend to the
common opinion of all, except children and savages, we shall
find that no real value is attached to any decoration, except
upon this principle; on the contrary, it becomes contemptible
in proportion as it *affects to seem what it is not.*[*]

The idea of costliness in ornament is increased by its
rarity, or, rather, by its being used only where it is most conspicuous, and this sort of economy is observable even in the
works of nature; for instance, the most beautiful coloured
feathers of birds are on the surface, while those for use, rather
than for shew, are generally of a dirty brown; it may also be
observed, that those butterflies, or moths, whose wings are
ornamented on the under side, generally bear them erect;
while those which have the upper side most beautiful, generally spread them flat.   The same remark may be extended
to all the vegetable tribe; every flower, and every leaf, has one
side more ornamented, more glossy, more vivid, or more
highly finished than the other, and this is always the side
presented to the eye.   Hence we are taught, by the example
of nature, not to lavish decorations where they cannot generally be seen.[†]

---

[*] If a lady of high rank were to decorate her person with gauze and gilt
paper, with glass beads, and the feathers of common English birds, instead
of muslins and gold lace, diamonds, and the feathers of an ostrich, or a bird
of paradise, although she might be equally brilliant, and even dispose her
dress with grace and fancied taste, we should pronounce it *tromperie*, as
affecting to *seem what it is not.*

[†] Good taste can only be acquired by leisure and observation; it is not,
therefore, to be expected in men, whose time is fully employed in the more
important acquirement of wealth or fame; while on certain subjects of taste
the most elegant women often excel the most learned men; and although

While treating on the subject of ornaments and decorations, I must not omit to mention colours, since improper colouring may destroy the intended effect of the most correct design, and render ridiculous what would otherwise be beautiful.*

Both the form and the colour of a small house in LANGLEY PARK [fig. 95], rendered it an object unworthy of its situation; yet, from peculiar circumstances, it was not deemed

[Fig. 95. House in Langley Park, the architecture of which is unworthy of its situation.]

advisable, either to remove it, or to hide it by plantations. I therefore recommended a Doric portico to cover the front; and thus a building formerly unsightly, because out of character with the park, became its brightest ornament, doing honour to the taste and feelings of the noble proprietor, who preserved the house for having been a favourite retreat of his

they may not have investigated the causes of the pleasure they either derive or communicate, yet they are more exquisitely sensible to both. This, if it were necessary, might be used as an apology for occasionally introducing allusions more familiar than the philosophic reader may deem conformable to the nature of a didactic work.

* I cannot help mentioning, that, from the obstinacy and bad taste of the Bristol mason who executed the design, page 251, I was mortified to find that Gothic entrance built of a dark blue stone, with dressings of white Bath stone; and in another place, the intention of the design, page 254, was totally destroyed, by painting all the wood-work of this cottage of a bright pea-green. Such, alas! is the mortifying difference betwixt the design of the artist, and the execution of the artificer.

mother, and which, thus ornamented, may be considered as a
temple sacred to filial piety [see fig. 96].

[Fig. 96. House at Langley Park as altered.]

In the following instances there is something more than
harmony of colours, there is an association from habit, which
causes part of our pleasure or disgust.

A compact red house displeases from the meanness of its
materials, because we suppose it to be of common red bricks,
although it may, perhaps, be of the red stone of Herefordshire.

On the contrary, a large pile of red buildings is not so displeasing ; witness the houses of COBHAM, GLEMHAM, &c., and
the royal palaces of St. James's, Hampton Court, Kensington,
&c. ; but, perhaps, the weather-stains of time may have contributed more than the quantity to reconcile us to the colour
of these large masses.

Lime-whited houses offend the eye, partly from the violent glare, and partly from the associated meanness of a lath
and plaster building ; but if a little black and yellow be mixed
with the lime, the resemblance to the colour of stone satisfies
the eye almost as much as if it were built of the most costly
materials, witness WOODLEY, BABWORTH, TAPLOW, &c.

To produce effect by difference of colour in buildings, such
as red and yellow bricks, black and white flints, or even edging brick-work with dressings of stone, is the poor expedient
of the mere bricklayer ; the same may be observed of that

paltry taste for pointing the joints of brick-work to render them more conspicuous, and, of course, more offensive.

As a general principle, I should assert, that no external effect, or light and shade on a building, ought to be attempted, except by such projections or recesses as will naturally produce them, since every effect produced by colour is a trick, or sham expedient; and on the same principle a recess in the wall is preferable to a painted window, unless it is actually glazed.

With respect to the colour of sashes and window-frames, I think they may be thus determined with propriety, first observing that, from the inside of the room, the landscape looks better through bars of a dark colour; but on the outside, in small cottages, they may be green, because it is a degree of ornament not incompatible with the circumstances of the persons supposed to inhabit them, and even in such small houses as may be deemed cottages, the same colour may be proper; but in proportion as it approaches to a mansion, it should not derive its decoration from so insignificant an expedient as colour, and, therefore, to a gentleman's house the outside of the sashes should be white, whether they be of mahogany, of oak, or of deal, because, externally, the glass is fastened by a substance which must be painted, and the modern sash-frames are so light, that, unless we see the bars, the houses appear at a distance unfinished, and as having no windows. In palaces, or houses of the highest description, the sash-frames should be gilt, as at HOLKHAM, WENTWORTH, &c. The effect of gold in such situations can hardly be imagined by those who have never observed it; and even at THORESBY, where the house is of red brick, the gilding of the sashes has wonderfully improved its importance.

There is a circumstance with respect to gold and gilding, of which few are aware who have not studied the subject. The colour of gold, like its material, seems to remove all difficulties, and makes everything pleasing; this is evident on viewing a finely coloured picture on a crimson hanging, with or without a gold frame; two discordant colours may be rendered more harmonious by the intervention of gilding, it is never tawdry or glaring, the yellow light catches on a very small part of its surface, while the brown shadows melt into the adjoining colours, and form a quiet tint, never offensive:

gold ornament may be applied to every colour, and every shade, and is equally brilliant, whether in contact with black or white.

All ornaments of gold should be more plain and simple than those of silver; not only because the costliness of the material renders the costliness of workmanship less necessary, but because the carved or enriched parts reflect very little light or brilliancy, compared with those that are plain.

On the contrary, in silver ornaments, if the surface be too plain, we annex the ideas of tin or pewter, and it is only by the richness, or the embossing, that its intrinsic value becomes apparent.

These remarks are applicable to gold and silver plate,* as well as to every species of ornament, in which those metals can be used.

Since the improvement in the manufactory of cast-iron has brought that material into more frequent use, it may not be improper to mention something concerning the colour it ought to be painted. Its natural colour, after it is exposed to wet, is that of rusty iron, and the colour of rust indicates decay; when painted of a slate colour it resembles lead, which is an inferior metal to iron; and if white or green, it resembles wood: but if we wish it to resemble metal, and not appear of an inferior kind, a powdering of copper or gold dust on a green ground, makes a bronze, and perhaps it is the best colour for all ornamental rails of iron. In a cast-iron bridge at WHITTON, the effect of this bronze colour, mixed with gilding,† is admirable; and for the hand-rails of staircases it is peculiarly appropriate.

---

* Lest it should be objected that I am going beyond the precise boundaries of my profession, either as a landscape gardener, or as an architect, I shall observe, that the professor of taste in those arts must necessarily have a competent knowledge of every art in which taste may be exercised. I have frequently given designs for furniture to the upholsterer, for monuments to the statuary, and to the goldsmith I gave a design for one of the most sumptuous presents of gold plate which was ever executed in this country: it consisted of a basin, in the form of a broad flat vase, and pedestal, round which were the figures of Faith, Hope, and Charity; the former spreading her hand over the water, as in the act of benediction; and the two latter supporting the vase, which resembled a baptismal font: the whole was executed in gold, and was the present of a noble duke to his son, on the birth of his first child.

† Those who have seen the gilded domes of Constantinople, mention

With respect to wooden fences, or rails, it is hardly necessary to say, that the less they are seen the better; and therefore a dark, or, as it is called, an invisible green, for those intended to be concealed, is the proper colour; perhaps there can hardly be produced a more striking example of the truth, " that *whatever is cheap*, is *improper for decorations*," than the garish ostentation of white paint, with which, for a few shillings, a whole country may be disfigured, by milk-white gates, posts, and rails.

them with admiration; and from the observations I have made on the effect of external gilding in large masses, I have often considered gilding the dome of St. Paul's as a subject worthy of this nation's wealth and glory. This idea will, I doubt not, excite ridicule from those who have never observed or studied the wonderful, the pleasing, the unexpected, and harmonious effect of gilding on smooth surfaces.

CHAPTER XII.

Architecture and Gardening inseparable—Some Inquiry into the Forms
and Arrangements of different Eras—Situation and Arrangement of
MICHEL GROVE—Singular Character of the House—Change in Customs
and Manners alters Uses of Rooms—An extended Plan—Example,
GARNONS—A contracted Plan—Example, BRENTRY HILL, &c.

IT has been objected to my predecessor, Mr. Brown, that
he fancied himself an architect.   The many good houses built
under his direction, prove him to have been no mean pro-
ficient in an art, the practice of which he found, from experience, to be inseparable from landscape gardening: he had not
early studied those necessary, but inferior branches of architecture, better known, perhaps, to the practical carpenter than
to Palladio himself: yet, from his access to the principal
palaces of this country, and his intercourse with men of
genius and science, added to his natural quickness of perception, and his habitual correctness of observation, he became
acquainted with the higher requisites of the art, relating to
form, to *proportion*, to *character*, and, above all, to *arrangement.**

* Mr. Brown's fame as an architect seems to have been eclipsed by his
celebrity as a landscape gardener, he being the only professor of one art,
while he had many jealous competitors in the other.   But when I consider
the number of excellent works in architecture designed and executed by
him, it becomes an act of injustice to his memory to record, that, if he was
superior to all in what related to his own peculiar profession, he was inferior to none in what related to the comfort, convenience, taste, and
propriety of design, in the several mansions and other buildings which he
planned.   Having occasionally visited and admired many of them, I was
induced to make some inquiries concerning his works *as an architect*, and,
with the permission of Mr. Holland, to whom, at his decease, he left his
drawings, I insert the following list:—

   For the Earl of Coventry.   Croome, house, offices, lodges, church, &c.,
        1751.
   The same.   Spring Hill, a new place.
   Earl of Donegal.   Fisherwick, house, offices, and bridge.
   Earl of Exeter.   Burleigh, addition to the house, new offices, &c.
   Ralph Allen, Esq. near Bath, additional building, 1765.
   Lord Viscount Palmerston.   Broadland, considerable additions.
   Lord Craven.   Benham, a new house.

These branches of architecture are attainable without much early practice, as we have seen exemplified in the designs of certain noblemen, who, like Lord Burlington, had given their attention to this study. A knowledge of *arrangement*, or *disposition*, is, of all others, the most useful: and this must extend to external appendages as well as to internal accommodation.

Robert Drummond, Esq. Cadlands, a new house, offices, farm buildings, &c.
Earl of Bute. Christ Church, a bathing-place.
Paul Methuen, Esq. Corsham, the picture gallery, &c.
Marquis of Stafford. Trentham Hall, considerable alterations.
Earl of Newbury. House, offices, &c., 1762.
Rowland Holt, Esq. Redgrave, large new house, 1765.
Lord Willoughby de Broke. Compton, a new chapel.
Marquis of Bute. Cardiff Castle, large additions.
Earl Harcourt. Nuneham, alterations and new offices.
Lord Clive. Clermont, a large new house.
Earl of Warwick. Warwick Castle, added to the entrance.
Lord Cobham. Stowe, several of the buildings in the garden.
Lord Clifford. Ugbrooke, a new house.

To this list Mr. Holland added: " I cannot be indifferent to the fame " and character of so great a genius, and am only afraid lest, in giving the " annexed account, I should not do him justice. No man that I ever met " with understood so well what was necessary for the habitation of all " ranks and degrees of society ; no one disposed his offices so well, set his " buildings on such good levels, designed such good rooms, or so well " provided for the approach, for the drainage, and for the comfort and " conveniences of every part of a place he was concerned in. This he did " without ever having had one single difference or dispute with any of his " employers. He left them pleased, and they remained so as long as he " lived ; and when he died, his friend, Lord Coventry, for whom he had " done so much, raised a monument at Croome to his memory."

Such is the testimony of one of the most eminent and experienced architects of the present time ; and, in a letter to me from the Earl of Coventry, written at Spring Hill, his lordship thus mentions Mr. Brown :—

" I certainly held him very high as an artist, and esteemed him as a " most sincere friend. In spite of detraction, his works will ever speak for " him. I write from a house which he built for me, which, without any " pretension to architecture, is, perhaps, a model for every internal and " domestic convenience. I may be partial to my place at Croome, which " was entirely his creation, and, I believe, originally, as hopeless a spot as " any in the island."

I will conclude this tribute to the memory of my predecessor, by transcribing the last stanza of his epitaph, written by Mr. Mason, and which records, with more truth than most epitaphs, the private character of this truly great man :—

" But know that more than genius slumbers here :
  Virtues were his which art's best powers transcend ;
  Come, ye superior train, who these revere,
  And weep the christian, husband, father, friend."

This knowledge cannot be acquired without observing and comparing various houses under various circumstances; not occasionally only, but the architect must be in the habit of living much in the country, and with the persons for whom he is to build; by which alone he can know their various wants with respect to comfort as well as to appearance, otherwise he will, like an ordinary builder, be satisfied in shewing his skill, by compressing the whole of his house and offices under one compact roof, without considering *aspect, views, approaches, gardens,* or even the *shape of the ground* on which the house is to be built.

It is impossible to fix or describe the situation applicable to a house, without, at the same time, describing the sort of house applicable to the situation.

This is so evident, that it scarcely requires to be pointed out; yet I have often witnessed the absurdity of designs for a house where the builder had never seen the situation; I have, therefore, long been compelled to make architecture a branch of my own profession.*

Having occasionally observed the various modes by which large houses and their appendages have been connected, at various periods, it may not be uninteresting if I attempt to describe them, by reference to the annexed plans.

No. 1 [fig. 97]. The earliest form of houses, or, rather, of palaces, in the country, prior to the reign of Elizabeth, consisted of apartments built round a large square court. These were formerly either castles or abbeys, and often received all their light from the inner courts; but, when afterwards converted into habitations, windows were opened on the outside of the building. The views from a window were of little

---

* Before I had the advantage of my eldest son's assistance in this department, I met with continual difficulties. I will mention one instance only, which occurred to me some years ago. Having been consulted respecting the situation for a villa, to be built near the metropolis, I fixed the precise spot, and marked the four corners of the house with stakes upon the ground, proposing that the best rooms should command the best views, and most suitable aspects; but, not having any consultation with the architect, I was afterwards surprised to find my position of the four corners of the house strictly observed; but, to accommodate the site to his previously settled plan on paper, the chimneys were placed where I had supposed the windows should be, to command the finest views, and the windows, alas! looked into a stable court.

consequence at a time when glass was hardly transparent, and in many of the ancient castles the small lozenge panes were glazed with coloured glass, or painted with the armorial bearings, which admitted light without any prospect. Perhaps there is no form better calculated for convenience of habitation, than a house consisting of one or more of these courts, provided the dimensions are such as to admit free

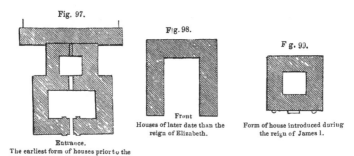

Fig. 97.

Fig. 98.

Fig. 99.

Front

Entrance.

The earliest form of houses prior to the reign of Elizabeth.

Houses of later date than the reign of Elizabeth.

Form of house introduced during the reign of James I.

circulation of air, because, in such a house, the apartments are all easily connected with each other, and may have a passage of communication for servants from every part. Of this kind are the old palaces at Hampton Court and St. James's, of Penshurst and Knowle in Kent, Warwick Castle, and various other ancient mansions.

No. 2 [fig. 98]. Houses of the next form I consider as of later date, although, from the various subsequent alterations, it is difficult to define their original shapes: they seem to have had one side of the quadrangle opened, and thus the line of communication being cut off, this sort of house becomes less commodious in proportion to the length of its projecting sides. Of this description were COBHAM HALL and CASHIOBURY, to both which have been judiciously added square courts of offices, under the direction of Mr. James Wyatt.

No. 3 [fig. 99], is a form introduced in the reign of James I., with the quadrangle so small, that it is often damp and dark; of this kind are CREWE, HILL HALL, GAYHURST, and CULFORD; although the latter has been modernised and changed to the form, No. 7 [fig. 103]. Houses of this shape may sometimes be greatly improved by covering the inner court entirely, and converting it into a hall of communication; this I advised at SARSDEN, a house of later date. The offices are generally attached to the side of these houses. In mansions of the foregoing three descriptions, a mixture of Grecian with Gothic is often observed, particularly in those repaired by Inigo Jones.

No. 4 [fig. 100], the form next in succession, was of the date of William III. and George I., and has been commonly called an H, or half H. This kind of house is often rendered

Fig. 100.

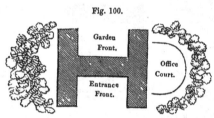

Form, of the date of William III. and George I.

very inconvenient by the centre being one great hall, which breaks the connexion of apartments above stairs. It is also further objectionable, because it is a mere single house in the centre, and must have offices attached on one side: of this description are STOKE PARK, LANGLEYS, GLEMHAM HALL, DULLINGHAM, and CONDOVER.

No. 5 [fig. 101]. When the Italian or Grecian architecture became more general, a greater display of façade was introduced than the body of the house required; the offices

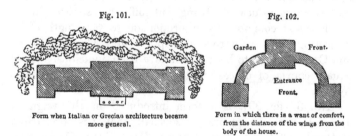

Fig. 101.

Form when Italian or Grecian architecture became more general.

Fig. 102.

Form in which there is a want of comfort, from the distance of the wings from the body of the house.

and appendages were, therefore, made in wings to extend the design, as at WENTWORTH HOUSE, WIMPOLE, ATTINGHAM, DRYHAM PARK, and numerous others.

A house on this plan, if it commands only one view, may be less objectionable; but when applied to situations where the windows are to look in opposite directions, it becomes very inconvenient, because the offices want that uninterrupted communication which is absolutely necessary to the comfort of a dwelling. After the views from the windows became an object of consideration, it was not deemed sufficient to preserve the views to the north and to the south, but even the views to the east and to the west were attempted to be preserved, and this introduced the plan, No. 6 [fig. 102].

No. 6 [fig. 102], has wings, not in the same line with the house, but receding from it, which, of course, destroy the symmetry proposed by wings, unless the whole be viewed from one particular point in the centre; of this form are

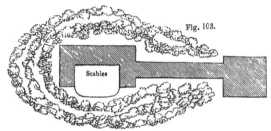

Fig. 103.

Stables

Form generally adopted in modern houses.

MERLEY, NEWTON PARK, NORMANTON, LATHOM HOUSE, &c. The houses built by Paine and Leadbeter are frequent instances of want of comfort in the two latter forms.

No. 7 [fig. 103], is a form so generally adopted in modern houses, that I will not mention any particular instances, especially as they are the works of living architects; yet I hope I shall be pardoned in also making some observations on their construction.

This last invented form consists in a compact square house, with three fronts, and to the back are attached offices, forming a very long range of buildings, courts, walls, &c., supposed to be hid by plantation. These * I have been often required to

* Such is the horror of seeing any building belonging to the offices, that, in one instance, I was desired by the architect to plant a wood of trees on the earth which had been laid over the copper roofs of the kitchen offices, and which extended 300 feet in length from the house.

hide by planting, while, in fact, during the lives of the architect and the proprietor, the buildings can never be concealed, and in the lives of their successors the trees must be cut down to give a free circulation of air to the buildings.

Notwithstanding the danger of giving offence, when I am obliged to speak of the works of living artists, I shall venture to point out some objections to the compact form, No. 7 [fig. 103], as applied to a large mansion, which have not an equal weight when applied to a villa or a house near the city, where land is valued by the foot, and not by the acre; for, however ingenious it may be, in such places, to compress a large house within a small compass, or to cover under the same roof a great number of rooms, yet a mansion in a park does not require such management, or warrant such economy of space.

Of all the forms which can be adopted, there is none so insignificant as a cube; because, however large it may be, the eye can never be struck with its length, its depth, or its height, these being all equal; and the same quantity of building which is often sunk under ground, raised in the air, or concealed in plantation, might have been extended, to appear four times as large, with less expense and more internal convenience.

A house in the country is so different from a house in town, that I never could see any good reason for disposing the living rooms above stairs: it may, perhaps, be said, that the views are more perfect from the higher level; but the same degree of elevation may be obtained by building the cellars above ground, and afterwards raising the earth above them, as I advised at DONNINGTON and BLAIZE CASTLE; and surely the inconvenience of an external staircase can scarcely be compensated by any improvement of the views. To counteract this error in modern houses, I have, in some instances, raised the earth to the principal floor; and, in others, where the architecture would not allow this expedient, I have advised a gallery to be added, as at HOOTON and HIGHAM HILL.

Few subjects having occurred in which I have so fully discussed the proper situation for a house, and all its appen-

dages, as that of MICHEL GROVE,* I shall subjoin the following extract from that *Red Book* :—

"There is no circumstance connected with my profession,
in which I find more error of judgment, than in selecting the
situation for a house, yet it is a subject every one fancies
easy to determine. Not only visitors and men of taste fall
into this error, but the carpenter, the land-steward, or the
nurseryman, feels himself equally competent to pronounce on
this subject. No sooner has he discovered a spot commanding
an extensive prospect, than he immediately pronounces that
spot the true situation for a house ; as if the only use of a
mansion, like that of a prospect-tower, was to look out of the
windows.†

After long experiencing the many inconveniences to which
lofty situations are exposed; after frequently witnessing the
repentance and vexation of those who have hastily made choice
of such situations, under the flattering circumstances of a clear
atmosphere and brilliant sky ; after observing how willingly
they would exchange prospect for shade and shelter, and, after
vainly looking forward to the effect of future groves, I am
convinced that it is better to decide the situation of a house
when the weather is unfavourable to distant prospects, and
when the judgment may be able to give its due weight to

* The plate of MICHEL GROVE HOUSE had been engraved when the
death of its late possessor put a stop, for the present, to these extensive
plans of improvement, which, from his perfect approbation and decisive
rapidity, would, probably, by this time have been completed. Whatever
disappointment I may feel from this melancholy interruption in my most
favourite plan, I must still more keenly regret the loss of a valuable friend,
and a man of true taste; for he had more celerity of conception, more
method in decision, and more punctuality and liberality in execution, than
any person I ever knew.

† The want of comfort, inseparable from a house in an exposed situation,
even in the climate of Italy, is well illustrated by Catullus.

> " Furi! villula nostra, non ad Austri
> " Flatus opposita est, nec ad Favoni,
> " Nec sævi Boreæ, aut Apeliotæ ;
> " Verum ad millia quindecim et ducentos.
> " Oh ventum horribilem ! atque pestilentem !"
>
> CATULLUS, Ode 26.

[My cottage, Furius, is not exposed to the blasts of the South, nor to
those of the West, nor to the raging North, nor to the South-east ; but to
fifteen thousand two hundred blasts. Oh, that horrible and pestilent
wind !]

every circumstance which ought to be considered in so material an object; that the comforts of habitation may not be sacrificed to the fascinating glare of a summer's day.

From these considerations, I do not hesitate to assert, that, if no house existed at MICHEL GROVE, the sheltered situation of the present magnificent and singular mansion [see figs. 104 and 105] is greatly to be preferred to any spot that could be found on the hill, every part of which is more or less exposed to the force of the winds from the south-west. I shall, therefore, inquire into the character of the present house, and consider how far the old mansion may be rendered convenient and adapted to modern comforts.

There are few old mansions in England which have not been either castles or monasteries altered into houses, but there is no trace of this house ever having been either; and, indeed, its situation in a dry valley is unlike that of any abbey, and it is so immediately commanded by the surrounding hills, that it never could have been a castle or place of defence.

The proposed addition of a drawing-room, an anti-room, and an eating-room of large dimensions, will alter those relative proportions, now so pleasing. It is not, therefore, with a view of improving, but with that of doing as little injury as possible to its appearance, that I venture to suggest the additions in the annexed sketch; because the terrace will tend to preserve the apparent height, which the additions to the east tend to destroy.*

The present style of living in the country is so different from that of former times, that there are few houses of ancient date which would be habitable, without great alterations and additions. Such, indeed, is the constant fluctuation in the habits and customs of mankind, and so great the change in the luxuries, the comforts, and even the wants of a more refined people, that it is, in these times, impossible to live in the baronial castle, the secularized abbey, or even in the more modern palaces, built in the reign of Queen Elizabeth, preserving all the apartments to their original uses.

* This house is said to have been built by a Knight of Malta, in the reign of Henry VIII., in imitation of a Morisco palace which he had seen in Spain; if this be true, it accounts for the singular style of architecture.

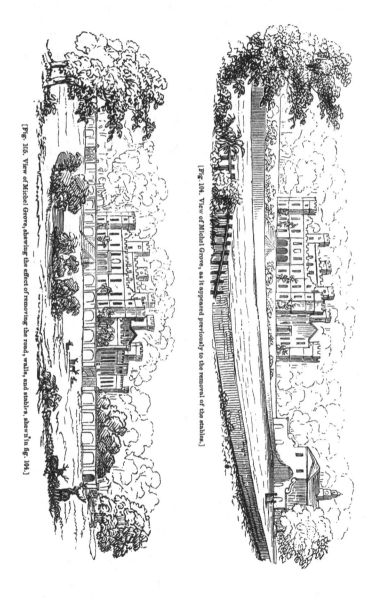

[Fig. 103. View of Michel Grove, shewing the effect of removing the road, walls, and stables, shewn in fig. 104.]

[Fig. 104. View of Michel Grove, as it appeared previously to the removal of the stables.]

The chief rooms formerly required in a house of that date were,—

The *Hall*, for the entertainment of friends and vassals; a large and lofty room, having the floor at one end raised above the common level, as at present in the halls of our colleges; this was to mark some distinction in the different ranks of the guests.

The next large room required was a *Gallery*, for the reception of company in a morning, for dancing in the evening, and for the exercise of the family within doors. Very few books were then in use; and, instead of the newspapers and pamphlets of the present day, the general information was collected in conversations held in those long galleries, which had large recesses, or bays,* sometimes called bowre-windows, and now bow-windows; into which some of the company would occasionally withdraw, for conversation of a more private nature, as we frequently read in the Memoires de Sully, &c.

But the apartment, of all others, which was deemed indispensable in former times, and in which the magnificence of the proprietor was greatly displayed, was the *Chapel.*

The other apartments were one or more *small parlours*, for the use of the ladies and their female attendants, in which they carried on their various works of embroidery, &c., and, instead of the present dressing-room and sitting-rooms, which are added to each modern bed-room, there was, generally,—

A *small closet* to each, with, perhaps, an oriel window for private morning devotions.

After thus mentioning the uses of ancient apartments, it is necessary to enumerate those additions which modern life requires. 1st. The *Eating-room*, which does not exactly correspond with the ancient hall, because it is no longer the fashion to dine in public. 2nd. The *Library*, into which the gallery may sometimes be changed with propriety. 3rd. The *Drawing-room*, or saloon. 4th. The *Music-room*. 5th. The *Billiard-room*. 6th. The *Conservatory* attached to the house;

---

* " If this law hold in Vienna ten years, I'll rent the fairest house in it after three-pence a bay."—*Measure for Measure*, Act II. Sc. I.

" The fashion of building, in our author's time, was to have two or three " juttings out in front, which we still see in old houses, where the windows " were placed, and these projections were called bays, as the windows were " from thence called bay-windows."—*Theobald, ibid.*

These projections answer to the Exhedra of the Greeks and Romans.

and, lastly, the *Boudoirs*, wardrobes, hot and cold baths, &c., which are all modern appendages, unknown in Queen Elizabeth's days. Under these circumstances, it is difficult to preserve the ancient style of a mansion without considerable additions. For this reason, we see few specimens of Gothic buildings which have not been mixed and corrupted with the architecture of various dates; and whilst every casual observer may be struck with the incongruity of mixing the Grecian with the Gothic styles, yet the nice antiquarian alone discovers, by the contour of a moulding, or the shape of a battlement, that mixture of the castle and abbey Gothic, which is equally incorrect with respect to their different dates and purposes.

The view of this house [fig. 105] will, I hope, justify my anxiety to preserve it, as far as may be consistent with modern habitation : for although it can neither be deemed a castle, an abbey, or a house of any Gothic character with which we are acquainted, yet its form is singularly picturesque, and the *slide* [our fig. 104] shews the effect of removing the present road, walls, and stables, which would obstruct the view from the new apartments.

In determining the situation for a large house in the country, there are other circumstances to be considered besides the offices and appendages immediately contiguous. These have so often occurred, that I have established, in imagination, certain positions for each, which I have never found so capable of being realized as at MICHEL GROVE.

I would place the *house* with its principal front towards the south or south-east.

I would build the *offices* behind the house ; but, as they occupy much more space, they will, of course, spread wider than the front.

I would place the *stables* near the offices.

I would place the *kitchen-garden* near the stables.

I would put the *home farm buildings* at rather a greater distance from the house ; but these several objects should be so connected by *back-roads* as to be easily accessible.

I would bring the *park* to the very front of the house.

I would keep the farm, *or land in tillage*, whether for use or for experiment, behind the house.

I would make the dressed *pleasure-ground*, to the right and left of the house, in plantations, which would screen the unsightly appendages, and form the natural division between the park and the farm, with walks communicating to the garden and the farm.

It will be found that these are exactly the positions of all the appendages at MICHEL GROVE.   But, in support of my opinion, it may be proper to give some reasons for the choice of these general positions.

1. The *aspect* of a house requires the first consideration, since no beauty of prospect can compensate for the cold exposure to the north, the glaring blaze of a setting sun, or the frequent boisterous winds and rains from the west and south-west; while, in a southern aspect, the sun is too high to be troublesome in summer, and, during the winter, it is seldom an unwelcome visitant in the climate of England.

2, 3. It can hardly be necessary to enumerate the advantages of placing the offices near, and stables at no great distance from the house.

4. The many interesting circumstances that lead us into a *kitchen-garden*, the many inconveniences which I have witnessed from the removal of old gardens to a distance, and the many instances in which I have been desired to bring them back to their original situations, have led me to conclude, that a kitchen-garden cannot be too near, if it be not seen from the house.

5. So much of the comfort of a country residence depends on the produce of its *home farm*, that even if the proprietor of the mansion should have no pleasure in the fashionable experiments in husbandry, yet a farm, with all its appendages, is indispensable : but, when this is considered as an object of *profit*, the gentleman-farmer commonly mistakes his aim ; and, as an object of *ornament*, I hope the good taste of the country will never confound the character of a park with that of a farm.

To every dwelling there must belong certain unsightly premises, which can never be properly ornamental; such as yards for coal, wood, linen, &c., and these are more than doubled when the farm-house is contiguous ; for this reason I am of opinion, that the farming premises should be at a greater distance than the kitchen-garden or the stables, which have a more natural connexion with each other.

The small pool in front of the house has been purposely left; not as an object of beauty in itself, but as the source of great beauty to the scenery; for, in the dry valleys of Sussex, such a pond, however small, will invite the deer and cattle to frequent the lawn in front of the house, and add to the view, motion and animation.

Those who only remember the former approaches to this house, over lofty downs, with a dangerous road to descend, will hardly believe that this venerable mansion is not situated in the bottom, but at the extremity of a valley; for, in reality, the house is on the side of a hill, and, by the proposed line of approach, it will appear that it actually stands on a considerable eminence, the road ascending along the whole course of the valley, for more than a mile." *

A house extended in length may be objectionable in many situations, but when built on the side of a hill, if the ground rises boldly behind it, the objection to it, as a single house, is removed.

Where a house, like that at GARNONS, by its situation and southern aspect, will constantly be a marked feature from the surrounding country, presenting only one front embosomed in wood, that front should be so extended as to distinguish the site of the mansion with adequate importance.

In such a situation, it would be difficult to produce the same greatness of character by a regular Grecian edifice, that will be effected by the irregularity of outline in the proposed house, offices, and stables; and, in defence of this picturesque style, I shall take the liberty to transcribe, in a note,† the

[* In 1832, the property on which MICHEL GROVE stood was purchased by the Duke of Norfolk, and added to the domain of ARUNDEL CASTLE. The house was pulled down, and the materials sold.   J. C. L.]

† " C'est par une suite de cet usage de voir et d'entendre par les yeux " et les oreilles de l'habitude, sans se rendre raison de rien, que s'est établie " cette manière de couper sur le *même patron* la droite et la gauche d'un " bâtiment.   On appelle cela de la symetrie ; Le Nôtre l'a introduite dans " les jardins, et Mansard dans les bâtiments, et cequ'il y a de curieux, c'est " que lorsqu'on demande à quoi bon ? aucun expert Juré, ne peut le dire ;

following very judicious remarks of R. L. Girardin Viscomte
d'Ermenonville.

" car cette sacrée symetrie ne contribue en rien à la solidité, ni à la com-
" modité des bâtiments, et loin qu'elle contribue à leur agrément, il n'y a si
" habile peintre, qui puisse rendre supportable dans un tableau un bâtiment
" tout plattement symetrique.   Or, il est plus que vraisemblable que si la
" copie est ressemblante et mauvaise, l'original ne vaut guères mieux,
" d'autant qu'en general tous les desseins de fabriques font plus d'effet en
" peinture qu'en nature."
    " C'est donc l'effet pittoresque qu'il faut principalement chercher, pour
" donner aux bâtiments le charme par lequel ils peuvent séduire et fixer les
" yeux. Pour y parvenir, il faut d'abord choisir le meilleur point de vue pour
" developper les objets ; et tacher, autant qu'il est possible, d'en presenter
" plusieurs faces."
    " C'est à donner de la saillie, et du relief à toutes les formes, par l'oppo-
" sition des renfoncemens, et par un beau contraste d'ombre et de lumière,
" c'est dans un juste rapport des proportions, et de la convenance avec tous
" les objets environnans, qui doivent se presenter sous le même coup d'œil ;
" c'est a bien disposer tous les objets sur différens plans, de manière que
" l'effet de la perspective semble donner du movement aux differentes parties
" dont les une paroissent eclairées, les autres dans l'ombre ; dont les unes
" paroissent venir en avant, tandis que les autres semblent fuir ; enfin c'est
" à la composer de belles masses dont les ornéments et les details ne com-
" battent jamais l'effet principal, que doit s'attacher essentiellement l'archi-
" tecture."
    " Les anciens l'avoient si bien senti, qu'ils ne se sont jamais occupées
" dans leur constructions, que de la grande masse, de manière que les plus
" precieux ornements sembloient se confondre dans l'effet general, et ne
" contrarioient jamais l'objet principal de l'ensemble, qui annonçoit toujours
" au premier coup d'œil, par son genre et ses proportions, le caractère et la
" destination de leur edifices."
    ["It is in consequence of this habit, of seeing and hearing with the eyes
and ears of custom and prejudice, without considering the reason of anything, that the practice of designing the right and left of a building to the
same pattern has arisen.   This is called symmetry ; Le Nôtre introduced it
in gardens, and Mansard in buildings ; and what is singular is, that if any
one asks to what purpose is it so? no adept in the art can tell ; for this detest-
able symmetry contributes, in no degree, either to the solidity or convenience
of the buildings : and so far is it from contributing to their beauty, that
there is no painter, however skilful he may be, who can render a building,
insipidly symmetrical, tolerable in a picture. Now, it is more than probable,
that if the copy, though a good likeness, be bad, the original is no better,—
inasmuch as, in general, all drawings of buildings have more effect in a
painting than in nature.
    It is picturesque effect that must principally be sought for, in order to
give to buildings the charm necessary to attract and rivet the eye.   For
this purpose a point of view should be chosen, which appears the best for
shewing all the objects ; and the building should be so contrived as to
present as many sides as possible at once.
    It is in giving prominence and relief to the principal forms, by the
opposition afforded by the others, and by a fine contrast of shade and light ;
it is in an accurate adjustment of the proportions of the buildings to those
of the surrounding objects, which will be seen in the same *coup d'œil ;* it is
in placing the objects on different levels, so that the effect of the perspective

A plan of the house proposed for this situation is added,
to shew how conveniently the comforts of modern habitations

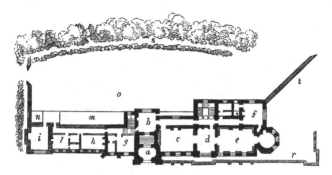

[Fig. 106. Example of a plan for an extended front on the steep side of a hill.]

References.

a Porch	g Butler's pantry, connected with	n Scullery
b Hall	bread-closet and sleeping-room,	o Office-court, with out-houses,
c Eating-room	and with a private stair to the	sheds, &c. &c.
d Breakfast or anti-room	cellar below	r Terrace in front of the living-
e Library and living-room	h Servants' hall	rooms
f Business-room of the master, with	i Kitchen	s A very steep bank, with rock and
strong-closet and water-closet	l Housekeeper's room	plantation
adjoining	m Larder, bakehouse, and other	t Situation for a conservatory.
	conveniences	

may be adapted to ancient magnificence; and I rejoice in
observing that many large houses are at this time building, or
altering, in this irregular style, under the direction of one of
our most eminent architects. I may mention those of CASHI-
OBURY and WICKHAM MARKET, which disdain the spruce
affectation of symmetry so fatal to the Gothic character.

When a house, as in the foregoing instance, is to be built
on the side of a hill, or on an inclined plane, it is hardly
possible to dispose it in any other form than that of an
extended front: but this supposes a certain degree of property
to belong to the house, or it is apt to appear too large for the

may seem almost to give movement to the different parts, of which some will
appear in strong light, and others in the shade, some will be brought pro-
minently forward, and others seem as though retiring; in short, it is in
composing beautiful masses, of which the ornaments and details never
interfere with the principal effect, that the great art of architecture consists.

The ancients understood this so well, that, in their buildings, the general
mass only was taken into consideration, so that the most costly ornaments
seemed to be absorbed in the general effect, and were never at variance
with the principal object of the whole, which always announced, at first
sight, by its style and proportions, the character and destination of their
edifices."]

2 o

annexed estate; this objection is, however, less forcible in a
villa than in a mansion; yet even a villa, which covers too
much of its own field, or lawn, partakes more of ostentation
than good taste.

[Fig. 107. Villa at Brentry Hill, near Bristol, given as a specimen of economy united with compactness, and
adapted to its situation, character, and uses.]

A field of a few acres, called BRENTRY HILL, near Bristol,
commands a most pleasing and extensive view.   In the fore-

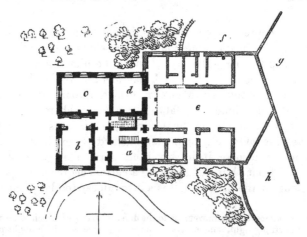

[Fig. 108. Ground-plan of the Villa at Brentry Hill.]

References.

a Breakfast-room	c Eating-room	f Drying ground
b Drawing-room, opening, with	d Kitchen	g Part of the kitchen-garden
folding doors, to a small library	e Kitchen court	h Stable court.

ground are the rich woods of King's Weston, and Blaize
Castle, with the picturesque assemblage of gardens and villas
in Henbury and Westbury; beyond which are the Severn and

Bristol Channel, and the prospect is bounded by the mountains of South Wales. This view is towards the west, and I have generally observed, that the finest prospects in England are all towards this point.* Yet this, of all aspects, is the most un-pleasant for a house; it was not, therefore, advisable to give an extended front in this direction, yet it would have been unpardonable not to have taken advantage of so fine a prospect.

A compact plan often demands more trouble and con-trivance than a design for a palace, in which the rooms may be so numerous, that different apartments may be provided for summer and for winter use; but where compactness and economy are studied, some contrivance is necessary to avail ourselves of views and aspects, without sacrificing convenience and *relative fitness* to the beauty of the prospect.

Under this restraint, perhaps, few houses have been built with more attention to the situation and circumstances of the place, than the villa at Brentry [figs. 107 and 108]. The eat-ing-room is to the north, with one window towards the pros-pect, which may be opened or shut out by Venetian blinds at pleasure. The breakfast-room is towards the south, and the drawing-room towards the prospect [see the plan, fig. 108].

Modern habits have altered the uses of a drawing-room: formerly, the best room in the house was opened only a few days in each year, where the guests sat in a formal circle; but now, the largest and best room in a gentleman's house is that most frequented and inhabited: it is filled with books, musical instruments, tables of every description, and whatever can contribute to the comfort or amusement of the guests, who form themselves into groups, at different parts of the room; and in winter, by the help of two fire-places, the re-straint and formality of the circle is done away.

This has been often happily effected in old houses by laying two rooms together, preserving the fire-places in their

* This remark concerning our finest prospects being towards the west, has been so often confirmed by repeated observations, that I have endeavoured to discover some natural cause for its general prevalence; and perhaps it may, in some degree, be accounted for from the general position of the strata in all rocky countries, which appear to dip towards the east and rise towards the west; in one direction, the view is along an inclined plane; in the other, it is taken from the edge of a cliff, or some bold promontory overlooking the country towards the west.

original situations, without regard to correspondence in size
or place; but two fires not being wanted in summer, a provision is made in this villa to preserve an additional window
towards the fine prospect at that season of the year; and the
pannel, which ornaments the end of the room, may be removed in winter, when the window will be less desirable than
a fire-place; thus the same room will preserve, in every season, its advantages of aspects and of views, while its elegance
may be retained without increasing the number of rooms for
different purposes.*

* This attention to the wants of different seasons has been too little
studied in this country, whilst in France almost every large house has its
Garçon tapissier, whose business it is to change the furniture of the apartments for summer and winter. Those who have compared the fitting up of
rooms in France, with that of any other country of Europe, must, doubtless,
give the preference to French taste, as far as it relates to the union of internal magnificence and comfort; but those architects who copy both the
inside and outside of Italian houses, should at least provide for such occasional alterations as our climate may require.

Another circumstance may be mentioned, in which economy has been
consulted at this small villa. More rooms are generally required on the
chamber than on the ground floor; yet, except the kitchen, there is no
part of a house which ought properly to be so lofty as the principal rooms;
instead, therefore, of increasing the quantity of offices, by what a witty
author calls, "turning the kitchen out of doors for smelling of victuals,"
this offence is here avoided by the external passage of communication.

The operations of landscape gardening have often been classed under the
general term of *improvement;* but there are three distinct *species.* The
first relates to places where the grounds are altered, and adapted to a house
already existing; the *second* to those where the houses, by additions, having
changed their original character, or aspect, renders it necessary to make
alterations in the ground also; the *third* includes those places where no
house previously exists, and where the entire plan of the house, appendages,
and grounds, has sometimes been called a *Creation.* Of the first kind it is
needless to enumerate examples. Among the second may be mentioned
those, in which the entrance of the house being changed, new rooms added,
or barns, stables, and kitchen-gardens removed, new arrangements have
taken place, as at ABINGTON HALL, CLAYBERRY, WALLHALL, WEST-COKER,
BETCHWORTH, HIGHLANDS, BRANDSBURY, HOLWOOD, &c. Of those places
which may be called *Creations,* the number is necessarily small, yet I may
refer to the following examples. In some, where new houses were built,
I was consulted by the respective architects on the situation and appendages; as at BRACONDALE, MILTON HOUSE, DONNINGTON, BUCKMINSTER,
COURTEEN HALL, BANK FARM, CHILTON LODGE, DULWICH CASINA, HOLME
PARK, STREATHAM, THE GROVE, SOUTHGATE, LUSCOMBE, &c. In others, I
gave general plans for the *whole,* with the assistance of my son only in
the architectural department, as at BRENTRY HILL, COTHAM BANK,
ORGAN HALL, STAPLETON, STRATTON PARK, SCARRISBRICK, PANSHANGER,
BAYHAM, &c.

CHAPTER XIII.

Ancient Mansions—Danger of modernizing—Three Characters of Gothic Architecture—for Castles, Churches, and Houses—CORSHAM HOUSE— Mixing Characters, how far allowable—PORT ELIOT—Remarks on Grecian and Gothic Architecture, extracted from the *Red Book* in the Library of MAGDALEN COLLEGE, Oxford—Example of Additions to the Gothic Mansion of ASHTON COURT.

THE following extract from the *Red Book* of CORSHAM, may serve to exemplify the impropriety of improving the grounds without previous attention to the style, character, and situation of the house.

At the time CORSHAM HOUSE [figs. 109 and 110] was

End added by Brown.   Original South Front.   End added by Brown.
[Fig. 109. Corsham House.]

erected, instead of the modern houses now placed in the centre of parks, distant from every other habitation, it was the glory and pride of an English baron to live in, or near, the town or village which conferred its title on his palace, and often on himself. Nor was the proximity of the village attended with any inconvenience, so long as the house was disjoined from it by ample court-yards, or massive gates; some of its fronts might look into a garden, lawn, or park, where the neighbours could not intrude. Yet even these views, in some instances, were confined, formal, and dull, by lofty walls and clipped hedges.

In determining the situation for a new house, it may often be advisable to place it at a distance from other habitations, that the modern taste for freedom and extent may be gratified ; but, in accommodating plans of improvement to houses

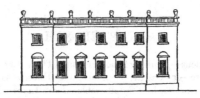

[Fig. 110, Corsham House.—East Front added by Brown.]

already built, it requires due consideration how far such taste should be indulged, otherwise we may be involved in difficulties and absurdities; for it is not uncommon to begin by removing walls which conceal objects far more offensive than themselves.

When additions or alterations are made to an old house, internal convenience and improvement should, certainly, be the first objects of consideration; yet the external appearance and character must not be neglected.   This is a circumstance which our ancestors seem to have little regarded, for we frequently distinguish the dates of additions to buildings by the

[Fig. 111, Corsham House.—East Front as altered.]

different styles of architecture ; and hence it often happens, that a large old house consists of discordant parts mixed together, without any attempt at unity either in date or character of building.

This was of less consequence, when each front, surrounded by its court, or *parterre*, became a separate and entire object; but since modern gardening, by removing those separations, has enabled us to view a house at the angle, and at once to see two fronts in perspective, we become disgusted by any want of unity in the design.

The south front of CORSHAM [fig. 109], is of the style called Queen Elizabeth's Gothic, although rather of the date of King James. The north front is of Grecian architecture.

[Fig 112. Corsham House.—The New Front to the North.]

The east front [fig. 110], is in a correct, but heavy style of regular architecture; and to alter the old south front in conformity to it, would not only require the whole to be entirely rebuilt, but make an alteration of every room in that part of the house unavoidable. This not according with the intention of the proprietor of CORSHAM HOUSE, the original south front becomes the most proper object for imitation. [Fig. 111 shews the east front as altered; and fig. 112, the new front to the north.]

A house of Grecian architecture, built in a town, and separated from it only by a court-yard, always implies the want of landed property; because, being evidently of recent erection, the taste of the present day would have placed the house in the midst of a lawn or park, if there had been sufficient land adjoining: while the mansions built in the Gothic character of Henry VIII., Elizabeth, and James, being generally annexed to towns, or villages, far from impressing the mind with the want of territory, their size and grandeur, compared with other houses in the town, imply that the owner is not only the lord of the surrounding country, but of the town also.

The valuable and celebrated collection of pictures at COR-SHAM HOUSE, in a modern Grecian edifice, might appear recent, and not the old inhabitants of an ancient mansion, belonging to a still more ancient family : and although Grecian architecture may be more regular, there is a stateliness and grandeur in the lofty towers, the rich and splendid assemblage of turrets, battlements, and pinnacles, the bold depth of shadow produced by projecting buttresses, and the irregularity of outline in a large Gothic building, unknown to the most perfect Grecian edifice.

Gothic structures may be classed under three heads, *viz.*, The *Castle* Gothic, the *Church* Gothic, or the *House* Gothic: let us consider which is the best adapted to the purposes of a dwelling.

The *Castle Gothic*, with few small apertures and large masses of wall, might be well calculated for defence, but the apartments are rendered so gloomy, that it can only be made habitable by enlarging and increasing these apertures, and, in some degree, sacrificing the original character to modern comfort.

The more elegant *Church Gothic* consists in very large apertures with small masses, or piers: here, the too great quantity of light requires to be subdued by painted glass; and, however beautiful this may be in churches, or the chapels and halls of colleges, it is seldom applicable to a house, without such violence and mutilation as to destroy its general character : therefore, a Gothic house of this style would have too much the appearance of a church; for, I believe, there are no large *houses* extant of earlier date than Henry VIII., or Elizabeth, all others being either the remains of baronial castles or conventual edifices.

At the dissolution of the monasteries by Henry VIII., a new species of architecture was adopted, and most of the old mansions now remaining in England, were either built, or repaired, about the end of that reign, or in the reign of Queen Elizabeth : hence, it has acquired, in our days, the name of *Elizabeth's Gothic*; and although, in the latter part of that reign, and in the unsettled times which followed, bad taste had corrupted the original purity of its character, by introducing fragments of Grecian architecture in its ornaments,

yet, the general character and effect of those houses is perfectly Gothic; and the bold projections, the broad masses,
the richness of their windows, and the irregular outline of
their roofs, turrets, and tall chimneys, produce a play of light
and shadow wonderfully picturesque, and, in a painter's eye,
amply compensating for those occasional inaccuracies urged
against them as specimens of regular architecture.

Although the old south front should be the standard of
character for the new elevations of CORSHAM HOUSE, yet I
hold it not only justifiable, but judicious, in the imitation of
any building, to omit whatever is spurious and foreign to its
character, and supply the places of such incongruities from
the purest examples of the same age.   For this reason, in the
plans delivered, the Grecian mouldings are omitted, which the
corrupt taste of King James's time had introduced, and the
true Gothic mouldings of Elizabeth's reign are introduced.

The turrets, chimney shafts, and oriels, will be found in
the examples of BURLEIGH, BLICKLING, HAMPTON COURT,
HATFIELD, &c., or in most of the buildings of Henry VIII.
and Elizabeth.   The centre of the north front, although of
the same character, being in imitation of a building somewhat
earlier than Elizabeth, together with the peculiarity of its
form, it is necessary to describe why it has been adopted.
Here another principle arises, *viz.*, that in designing any
Gothic building, it is presumed that some fragments exist of
the style we propose to imitate; otherwise it ceases to be an
imitation.

In pursuance of this principle, *we** looked for an instance
of an octangular room projecting beyond the general line of
the wall, in some building of that date.   The chapel of Henry
VII., at Westminster, though not an octagon, was the only

* In speaking of this house, I use the plural number, because the plans
were the joint effort of a connexion and confidence which then so intimately
existed between me and another professional person, that it is hardly possible to ascertain to whom belongs the chief merit of the design.   Yet I
claim to myself all that relates to the reasoning and principles on which the
character of the house was adopted : to my son's knowledge and early study
of the antiquities of England, may justly be attributed a full share of the
general effect and proportions of the buildings; but as we did not direct the
execution of the work, the preceding elevations are on so small a scale, as to
describe only the general outline proposed, without copying the detail of
what has been executed.

[Fig. 113. Port Eliot, St. Germain's, before the Abbey was united with the mansion.]

[Fig. 114. Port Eliot, as it will appear when the mansion is united with the Abbey by the cloister shewn at b in fig. 115.

projecting regular polygon : this, therefore, became our model
for the centre room of the north front, and this example not
only furnished a precedent for a projecting room, but other
parts of its composition peculiarly suited our situation.

In the modern rage for removing to a distance all those
objects which were deemed appendages to the ancient style of
gardening, such as terraces, lofty walls, almshouses, quadrangular courts, &c.; a mistaken idea has prevailed, that the
house should stand detached from every surrounding object :
this injudicious taste has, in many parts of the kingdom,
destroyed towns and villages, to give solitary importance to
the insulated mansion.

" The situation of PORT ELIOT [figs. 113, 114, and 115],
is apparently oppressed by its vicinity to St. Germain's and
its stupendous cathedral [fig. 115], whose magnitude and

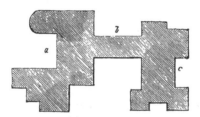

[Fig. 115. Ground plan of Port Eliot, and the Abbey of St. Germain's.
References. *a*, Mansion and offices. *b*, Proposed cloister to join
the mansion to the Abbey. *c*, The Abbey.]

lofty situation forbid its being made subordinate to the mansion. Under such circumstances, instead of shrinking from
this powerful neighbour, it will rather be advisable to attempt
such an union as may extend the influence of this venerable
pile to every part of the mansion.

This I purpose to effect by a narrow building, or cloister
[fig. 115, *b*], to connect the house with the Abbey, as shewn
in the view [fig. 114]. The plan [fig. 115] is introduced to
shew how inconsiderable in proportion to the present buildings would be such addition, although it appears to be a work
of great magnitude; and this being a deception arising from
perspective, I shall explain its cause.

The south front of the house being only about fourscore feet distant from the Abbey, it is impossible to view it, except in such perspective as must shew it very much foreshortened. For this reason, as it appears by the drawing, the west end of the house, though containing only two windows, is more conspicuous than the whole south front, in which there are twenty-six; it is therefore the more necessary that this small part of the building which faces the west should be enriched by such ornaments as may be in harmony with the Gothic character of the Abbey: the Venetian window, and the paladian window over it, may be externally united into one Gothic window, which, by its size and character, will extend the importance of the Abbey to the whole of the mansion.*

A large window is necessary, because a number of small parts will never constitute one great whole; but if a few large parts, such as the window here mentioned, the gateway, and another large window in the cloister, be properly introduced, they will extend the impression of greatness, and overpower all the lesser parts of the building in the same manner that the great west entrance of the Abbey takes off the attention from the smaller windows in the same massive pile.

It may, perhaps, be observed, that, in the cloister proposed, I have not strictly followed the architecture of the Abbey, which is either Saxon or Norman (a distinction in which very learned antiquarians have differed in opinion). It is certainly of a style anterior to the kind of Gothic distinguished by pointed arches and pinnacles. But I conceive there is no incongruity in mixing these different species of Gothic, because we see it done in every cathedral in the kingdom; indeed the greatest part of this Abbey itself is of the date and style which I have adopted."

———————

The following remarks on the improvement of MAGDALEN COLLEGE, OXFORD, were accompanied with many drawings,

---

* A beautiful specimen of thus uniting two floors by one window, may be seen at SHEFFIELD PLACE, where, I believe, it was first introduced by Mr. James Wyatt.

on a scale too large for this work; but as the book is in the library of that college, I suppose such of my readers as are interested in these observations concerning Grecian and Gothic architecture, may have access to the original designs, if they wish farther to consider the subject; at the same time their inquiries will be facilitated by having previously perused the following extract from that manuscript.

" The love of novelty and variety, natural to man, is alone sufficient to account for the various styles of building with which our universities abound.

When Grecian architecture was first introduced into this country, it was natural to adopt the new style, without considering how far its uses or general character might accord with the buildings to which it was applied; and, without recollecting the climate from whence it was imported, every other consideration was sacrificed, or made subservient to the external ornaments of Greece and Rome.*   On a more exact inquiry, we shall find, it was not the habitable buildings of ancient Greece or Rome which formed our models: the splendid and magnificent remains of Athens, of Palmira, of Balbec, of Pæstum, or of Rome herself, supply only temples with columns, entablatures, and porticos, but without windows or chimneys, or internal subdivisions by floors for apartments, indispensable in our English habitations, and even to our public buildings.

In this climate we should seldom visit a hall, or a chapel, where all the light admitted was from the entrance, or from an uncovered aperture in the roof; and yet on such plans were constructed all the temples of the ancients.

Our students in architecture, who have visited southern climates, were therefore obliged to copy the works of more modern artists, who, by various expedients, had endeavoured to make their buildings habitable ; and from the *modern*

---

* Among the conveniences observable in Gothic colleges, may be mentioned the uninterrupted communication ; this was formerly provided for by cloisters, that each member of the society might at all times, in all weather, walk under cover from his respective apartment to the hall, the chapel, the library, or to the apartment of any other member. Such cloisters also yielded a dry and airy walk when the uncertainty of our climate would otherwise have prevented that sort of moderate exercise necessary to the sedentary occupations of the learned.

Italian, rather than from the buildings of *ancient Rome*, have been introduced floors intersecting the shaft of a lofty column, or, what is still more offensive, columns of various orders, built over each other; while the whole face of the building is cut into minute parts by ranges of square apertures. Having at length discovered how seldom a very lofty portico* can be useful in this climate, where we have little perpendicular sun, the portico itself is filled up with building, and the columns are nearly half buried in the walls: this is the origin of that unmeaning ornament called a *three-quarter column*.

By degrees these columns were discovered to be totally useless, and were at length entirely omitted; yet the *skeleton* of the portico and its architectural proportions still remain, as we frequently observe in the entablature and pediment of what is called a Grecian building.

This is all that remains of Grecian architecture in the present new building at Magdalen College; yet, from its simplicity, we are still pleased with it, and more from its utility, because it evidently appears to be a succession of similar apartments for the separate habitations of a number of members of the same society, equal in their rank and in their accommodations, and only claiming that choice of aspect or situation which seniority or priority confers.

It has been observed, that the age of every manuscript is as well known to the learned antiquarian, from the letters or characters, as if the actual date were affixed. The same rule obtains in architecture. And even while we profess to copy the models of a certain era, we add those improvements or conveniences which modern wants suggest; and thus, in after ages the dates will never be confounded.

In Gothic, which is the style of architecture most congenial to the uses and to the character of a college, we are to

---

* I have frequently smiled at the incongruity of Grecian architecture applied to buildings in this country, whenever I have passed the beautiful Corinthian portico to the north of the Mansion House, and observed, that on all public occasions it becomes necessary to erect a temporary awning of wood and canvas to guard against the inclemency of the weather. In southern climates, this portico, if placed towards the south, would have afforded shade from the vertical rays of the sun; but in our cold and rainy atmosphere, such a portico towards the north, is a striking instance of the false application of a beautiful model.

study first, the general and leading principles, and afterwards that detail of which we can collect the best specimens from buildings of the date we mean to imitate.

The leading principles of all Gothic buildings were these:

1. The *uses* of a building were considered before its *ornaments*.

This principle is obvious in the staircases of towers, which were generally made in a turret at one corner, larger than the other three, and often carried up higher to give access to the roof of the building. Small turrets and pinnacles, or fineals, will be considered only as ornaments by the careless observer, but the mathematician discovers that such projections above the roof, form part of its construction; because they add weight and solidity to those abutments which support the Gothic arch.

2. The ornaments prevailed most where they would be most conspicuous.

The richest ornaments of Gothic architecture are the turrets, pinnacles, or open battlements at the top of the building. These were seen from all parts, and in the beautiful tower at Magdalen, it may be observed, that the enrichment ceases below, where it would not be so much seen. The gates and entrances are highly ornamented, because they are immediately subject to the eye; but the walls are frequently without any decoration. This economy in ornaments is confirmed by the laws of nature. See page 260.

3. The several principal parts of the building were marked by some conspicuous and distinguishing character.

As the chapel, the hall, the chapter-room, and the bishop's, abbot's, or president's habitation, &c. The dormitories were not less distinguished as a suite of similar apartments. But where, in conformity to the modern habits of symmetry, it is necessary to build two parts exactly similar, it is difficult for a stranger to distinguish their separate uses.

4. Some degree of symmetry, or correspondence of parts, was preserved, without actually confining the design to such regularity as involved unnecessary or useless buildings.

This irregularity, which has been already noticed in speaking of the towers for staircases, is carried still farther in those projections, by which an apparent centre is marked: for if any ancient Gothic building be attentively examined, it will be found that the *apparent* centre is seldom in the middle. Thus in the beautiful cloister of Magdalen, the gateway is not in the centre of the west, nor the large window of the hall in the centre of the south side of the quadrangle; yet the general symmetry is not injured, and the dimensions are, perhaps, enlarged by this irregularity.

5. This degree of irregularity seems often to have been studied, in order to produce increased grandeur by an intricacy

and variety of parts. A perfect correspondence of two sides
assists the mind in grasping the whole of a design on viewing
only one-half; it therefore, in fact, lessens the apparent
magnitude, while the difficulty with which dissimilar parts are
viewed at once, increases the apparent dimensions, provided
the eye be not distracted by too much variety.

The frequency of Gothic towers having been placed at a different angle
with the walls of the chapel, must have been more than accident. The position of the tower at Magdalen, with respect to the chapel, is a circumstance
of great beauty, when seen from the centre of the cloisters, because two
sides are shewn in perspective. And, upon actual measurement, it will be
discovered that few quadrangular areas are correctly at right angles.

And, lastly, the effect of perspective, and of viewing the
parts of a building in succession, was either studied, or
chance has given it a degree of interest that makes it worthy
to be studied; since every part of a building is best seen from
certain points of view, and under certain relative circumstances of light, of aspect, of distance, or of comparative size.

The great scale on which Gothic architecture was generally executed, is
one source of the grand impression it makes on the mind, since the most
correct model of a cathedral would convey no idea of its grandeur. The
false Gothic attempts of our modern villas, offend as much by their littleness as by the general incorrectness of detail."

The *Red Book* in Magdalen College contains such examples and remarks, concerning the detail of Gothic architecture, as might be curious to the antiquarian; but which
can only be understood by the numerous drawings with which
the subject was elucidated.

Having assigned as a reason for writing in the plural number in the
Red Book of CORSHAM, that a third person was there consulted, it may
perhaps be proper to mention, that, in the architectural part of the plans for
Magdalen College, and all the other buildings described in this volume, I
have been assisted by my son *only*.

———————

The annexed plate of ASHTON COURT [fig. 116], furnishes an example of making considerable additions to a very
ancient mansion, without neglecting the comforts of modern
life, and without mutilating its original style and character.

This house was built about the reign of Henry VI., and
originally consisted of many different courts, surrounded by

[Fig 116. Ashton Court.]

The old part built in the reign of Henry VI.

The new part added in the reign of George III.

building, of which three are still remaining; in all these the
Gothic windows, battlements, and projecting buttresses, have
been preserved; but the front towards the south, 150 feet
in length, was built by Inigo Jones, in a heavy Grecian style;
this front was designed to form one side of a large quadrangle,
but, from the unsettled state of public affairs, the other three
sides were never added, and the present long front was never
intended to be seen from a distance: this building consists of
a very fine gallery, which has been shortened to make such
rooms as modern habits require; but it is now proposed to
restore this gallery to its original character, and to add in the
new part, a library, drawing-room, eating-room, billiard-room,
with bed-rooms, dressing-rooms, and a family apartment, for
which there is no provision in the old part of the mansion.
It is also proposed to take down all the ruinous offices, and
rebuild them with the appearance of antiquity, and the conveniences of modern improvement.

If, in conformity to buildings of this date, the courts were
all to be preserved, and surrounded with buildings, or lofty
walls, the damp and gloom, as well as the grandeur of former
times, would be recalled; but by opening the side of these
courts to the park with an iron rail, cheerful landscapes will
be admitted; and by keeping the buildings in some parts low,
a free circulation of air will be encouraged, and the more
lofty buildings, rising above these subordinate ones, will produce that degree of grandeur and intricacy exemplified in the
east view of ASHTON COURT.

The old part (as distinguished in the plate) [fig. 116],
consists of the hall, the chapel, and the two turrets; but no
part of the gallery, added by Inigo Jones, is visible, except
the chimneys in perspective. The new part consists of the
entrance porch and cloister, which supplies a covered way to
the great hall, and forms one side of the quadrangle.*

Over this low range of offices the more lofty range of new

---

* The idea of an octagon kitchen is taken from that still remaining
among the ruins of Glastonbury Abbey: I mentioned it to the architect engaged at KENWOOD many years ago, and I have since observed it is introduced at CASHIOBURY, with admirable effect, by Mr. James Wyatt, under
whose direction that ancient Abbey has been lately altered with such good
taste and contrivance, that I shall beg leave to refer to it as a specimen of
adapting ancient buildings to modern purposes.

building appears, consisting of a large square tower, which will also be seen rising above the long south front. In that part which joins the new to the old buildings, are a dressing-room and *boudoir*, lighted by a bow-window, placed at the angle in such direction as to command an interesting view of Bristol, and the river Avon, with its busy scene of shipping. To take advantage of this view, from a house in the country, may appear objectionable to some; but I consider it among the most interesting circumstances belonging to the situation of Ashton Court. To the wealthy mechanic, or the more opulent merchant, perhaps the view of a great city may recall ideas of labour, of business, of difficulty, and dangers, which he would wish to forget in the serenity of the country; but the country gentleman, who never visits the city but to partake in its amusements, has very different sensations from the *distant* view of a place which, by its neighbourhood, increases the value and the enjoyment of his estate.

A general idea prevails, that, in most cases, it is better to rebuild than repair a very old house; and the architect often finds less difficulty in making an entire new plan, than in adapting judicious alterations: but if a single fragment remains of the grandeur of former times, whether of a castle, an abbey, or even a house, of the date of Queen Elizabeth, I cannot too strongly enforce the propriety of preserving the original character of such antiquity, lest every hereditary mansion in the kingdom should dwindle into the insignificance of a modern villa.*

* There is not more false taste in adding pointed arches and wooden battlements to a modern building, than in cutting off the projections, filling up the recesses, and mutilating the picturesque appendages of a true Gothic structure.

CHAPTER XIV.

Application of Gardening and Architecture united, in the Formation of a
new Place—Example, from BAYHAM—River—Lake—The House—
Character—Observations on Grecian Houses—Characteristic Architecture—External Gothic not incompatible with Comfort—How far it
should prevail internally.

THE necessity of uniting architecture and landscape-gar-
dening is so strongly elucidated in the *Red Book* of BAYHAM,
that I gladly avail myself of the permission of its noble possessor to insert the following observations; but as the ruins of
Bayham Abbey are generally known to those who frequent
Tunbridge Wells, it is necessary to premise that the situation
proposed for a new house is very different from that of the
Abbey.

" No place, concerning which I have had the honour to be
consulted, possesses greater variety of water, with such difference of character as seldom occurs within the limits of the
same estate.

The water near the Abbey, now intersecting the meadow
in various channels, should be brought together into one
river, winding through the valley in a natural course: this
may be so managed as to drain the land while it improves the
scenery; and I suppose the whole of this valley to be a more
highly dressed lawn, fed by sheep and cattle, but without
deer.

Above this natural division the water will assume a bolder
character; that of a lake, or a broad river, filling the entire
bottom of the valley, between two wooded shores, and dashing
the foot of that steep bank on which the mansion is proposed
to be erected. This valley is so formed by nature, that an
inconsiderable dam will cause a lake, or, rather, broad river,
of great apparent extent: for, when I describe water, I never
estimate its effects by the number of acres it may cover, but
by its form, its continuity, and the facility with which its
termination is concealed.

Where a place is rather to be *formed* than *improved*, that

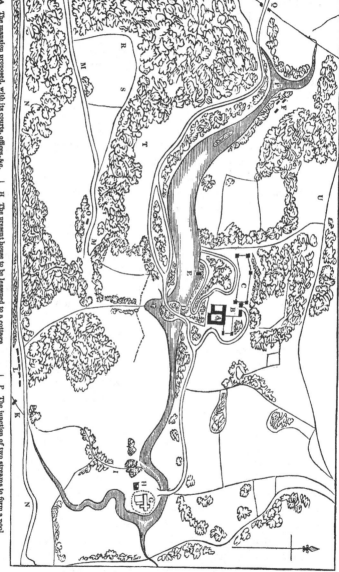

A The mansion proposed, with its courts, offices, &c.
B Stables, barn, wood and carpenter's yards, &c.
C Kitchen garden, fruit walls, gardener's house, &c.
D Bridge, wier, and engine-house to supply water
E Boat-house, cold bath, landing quay, &c.
F Farm-house at Tolsely for poultry, &c.
G Ruins of Bayham Abbey

H The present house to be lessened to a cottage
I The present water-mill to be removed to K
L Situation for blacksmith, wheelwright cottages, &c.
M The course of the old high-road changed to
N The new high-road and boundary of the park to the

O Little Bayham farm-house, to be a keeper's house

P The junction of two streams to form a pool
Q Entrance to the park from the principal approach
R The summit of the hill from whence the ground
S falls every way
T Terrace field
U A grade in the wood to be opened
U To be planted.

[Fig. 117. Map of Bayham Park.]

is, where no mansion already exists, the choice of situation for
the house will, in some measure, depend on the purpose for
which it is intended, and the character it ought to assume:
thus a *mansion*, a *villa*, and a *sporting seat*, require very
different adaptation of the same principles, if not a variation
in the principles themselves. The purpose for which the
house at BAYHAM is intended, must decide its character: it is
not to be considered as a small villa, liable to change its
proprietor, as good or ill success prevails, but as the established mansion of an English nobleman's family. Its character, therefore, should be that of greatness and of durability.
The park should be a forest, the estate a domain, the house a
palace. Now, since magnificence and compactness are as
diametrically opposite to each other as extension and contrac-
tion, so neither the extended scale of the country, nor the
style, nor the character of the place, will admit of a compact
house.

In determining effects, it is not sufficient to consider
merely the size of the building; but as all objects appear
great or small only by comparison, it is also necessary to consider the size and character of those by which this mansion
will be accompanied.

The surrounding scenery of BAYHAM must influence the
character of the house; we must, therefore, consider what
style of architecture will here be most appropriate. There
has ever appeared to me something wrong, or misunderstood,
in the manner of adapting Grecian architecture to our large
mansions in the country: our professors having studied from
models in a different climate, often forget the difference of
circumstances, and shew their classic taste, like those who
correctly quote the words, but misapply the sense, of an
author. The most striking feature of Grecian architecture is
a portico, and this, when it forms part of a temple, or a
church, may be applied with propriety and grandeur; but
when added to a large house, and intersected by two or three
rows of windows, it is evidently what, in French, is called an
Appliqué, something added, an after-thought; and it has but
too often the appearance of a Grecian temple affixed to an
English cotton-mill.

There is, also, another circumstance belonging to Grecian

architecture, *viz.*, *symmetry*, or an exact correspondence of the sides with each other. *Symmetry* appears to constitute a part of that love of order so natural to man; the first idea of a child, in drawing a house, is to make the windows correspond, and, perhaps, to add two correspondent wings.

There are, however, some situations, where great magnificence and convenience are the result of a building of this description; yet it can only be the case where the house is so large, that one of the wings may contain a complete suite of private apartments, connected with the house by a gallery or library, while the other may consist of a conservatory, &c.

Every one who has observed the symmetrical elevations scattered round the metropolis, and the small houses with wings, in the neighbourhood of manufacturing towns, will allow, that symmetry so applied is apt to degenerate into *spruceness;* and of the inconvenience of a house, separated from its offices by a long passage (however dignified by the name of colonnade), there cannot surely be a question. There is yet another principle which applies materially to BAYHAM, *viz.*, that symmetry makes an extensive building look small, while irregularity will, on the contrary, make a small building appear large: a symmetrical house would, therefore, ill accord with the character of the surrounding country.

Having expressed these objections against the application of Grecian architecture, before I describe any other style of house, I shall introduce some remarks on a subject which has much engaged my attention, *viz.*, the adaptation of buildings not only to the situation, character, and circumstances of the scenery, but also to the purposes for which they are intended; this I shall call *characteristic architecture.*

Although it is obvious that every building ought ' to tell its own tale,' and not to look like anything else, yet this principle appears to have been lately too often violated: our hospitals resemble palaces, and our palaces may be mistaken for hospitals; our modern churches look like theatres, and our theatres appear like warehouses. In surveying the public buildings of the metropolis, we admire St. Luke's Hospital as a mad-house, and Newgate as a prison, because they both announce their purposes by their appropriate appearance, and no stranger has occasion to inquire for what uses they are intended.

From the palace to the cottage, this principle should be observed. Whether we take our models from a Grecian temple, or from a Gothic abbey, from a castle, or from a college, if the building does not look like a *house*, and the residence of a nobleman, it will be out of character at BAYHAM. It may, perhaps, be objected, that we must exactly follow the models of the style or date we mean to imitate, or else we make a *pasticcio*, or confusion of discordant parts. Shall we imitate the thing, and forget its application? No: let us rather observe how, in Warwick Castle, and in other great mansions of the same character, the proud baronial retreat ' of the times of old,' has been adapted to the purposes of modern habitation. Let us preserve the massive strength and durability of the castle, and discard the gloom which former tyranny and cruelty inspired; let us preserve the light elegance of Gothic abbeys in our chapels, but not in our houses, where such large and lofty windows are inadmissible; let us, in short, never forget that we are building a *house*, whether we admire and imitate the bold irregular outline of an ancient castle, the elegant tracery in the windows of a Gothic church, or the harmony of proportions, and the symmetrical beauty of a Grecian temple.

Of the three distinct characters, the *Castle*, the *Abbey*, and the *House-Gothic*, the former of these appears best calculated for BAYHAM [see figs. 118, 119, 120, and 121]. Yet, as the object is not to build a castle, but a house, it is surely allowable to blend with the magnificence of this character the advantages of the other two, as well as the elegance, the comfort, and the convenience of modern habitation. It may be urged, that the first purpose of a castle is defence; that of a house, habitation; but it will surely be allowed, that something more is required than the mere purposes of habitation. An ordinary carpenter may build a good room; a mechanic, rather more ingenious, may connect a *suite* of rooms together, and so arrange their several offices and appendages as to make a good house, that is, a house sufficient for all the purposes of habitation. But an architect will aim at something higher; he will add to the internal convenience, not merely external beauty, but external propriety and character; he will aim not only to make a design perfect in itself, but perfect in its application.

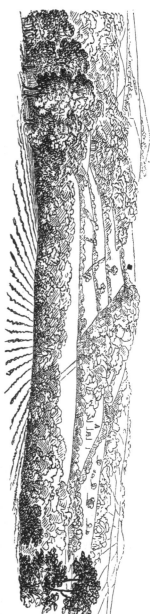

[Fig. 118. Panoramic view, shewing the situation intended for the mansion at Bayham. This view is taken from the hill in Terrace-field. On the field, A, a length of 150 feet was marked by poles, which, with a chaise left on the spot, served as a scale for the intended house.]

[Fig. 119. View, shewing Bayham Abbey as proposed to be built.]

[Fig. 120. A portion of the view fig. 119, on a larger scale.]

Where the lawn, the woods, the water, the whole place, and the general face of the surrounding country, are on so extensive a scale, the only means of preserving the same character is, by extending the plan of the house also. How can this be effected, unless we adopt the Gothic style of architecture? In Grecian or modern buildings, it has been considered an essential part of the plan to conceal all the subordinate appendages of the mansion, such as the stablès, the offices, the garden walls, &c.; and why? Because they neither do, nor can, partake of the character of the house; and the only method by which this extension of site is usually acquired in a Grecian building, is, by adding wings to the house. Thus the same mistaken principle obtains, and is considered material, for it is a part of the duty of these wings to conceal the offices. But, if continuity be an essential cause of the sublime, if extension be an essential cause of magnificence, whatever destroys continuity weakens the sublime, and whatever destroys extension lessens magnificence; therefore, as the offices and court-yards attached to a house are generally five times more extensive than the house itself, where magnificence is the object, why neglect the most effectual means of creating it? viz., continuity and extension, blended with unity of design and character; or, in other words, when it is desirable to take advantage of every part of the buildings, why conceal five parts in six of them?

If the truth of this principle be allowed, I trust the propriety of its application will be obvious; and, for its effect, I appeal to the following sketch, where both the actual size of the house, and its comparative proportion to the surrounding scenery, are correctly ascertained.

However pleasing these representations may appear, I should consider myself as having planned a ' castle in the air,' unless it should be proved that this design is not only practicable, but that it actually contains no more building than is absolutely necessary for the purposes of modern habitation. By the plan, it appears to contain,—

A Gothic hall, for the sake of ancient grandeur, but leading through a passage lower than the rooms, for the sake of not depressing their comparative height. The hall and passages should be rather dimly lighted by painted glass, to

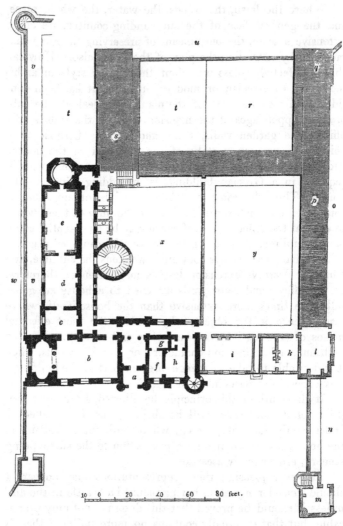

[Fig. 121. Plan of Bayham as proposed to be enlarged]

References:

a Porch	h Housekeeper's room	t t Flower gardens
b Eating-room	l Kitchen	u Shrubbery connecting the house
c Anti-room	m Brewhouse	with the kitchen garden
d Drawing-room	n Wood yard, &c.	v Terrace
e Library	o Stable yard	w Shrubbery on a steep bank hang
f Butler's bed-room	p Stables, &c.	ing down to the lake
g Plate-room	q Washhouse and laundry over	x Fountain court
h Butler's pantry	r Drying ground	y Kitchen court
Servants' hall	s Apartments for the family	

impress a degree of gloom essential to grandeur, and to render the entrance into the rooms more brilliant and cheerful.

This, it may be objected, is in character with those houses
which GRAY describes as having

'Windows that exclude the light,
And passages that lead to nothing.'

Yet I trust these passages will be found no less useful than
magnificent; they lead to the several rooms, which form
a complete suite of apartments, consisting of eating-room,
breakfast-room, drawing-room, and library. The rooms all
open by windows to the floor on a terrace, which may be enriched with orange-trees and odoriferous flowers, and will
form one of the greatest luxuries of modern, as well as one of
the most magnificent features of ancient habitation.

It now remains for me to shew that I have not suggested
a design more *expensive* than a house of any other character,
containing the same number of apartments. The chief difficulty of building arises from the want of materials: a house
of Portland stone would be very expensive; a red brick
house, as Mr. Brown used to say, ' puts the whole valley in
a fever;' a house of yellow brick is little better; and the
great Lord Mansfield often declared, that had the front of
Kenwood been originally covered with Parian marble, he
should have found it less expensive than stucco.   Yet one of
these must be used in any building except a castle; but for
this the rude stone of the country, lined with bricks, or faced
with battens, will answer every purpose ; because the enrichments are few, except to the battlements and the entrance
tower, which are surely far less expensive than a Grecian
portico.

The attached offices, forming a part of the front, are so
disposed as to lie perfectly convenient to the principal floor
and to the private apartments, while the detached offices, the
court-yards, and even the garden-walls, may be so constructed
and arranged, as to increase in dimensions the extent of the
castle.   This unity of design will be extended from the house
to the water, by the boat-house, the cold-bath, and the walls,
with steps leading to a bridge, near which the engine-house

may form a barbican, and contribute to the magnificent effect of the picture, as well as to the general congruity of character.

When we look back a few centuries, and compare the habits of former times with those of the present, we shall be apt to wonder at the presumption of any person who shall propose to build a house that may suit the next generation. Who, in the reign of Queen Elizabeth, would have planned a library, a music-room, a billiard-room, or a conservatory? Yet these are now deemed essential to comfort and magnificence: perhaps, in future ages, new rooms for new purposes will be deemed equally necessary. But to a house of perfect symmetry these can never be added: yet it is principally to these additions, during a long succession of years, that we are indebted for the magnificent irregularity, and splendid intricacy, observable in the neighbouring palaces of Knowle and Penshurst. Under these circumstances, that plan cannot be good which will admit of no alteration.

'Malum consilium est, quod non mutari potest.'

[It is a bad counsel which cannot be changed.]

But in a house of this irregular character, every subsequent addition will increase the importance: and if I have endeavoured to adopt some of the cumbrous magnificence of former times, I trust that no modern conveniences or elegances will be unprovided for."

It has been doubted how far a house, externally Gothic,* should internally preserve the same character; and the most ridiculous fancies have been occasionally introduced in libraries and eating-rooms, to make them appear of the same date

* It has occasionally been objected to Gothic houses, that the old form of windows is less comfortable than modern sliding sashes; not considering that the square top to a window is as much a Gothic form as a pointed arch, and that to introduce sash-frames, as at DONNINGTON, we have only to suppose the mullions may have been taken out without injuring the general effect of the building; while, in some rooms, the ancient form of window with large mullions may be preserved. Those who have noticed the cheerfulness and magnificence of plate-glass in the large Gothic windows of CASHIOBURY and COBHAM, will not regret the want of modern sashes in an ancient palace.

with the towers and battlements of a castle, without considering that such rooms are of modern invention, and, consequently, the attempt becomes an anachronism : perhaps the
only rooms of a house which can, with propriety, be Gothic,
are the hall, the chapel, and those long passages which lead
to the several apartments; and in these the most correct
detail should be observed. As a specimen of internal Gothic,
my son has inserted a design for a Gothic hall [omitted as
unnecessary], which is supposed to occupy two stories; yet the
comparative loftiness will not depress the height of the rooms,
because the gallery which preserves the connexion in the
chamber floor, marks a decided division in the height; and, as
this hall ought not to open into any room without an intermediate and lower passage, the several apartments will
appear more lofty and magnificent.

CHAPTER XV.

Conclusion—Concerning Colour—New Theory of Colours and Shadows, by
    Dr. Milner—Application of the same—Harmony—Discord—Contrast—
    Difficulty of comparisons between Art and Nature.

THE Art of Painting has been usually treated under four
distinct heads, *viz.*

Composition; Design, or *Drawing; Expression;* and *Colour-
    ing.* Each of which may, in some measure, be applied
    to landscape gardening, as it has been treated in this
    work.

Composition, includes those observations on utility, scale, per-
    spective, &c., contained in Chapters I. and II.
Design, may be considered as belonging to the remarks on
    water, wood, fences, lines, &c., contained in Chap-
    ters III. IV. V. VI. and VII.
Expression, includes all that relates to character, situation,
    arrangement, and the adaptation of works of art
    to the scenery of nature, which have been discussed
    in the remaining Chapters of this work; and, lastly,
Colouring, so far as it relates to certain artificial objects, has
    been mentioned in Chapter XI.

Having since been led to consider this subject more attentively, in consequence of a conversation with Mr. Wilberforce
concerning a new theory of colours and shadows, I have,
through his intervention, obtained permission to enrich my
work with the following curious remarks: and as Mr. Wilberforce, in his letter which enclosed them, observes of their
reverend and learned author, that " he is a man unequalled
" for the store of knowledge he possesses, for the clearness
" with which he views, and the happy perspicuity with which
" he communicates his conceptions," so I shall give this
theory in his own words.

THEORY OF COLOURS AND SHADOWS.

By the Rev. Dr. MILNER, F. R. S.

DEAN OF CARLISLE, AND PRESIDENT OF QUEEN'S COLLEGE, CAMBRIDGE.

SECT. 1. Several years ago, some curious questions concerning the colours of the shadows of bodies, were proposed to me by an ingenious and philosophical friend, who himself can paint very well, and is an excellent judge of colours. He first mentioned the following facts :—

2. Supposing a piece of writing paper to be weakly illumined by *white* light, and, at the same time, to have a strong *red* light thrown upon it by any contrivance, the shadow upon the paper, of a body placed in the said red light, will be *green*.

3. Or, *vice versâ*, if a strong *green* light be thrown upon the same paper, the shadow of a body placed in the green light will be *red*.

4. Under similar circumstances, the shadow of a body intercepting orange-coloured light will be blue, purple, or almost violet, according as the orange light contains more or less red ; and *vice versâ*.

5. And lastly, the shadow of a body which intercepts yellow light will be purple, and *vice versâ*.

6. The phenomena just mentioned may be exhibited in several ways. The weak white light may always be had in a dark room, either by admitting a small portion of daylight, or by means of a small lamp or wax taper, the light of which is sufficiently white for the purpose ; and in regard to the strong coloured lights, they are also easily procured, either by using transmitted or reflected light of the particular colour wanted. As candles and lamps are always at hand, and solar rays not so, I will here briefly describe the method of shewing any *one*, and, consequently, *all*, of these beautiful experiments by candle-light.

7. L, M, N, O, [in fig. 122], is a piece of white paper, illumined as in the figure ; D is a small cylinder of wood, as a black lead pencil, or even

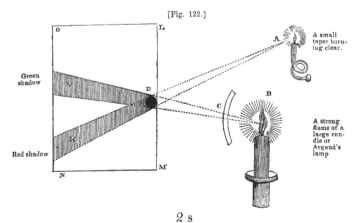

[Fig. 122.]

2 s

one's finger, in such a manner, as to produce the respective shadows D, V, and D, K; c, being a piece of red glass in this experiment.

8. If, instead of red glass, a piece of green glass be placed at c, then the shadow D, V, will no longer be green, but of a reddish cast; and so of the rest as mentioned above, at Sect. 3.

9. My friend was very desirous that I should endeavour to account for these beautiful and most extraordinary appearances ; with this view, I first observe, that the burning lights, A and B, when the experiments are made without daylight, may be reckoned nearly white, particularly if they are made to burn without smoke, though, in reality, they are yellowish, or even orange-coloured sometimes, as is very plain when they are compared with strong daylight.

10. Secondly, white light is well known to consist of several other colours, as red, orange, yellow, green, blue, purple, and violet; and, further, as violets and purples, with all their varieties, may be imitated by mixing blue and red in different proportions; and, as green also may be compounded in a similar way by mixing blue and yellow; and oranges by mixing red and yellow; we need not attend to more than the three primary colours, *red, yellow,* and *blue;* for, in fact, it is found that by mixing these three colours in certain proportions, a sort of white, or any colour may be formed; and there is reason to believe that, if we had colours equal in brilliancy to the prismatic colours, the white so formed would be perfect.

11. This last observation shews us, that white may be considered as made up even of two colours only, and we shall find it very convenient in the explanation of the phenomena in question, to consider white as so made up, namely, of red and green, of yellow and purple, or of blue and orange. These colours are called contrasts to each other respectively: their apparent brilliancy, when they are placed contiguous to each other, is promoted in a remarkable manner, but they cannot be mixed together without mutual destruction to their natural properties, and an approach to a white or a grey colour.

12. To understand the experiment above represented on the paper, we are first to consider the nature of the shadow D, V, green, as it is in appearance; that is, we are to consider what kind of light or lights can possibly come to this portion of the paper which we call the shadow D, V; and here it is plain, that this space, D, V, is illumined only by the *white light* * (I will call it) which comes from the small taper A, directly, and also by a small quantity of white light from B, not directly, but by reflection from the sides of the room, or from other objects. The direct red light coming from B, through the red glass c, is intercepted by D; and the small quantity of this red light, which can arrive at the space D, V, by reflection, is not worth mentioning; the green shadow D, V, therefore, is illumined by a small quantity of white light, and our business is to explain why it should appear green to the eye.

---

 * I call it white light because it is nearly so, and because it answers all the purposes of perfectly white light in such an experiment, supposed to be made in a room without daylight. When actually compared with daylight, is is found to be yellowish, or even orange-coloured.

13. Keep in mind that the idea of a perfect shadow excludes all light, and that the space D, V, is an imperfect shadow, illumined, as we have seen, with a small portion of white light. Let this small portion of white light be considered as made up of red light and green light, according to what has been stated above, in Sect. 12, and the reason of the phenomenon will be readily understood. For we must now attend to the strong red light which passes through the glass C, and covers the paper everywhere, except in the space D, V, where it is intercepted: the effect of this strong light coming up to the very boundaries of the shadow D, V, is such as to incapacitate the eye from seeing at the same time the weaker red light contained in the shadow D, V, which we have proved to be really of a weak dull white colour, but which, because its *red* light cannot be seen, appears *green* to the eye.

14. This effect of rendering the organs of perception insensible to weaker excitations, by strongly exciting those organs, is analogous to the constitution of the human frame in many instances. Accustom the eye either to much light, or to intense colours, and, for a time, it will hardly discern anything by a dull light, or by feeble colours, provided the feeble colours be of the *same* kind with the previous strong ones. Thus, after it has been excited by an intense red, for example, it will, for a time, be insensible to weak red colours, yet it will still easily perceive a weak green, or blue, &c., as in the instance before us respecting the shadow D, V, where the green part of the compound still affects the eye, after the red has ceased to produce any effect, owing to the previous excitation of a stronger red.*

15. Nor is this the case only with the eye, it is the same with every other sense; precise instances of this kind in regard to the taste, the smell, the touch, &c., will occur plentifully to every one.

16. I consider this solution of the appearances of the colours as perfectly satisfactory. Here it is applied only to one instance, but it is equally applicable to all the rest; and it appears to me to account for all the difficulties which seem to have embarrassed Count Rumford, in his very ingenious and entertaining paper, Phil. Trans. 1794, p. 107. Also in Dr. Priestley's History of Optics, p. 436, there is a curious Chapter, containing the observations of philosophers on blue and green shadows; the true cause of these shadows is not, I think, there mentioned; and it may be entertaining to read that Chapter with these principles in the mind.

17. When the sun has been near setting on a summer evening, I have often observed most beautiful blue shadows upon a white marble chimney-

---

* This distinction should always be kept in mind, for, unless the eye has been absolutely injured or weakened by excessive excitation, there is reason to believe that strong excitations of it, whether immediately preceding weaker ones, or contemporaneous with them, much *improve* its sensibility in regard to those weaker ones, provided only that they be of a *different* class. If the eye has been excited by a lively red colour, it will scarcely perceive a weak red, but it will perceive a weak green much better, on account of the previous excitation by the strong red; and the reason may be, that, in looking at a red colour, the eye wastes *none* of that nervous sensibility which is necessary for its seeing a green colour; and the same reasoning holds in all other cases where the colours are contrasts to each other. For such colours seem incapable of mixing with each other, in the proper sense of the word, as when red and yellow are mixed together, and produce a compound evidently partaking of the obvious properties of the two ingredients. When contrasts are mixed together, as red and green, these colours seem destructive of each other, and effect a compound approaching to whiteness. *Similar* observations may be made on the other senses.

piece. In this case, the weak white light of the evening, which illumines the shaded part of the marble, is to be considered as compounded of two colours, orange and blue. The direct orange rays of the sun at this time, render the orange part invisible, and leave the blue in perfection.

18. And in the same way is to be explained that beautiful and easy experiment mentioned by Count Rumford, p. 103, Phil. Trans. 1794, where a burning candle in the day-time produces two shadows, and one of them of a most beautiful blue colour. The experiment is the more valuable, as it may be made at any time of the day with a burning candle. Almost darken a room, and then by means of a lighted candle and a little daylight, produce two shadows of any small object, as of a pencil, &c., one from the candle, and another from the daylight received at a small opening of one of the window-shutters; the light of the candle will appear orange-coloured in the day-time, and so will that shadow of the body which belongs to, or is made by, the daylight; but the shadow of the body made by the candle, will surprise any person, by being of a fine blue.

19. More than once I have been agreeably struck with this appearance, produced unintentionally when I have been writing by candle-light on a winter's morning; upon the daylight being let in, the shadow of my pen and fingers in the orange-light of the candle, were beautifully blue.

20. I suppose there is such a thing as the harmony of colours, of which painters speak so much; according to the explanation here given, our key to the solution of every case of harmony and of contrast, is to consider what is the *other* colour, simple or compound, which, joined to a given one, simple or compound, will constitute *white*. Thus red, requires green; yellow, purple; blue, orange; and *vice versâ*, the mixtures in proper proportions will be white.

21. Sir Isaac Newton (Prop. 6. part 2, of book i. Optics,) has given a method for judging of the colour of the compound in any known mixture of primary colours, but it is not easy, even for mathematicians, to put his rules in practice. The gentleman who consulted me on this subject of shadows, has been accustomed, for a long time, to assist his memory, when he is painting, by the use of the simple diagram [fig. 123]. Let R, Y, B, represent the three uncompounded colours, red, yellow, blue; and let O, G, P, represent the compounds orange, green, and purple; it is evident that, to make a deeper orange, we must add more red; and to make a bluer green, we must add more blue; and to make the purple redder, we must add more red, and *vice versâ* : but besides this, the diagram puts us in mind that G is the contrast to R, and that, therefore, those two colours cannot be mixed without approaching to a dull whiteness, or greyness; and the same may be said of Y and P, and of B and O: these colours are also contrasts to each other; by mixture they destroy each other, and produce a whiteness, or greyness, according as they are more or less perfect; but when kept distinct, they are found to make each other look more brilliant by being brought close together: and all this is agreeable to what is said in Sect. 11, and in the note to Sect. 14.

22. Sir Isaac Newton observes, that he had never been able to produce

a perfect white by the mixture of only two primary colours, and seems to doubt whether such a white can be compounded even of three. He tells us, that one part of red lead, and five parts of verdigris, composed a dun colour, like that of a mouse; but there is nothing in all this which militates against the explanation here given of the cause of the coloured shadows of bodies; for even supposing that there did not exist in nature any two bodies of such colours as to form perfect whiteness by their mixture; or, to go still further, supposing that no two prismatic colours of the sun could form a compound perfectly white; still the facts and reasonings here stated respecting the mixtures of such colours as are called contrasts, are *so near* the truth, that they furnish a satisfactory account of the appearances of the colours of the shadows which we have been considering. The terms by which we are accustomed to denominate colours, have not a very accurate or precise meaning, and particularly those terms which denote colours that are known to be mixtures of others, as green, purple, and orange : neither the prismatic green, nor the colour of any known green body, may, perhaps, combine with red so as to make actually an accurate white, and yet the existence or composition of such a green may not be impossible. The philosophical reader will clearly perceive, that no argument of any weight can be drawn from considerations of this sort against this theory of coloured shadows.

23. Every one knows that red colours and yellow colours mixed together, in different proportions, produce orange colours of various kinds : also that reds and blues produce purples and violets; and, lastly, that blues and yellows produce greens in great variety; but it is not so generally known that green, purple, and orange colours, are, as it were, almost annihilated by mixture, and much improved by contiguity with red, yellow, and blue colours respectively.

The little diagram [fig. 123], suggests all these things to the memory, and a great many more of the same kind; and, therefore, must be extremely

[Fig. 123.]

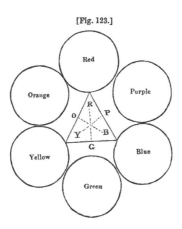

useful to the artist who is endeavouring to produce certain effects by contrast, harmony, &c., but it should always be carefully remembered, that it contributes nothing to *the proof* of any of the truths here advanced; the proof rests upon the reasons given for each of them respectively.

This curious and satisfactory theory demonstrates that the choice of colours which so often distinguishes good from bad taste in *manufactures, furniture, dress,* and in every circumstance where colour may be artificially introduced, is not the effect of chance, or fancy, but guided by certain general laws of NATURE.

Sir Isaac Newton discovered a wonderful coincidence between sound and colours, and proves, mathematically, that the *spaces* occupied by the colours in the prismatic spectrum correspond with the parts of *a musical chord,* when it is so divided as to sound the notes of an octave. So this resemblance may now be considered as extending further, for, as in music, so likewise in colours, it will be found that *harmony* consists in distance and contrast, not in similitude or approximation. Two notes near each other, are grating to the ear, and are called *discords;* in like manner, two colours very near each other, are unpleasing to the sight, and may be called *discordant;* this may be proved by covering all the colours in the preceding diagram except the two adjoining, which, in every part of the scale, will appear discordant; while, on the contrary, if the two sides be covered in any direction so as only to shew the two opposite colours, they will appear in perfect harmony with each other; and this experiment confirms the good taste of those who, in the choice of colours, oppose reds to greens, yellows to purples, and blues to oranges, &c. But instead of contrasting these colours, they are mixed, or so blended, as not to appear each distinctly, as in silks or linens where the stripes are too narrow; when seen at a little distance, instead of relieving, they will destroy each other. In the application of this theory to some familiar instances, particularly in the furniture of rooms, I have observed that two colours, here deemed *discordant,* may be used without offending the eye, as green and blue, or green and yellow; but I have always considered such assortment intolerable, unless one were very dark, and the other very light; and thus the effect is again produced by *contrast,* although on a different principle; it is the contrast, not between colours, but between *light* and *darkness*.

So far this theory is perfectly satisfactory with respect to works of *art*, but, when carried to those of *nature*, I confess my inability to reconcile a conviction of its truth, with certain appearances which seem to contradict it.

By the universal consent of all who have considered the harmony of colours, it is allowed that, in works of art, the juxta position of bright blues and greens is discordant to the eye, and the reason of this discordance has been shewn by the foregoing remarks. Yet these are the two prevailing colours in nature; and no person ever objected to the want of harmony in a natural landscape, because the sky was *blue*, and the surface of the earth covered with *greens*, except he viewed it with a painter's eye, and considered the difficulty, or even impossibility, of exciting the same pleasurable sensations by transferring these colours to his canvas; the only way in which I can solve this seeming paradox, is, by observing that the works of nature, and those of art, must ever be placed at an immeasurable distance from the different scale of their proportions; and whether we compare the greater efforts of man, with the system in which the world he inhabits forms but an inconsiderable speck; or the most exquisite miniature of mechanism with the organs of sense and motion in an insect, we must equally feel the deficiency of comparison, the incompetency of imitation, and the imperfection of all human system. Yet, while lost in wonder and amazement, the man of taste, and the true philosopher, will feel such agreement existing in the laws of nature as can only be the consequence of Infinite Wisdom and Design ; while, to the sceptic, whether in moral or in natural philosophy, the best answer will be in the words of the poet :—

> " All nature is but art unknown to thee ;
> All chance, direction which thou canst not see ;
> All *discord, harmony* not understood ;
> All partial evil, universal good."

INDEX

OF THE

SUBJECTS CONTAINED IN EACH CHAPTER.

AN INQUIRY

INTO THE

CHANGES OF TASTE

IN

LANDSCAPE GARDENING:

TO WHICH ARE ADDED

SOME OBSERVATIONS

ON

ITS THEORY AND PRACTICE,

INCLUDING

A DEFENCE OF THE ART.

BY

H. REPTON, Esq.

[*Originally published in 8vo. in 1806.*]

2 T

ADVERTISEMENT.

THE following " Inquiry into the Changes of Taste in Landscape Gardening " has been produced in consequence of a request from Professor Martyn, that I would furnish him with some matter for his elaborate edition of Miller's Gardener's Dictionary; in the Preface to which, he proposes to " take up the History of Landscape Gardening from the period when Mr. Walpole left off, and to trace it from Kent, through Brown, to the present time."

Having also been called on by my bookseller for a new edition of my " Observations on the Theory and Practice of Landscape Gardening," I have directed that work to be re-printed without any alteration or addition, considering it an act of justice to the original subscribers and purchasers of so expensive a volume, not to make the second edition more perfect than the first.

I have also been desired to publish a new edition of my first work, entitled, " Sketches and Hints on Landscape Gardening," of which, only 250 copies being published by Messrs. Boydells in 1794, the book is become so scarce, that above four times the original price has been paid for some copies. In compliment to the present possessors of that work, I have determined never to publish another costly edition of it with plates, but rather to extract from it such matter as may not interfere with the quarto volume ; incorporating it with such further observations on the Theory and Practice of the Art as have occurred from more recent practice : to which are added, answers to the attacks made on the art by some late publications.

Although I am aware of the utility of plates to exemplify many parts of this subject, and that one stroke of the *pencil* will often say more than a page with the *pen*, yet the enormous expense of engraving has hitherto so confined my opinions to a certain class of purchasers, that they have been either not generally known, or they have been repeated by some without acknowledgment, and misrepresented by others without sufficient quotation.

Harestreet, near Romford,
    1806.

[The greater part of this volume being a republication of " Hints and Sketches," we have, of course, confined ourselves to reprinting what has not before appeared, either in that work or in the *Observations.*—J. C. L.]

CONTENTS.

PART I.

HISTORICAL NOTICES.

Taste influenced by Fashion.—Every revolution in the taste of a country may be accounted for on the same principles with the revolutions in its laws, its customs, and opinions—the love of *change* or *novelty* in a few, and of *sameness* or *imitation* in the many. And however the pride of system may revolt at taste being influenced and liable to change with the fashions of the day, it is impossible to fix any standard for taste, that may not be shaken by the prevailing opinions of the public, whether right or wrong. Thus, in whatever relates to the amusements and pleasures of mankind, though an old fashion may be most rational, yet a received new fashion will be deemed in the best taste. This leads me to consider the origin of what is called fashion, and, by the multitude, generally considered as taste.

Origin of Fashion.—Although each individual may have the power of thinking, yet the mass of mankind act without thought, and, like sheep, follow a leader through the various paths of life. Without this natural propensity for imitation, every member of society would hold a different opinion, and the world would be at perpetual warfare. Indeed, every disagreement, from the enmity of nations to the petty squabbles of a parish, is caused and conducted by some leader, whom the multitude follow, imitate, and support.

This is the origin of changes, in customs or fashions, in every shape. Opinions are declared by one man, and followed by the many. If persons only of superior sense were the leaders, or if mankind always examined what they followed, fashion might, perhaps, be more reasonable : but this supposes mankind always to act like rational beings, which is contrary to every test of experience.

Therefore, whether in religion, in politics, in philosophy, in medicine, in language, in the arts, in dress, in equipage, in furniture, or in the most trifling concerns of life, we see thousands move in the way that some one has gone before : and if it be too great a stretch of thought to mark a new track, it is also too great to investigate whether the new track marked out by another be good or bad.

Changes, by whom made.—Changes in the fashion, or, in other words, in the customs of a country, become a source of wealth and commerce, and contribute to those daily occupations which make life preferable in civilized society. The clown and the savage require no change, no variety ; and the vulgar, who are one degree above them, slowly adopt the changes of others, although they insensibly slide into the fashion. On the contrary, the nice observer, the ' *elegantiæ formarum spectator*' [exact judge of beauty], eagerly seizes and imitates whatever appears new ; and, perhaps, without inquiring into its reasonableness or propriety. Thus, forms and fashions of one climate are often brought into another, without attending to their uses or original intentions.

Fashions in dress, in furniture, &c. are comparatively harmless ; they soon pass away, and become ridiculous, in proportion to the distance of their dates. Thus we laugh at the odd figures of our ancestors on canvas, and wonder at the bad taste of old worm-eaten furniture, without reflecting that, in a few years, our own taste will become no less obsolete.

But in the more lasting works of art, fashion should be guided by common sense, or we may perpetuate absurdities. Of this kind was the general rage for destroying those old English buildings called Gothic ; and for introducing the architecture of a hot country, ill adapted to a cold one—as the Grecian and Roman portico to the north front of an English house, or the Indian *verandah* as a shelter from the cold east winds of this climate.

In Gardening.—Fashion has had its full influence on gardening, as on architecture, importing models from foreign countries. The gardens in England have, at one time, imitated those of Italy, and, at another, those of Holland.

Italian style.—The Italian style of gardens consisted in ballustraded terraces of masonry, magnificent flights of steps, arcades, and architectural grottos, lofty clipped hedges, with niches and recesses, enriched by sculpture. This was too costly for general use; and where it was adopted, as at Non-such, and some other palaces, it was discovered to be in-applicable to the climate of England; and no traces now remain of it, except in some pictures of Italian artists. *

Dutch style.—To this succeeded the Dutch garden, intro-duced by King William III., and which prevailed in this country for half a century. It consisted of sloped terraces of grass, regular shapes of land and water, formed by art, and quaintly adorned with trees in pots, or planted alternately, and clipped, to preserve the most perfect regularity of shape. These were the kind of terraces, and not those of the grand Italian style, which Brown destroyed, by endeavouring to restore the ground to its original shape.

English style.—He observed that Nature, distorted by great labour and expense, had lost its power of pleasing, with the loss of its novelty; and that every place was now be-come nearly alike. He saw that more variety might be introduced by copying Nature, and by assisting her operations. Under his guidance, a total change in the fashion of gardens took place; and as the Dutch style had superseded the Italian, so the English garden became the universal fashion. Under the great leader, Brown, or rather those who patronised his discovery, we were taught that Nature was to be our only model. He lived to establish a fashion in gardening, which might have been expected to endure as long as Nature should exist.

Nature : Brown's model.—Nature is alike the model to the poet, the painter, and the gardener, who all profess to be her imitators: but how few have genius or taste to avoid becom-ing mannerists! Brown *copied Nature,* his illiterate followers *copied him ;* and, in such hands, without intending to injure his fame, or to depart from his principles, the fashion of English gardening was in danger of becoming more tiresome,

---

* Some mention of the French style of gardening may here be ex-pected; but as this was only a corruption of the Italian style, and was never generally adopted in England, it is purposely omitted; although, in prac-tice, I have occasionally availed myself of its more massive Trellis, *Boccages,* a nd *Cabinets de Verdure,* to enliven the scenery of a flower garden.

insipid, and unnatural, than the worst style of Italian or
Dutch examples.

Brown's style corrupted.—Mr. Brown, after his death, was
immediately succeeded by a numerous herd of his foremen and
working gardeners, who, from having executed his designs,
became *consulted*, as well as *employed*, in the several works
which he had entrusted them to superintend.   Among these,
one person had deservedly acquired great credit at Harewood,
at Holkham, and other places, by the execution of gravel
walks, the planting of shrubberies, and other details belonging
to pleasure grounds, which were generally divided from the
park by a sunk fence, or *ha! ha!* and happy would it have
been for the country, and the art, if he had confined his talents
within such boundary.   Unfortunately, without the same great
ideas, he fancied he might *improve* by *enlarging* his plans.
This introduced all that bad taste which has been attributed to
his great master, Brown.

Extent mistaken for Beauty.—Hence came the mistaken
notion, that greatness of dimensions would produce greatness
of character: hence proceeded the immeasurable extent of
naked lawn ; the tedious lengths of belts and drives; the useless breadth of meandering roads; the tiresome monotony of
shrubberies, and pleasure grounds; the naked expanse of
waters, unaccompanied by trees; and all the unpicturesque
features which disgrace modern gardening, and which have
brought on Brown's system the opprobrious epithets of bare
and bald.   Yet such is the fondness for what is *great by
measurement*, that the beauty of parks is estimated by the
acre, and the perfection of walks and drives computed by the
mile, although we look at them without interest, and fly from
them to farms and fields, even preferring a common or a heath,
to the dull round of a walk or drivé, without objects and without variety.

Park Scenery.—When, by this false taste for extent,
parks had become enlarged beyond all reasonable bounds of
prudence and economy in the occupation, it then became
advisable to allot large portions of land for the purposes of
agriculture, within the belt or outline of this useless and
extravagant inclosure ; and thus great part of the interior of
a park is become an arable farm.   Hence arises the necessity

of contracting that portion of an estate, in which beauty, rather than profit, is to be considered.

Garden Scenery. — Much of the controversy concerning modern gardening seems to have arisen from the want of precision in our language. *Gardening* is alike applied to the park, the lawn, the shrubbery, and the kitchen garden; and thus the scenery of one is blended with that of another, when there is as much difference between garden scenery, park scenery, and forest scenery, as between horticulture, agriculture, and uncultivated nature. The first is an artificial object, and has no other pretence to be natural, than what it derives from the growth of the plants which adorn it : their selection, their disposition, their culture, must all be the work of art; and instead of that invisible line, or hidden fence, which separates the mown turf from the lawn fed by cattle, it is more rational to shew that the two objects are separated, if the fence is not unsightly ; otherwise, we must either suppose that cattle are admitted to crop the flowers and shrubs, or that flowers and shrubs are absurdly planted in a pasture exposed to cattle, or, which is more frequently the case, we must banish flowers entirely from the windows of a house, and suppose it to stand on a naked grass field. *

By the avenues and symmetrical plantations of the last two centuries, the artificial garden was extended too far from the mansion ; but, in the modern gardening, the natural lawn is brought too near.

Example from Woburn. — As there are few palaces in England that can vie in magnificence with that of Woburn, it may furnish an example of greatness in variety and character in its garden scenery, without making its dimensions the standard of its greatness. The mansion is connected with its appendages, such as the stables, riding-house, tennis-court, orangery, Chinese-pavilion, game-larder, &c. &c., by a corridor, or covered passage, of considerable length, which is

---

* Fences are not objectionable when they mark a separation, and not a boundary of property. Thus a park-pale marks the precise limits of the park, and a hedge before a wood renders it liable to be mistaken for a wood belonging to some other person, and, therefore, acts as a boundary : but the hurdle, which makes a temporary division of a lawn, or a light open fence that divides the garden from the park, can only offend the fastidious critic, who objects to all fences, without knowing or assigning any reason.

enriched with flowers and creeping plants. This passage is proposed to be extended to the hot-houses in the forcing garden, which is to form a centre, for a series of different gardens, under the following heads :—

The terrace and parterre near the house.

The private garden, only used by the family.

The rosary, or dressed flower garden, in front of the green-house.

The American garden, for plants of that country only.

The Chinese garden, surrounding a pool in front of the great Chinese pavilion, to be decorated with plants from China.

The botanic garden, for scientific classing of plants.

The animated garden, or menagerie.

And, lastly, the English garden, or shrubbery walk, connecting the whole ; sometimes commanding views into each of these distinct objects, and sometimes into the park and distant country.

The word Gardening misapplied.—By a strange perversion of terms, what is called modern, or English *gardening*, seldom includes the *useful garden*, and has changed the name of the *ornamental garden* into *pleasure ground*. But it is not the name only that has been changed; the character of a garden is now lost in that of the surrounding park, and it is only on the map that they can be distinguished; while an invisible fence marks the separation between the *cheerful lawn* fed by cattle, and the *melancholy lawn* kept by the roller and the scythe. Although these lawns are actually divided by a barrier as impassable as the ancient garden wall, yet they are apparently united in the same landscape, and,

> ————— " wrapt all o'er in everlasting green,
> Make one dull, vapid, smooth and tranquil scene."
> R. P. KNIGHT.

Similitude between House and Gardens.—The gardens, or pleasure-grounds, near a house, may be considered as so many different apartments belonging to its state, its comfort, and its pleasure. The magnificence of a house depends on the number as well as the size of its rooms; and the similitude between the house and the garden may be justly extended to the mode of decoration. A large lawn, like a large room, when unfurnished, displeases more than a small one. If only in part, or meanly furnished, we shall soon leave it with disgust ; whether it be a room covered with the finest green baize, or a lawn kept with the most exquisite verdure, we look for carpets in one, and flowers in the other.

If, in its unfurnished state, there chance to be a looking-glass without a frame, it can only reflect the bare walls ; and thus a pool of water, without surrounding objects, reflects only the nakedness of the scene. This similitude might be extended to all the articles of furniture, for use or ornament, required in an apartment, comparing them with the seats, and buildings, and sculpture appropriate to a garden.

Its application.—Thus, the pleasure-ground at Woburn requires to be enriched and furnished like its palace, where good taste is everywhere conspicuous. It is not by the breadth or length of the walk, that greatness of character in garden scenery can ever be supported ; it is rather by its diversity, and the succession of interesting objects. In this part of a great place we may venture to extract pleasure from *variety*, from *contrast*, and even from *novelty*, without endangering the character of *greatness*.

Changes near the House.—In the middle of the last century, almost every mansion in the kingdom had its kitchen and fruit-gardens, surrounded by walls, in the front of the house. To improve the landscape from the windows, Brown was obliged to remove these gardens ; and not always being able to place them near the house, they were sometimes removed to a distance. This inconvenient part of his system has been most implicitly copied by his followers ; although I observe that at Croome, and some other places where he found it practicable, he attached the kitchen garden to the offices and stables, &c. behind the mansion, surrounding it with a shrubbery ; and, indeed, such an arrangement is the most natural and commodious.

Kitchen Garden.—The intimate connexion between the kitchen and the garden, for its produce, and between the stables and the garden, for its manure, is so obvious, that every one must see the propriety of bringing them as nearly together as possible, consistent with the views from the house ;

yet we find in many large parks that the fruit and vegetables are brought from the distance of a mile, or more, with all the care and trouble of packing for much longer carriage; and the park is continually cut up by dung-carts passing from the stables to the distant gardens.

Winter Garden.—To these considerations may be added, that the kitchen garden, even without hot-houses, is a different climate. There are many days in winter when a warm, dry, but secluded walk, under the shelter of a south wall, would be preferred to the most beautiful but exposed landscape; and in the spring, when

> " Reviving Nature seems again to breathe,
> As loosen'd from the cold embrace of death,"

on the south border of a walled garden some early flowers and vegetables may cheer the sight, although every plant is elsewhere pinched with the north-east winds peculiar to our climate in the months of March and April, when

> " Winter, still ling'ring on the verge of spring,
> Retires reluctant, and, from time to time,
> Looks back, while at his keen and chilling breath
> Fair Flora sickens."
> STILLINGFLEET.

Changes in Planting.—Straight Lines.—Quincunx.—Let us now trace the progress of change in the fashions of planting; by which I mean the various systems adopted at different periods for making trees artificial ornaments. The first was, doubtless, that of planting them in a single row at equal distances, which prevailed in the garden mentioned by Pliny. The next step was that of doubling these straight rows, to form shady walks, or, adding more rows, to make so many parallel lines. But fashion, not content with the simplicity of such avenues of trees placed opposite to each other, invented the quincunx, by which these straight lines were multiplied in three different directions. As the eagerness for adopting this fashion could not always wait the tedious growth of trees, where old woods existed, they were cut through in straight lines and vistas, and in the forms of stars and *pates d'oies* which prevailed at the beginning of the last century.

Regular Curves.—Fashion, tired of the dull uniformity of straight lines, was then driven to adopt something new; yet, still acting by geometric rules, it was changed to regular forms of circles and curves, in which the trees were always planted at equal distances. This introduced, also, the serpentine avenue for a road.

Platoons.—The next bold effort of fashion was that of departing from the equidistant spaces; and trees were planted in patches or clumps (called, in some old maps, platoons): these were either square or round, alternately shewing and hiding the view on each side of the road; and where no view was required, a screen, or double row of trees, entirely shut up one side, while, on the other, the view was occasionally admitted, but still at regular intervals: this prevails in the drives at Woburn.

Avenues ceased.—I perfectly remember, when I was about ten years old, that my father (a man of such general observation, that no innovation or novelty escaped him) remarked to me the change which was then taking place in ornamental planting; and then, although little supposing how much it would become the future study of my life, I recollect his observing the discovery made by some ingenious planter (perhaps Kent or Brown), that the straight line might be preserved in appearance from the ends of a vista, or avenue, without actually filling up all the sides; and thus alternate openings of views to the country might be obtained, without losing the grandeur of the straight line, which was then deemed indispensable. He also observed, that, perhaps, this would lead to the abolishing of avenues; and I believe few were planted after that date, *viz.*, the middle of the last century.

Natural Planting.—About this time a total change in the fashion took place. It was asserted, that nature must be our only model, and that nature abhorred a straight line; it was not, therefore, to be wondered at, that Brown's illiterate followers should have copied the means he used, and not the model he proposed: they saw him prefer curved lines to straight ones; and hence proceeded those meandering, serpentine, and undulating lines in all their works, which were, unfortunately, confirmed by Hogarth's recommendation of his *imaginary line of beauty.* Thus we see roads sweeping round, to avoid the direct line, to their object, and fences fancifully taking a longer course; and even belts and plantations in useless curves, with a drive meandering in parallel lines, which are full as much out of nature as a straight one. Thus has fashion converted the belt or screen of plantation, introduced by Brown, into a drive quite as monotonous, and more tedious, than an avenue, or vista, because a curved line is always longer than a straight one.

Brown's Belt.—Brown's belt consisted of a wood, through which a road might wind to various points of view, or scenery shewn under various circumstances of foreground; but the drive was only made among the trees, and under the shade of their branches.

The modern Belt.—The last fashion of belt, which Brown never made, is an open drive, so wide, that it never goes near the trees, and which admits such a current of air, that the front trees are generally the worst in the plantation: add to this, that two narrow slips of planting will neither grow so well, nor be such effectual harbours for game, as deeper masses, especially when the game are liable to be disturbed by a drive betwixt them. The belt may be useful as a screen, but, unless very deep, it should never be used as a drive, at least till after the trees have acquired their growth, when a drive may be cut through the wood to advantage.

Variety destroyed by its excess.—It is not only the line of the modern belt and drive that is objectionable, but also the manner in which the plantations are made, by the indiscriminate mixture of every kind of tree. In this system of planting all variety is destroyed by the excess of variety, whether it is adopted in belts or clumps, as they have been technically called; for example, if ten clumps be composed of ten different sorts of trees in each, they become so many things exactly similar; but if each clump consist of the same sort of trees, they become ten different things, of which one may hereafter furnish a group of oaks, another of elms, another of chestnuts, or of thorns, &c. In like manner, in the modern belt, the recurrence and monotony of the same mixture of trees, of all the different kinds, through a long drive, make it the more tedious, in proportion as it is long.

Woburn Evergreen Drive.—I must not here omit the full tribute of applause to that part of the drive at Woburn, in which evergreens alone prevail : it is a circumstance of grandeur, of variety, of novelty, and, I may add, of winter comfort, that I never saw adopted in any other place on so magnificent a scale. The contrast of passing from a wood of deciduous trees to a wood of evergreens, must be felt by the most heedless observer; and the same sort of pleasure, though in a weaker degree, would be felt in the course of a drive, if the trees of different kinds were collected in small groups or masses by themselves, instead of being blended indiscriminately.

Variety, how produced.—I do not mean to make separate groves or woods of different trees, although that has its beauty, but, in the course of the drive, to let oaks prevail in some places, beech in others, birch in a third; and, in some parts, to encourage such masses of thorns, hazel, and maple, hollies, or other brush-wood of low growth, as might best imitate the thickets of a forest.*

Forest Groups.—In that part of the forest near Laytonstone and Woodford, and, indeed, in all forest thickets, it will be seen that each mass of thorns, or brushwood, contains one or more young trees, to which it acts as nurse and protector; these trees require no other defence against the numerous herds of cattle, and they grow to a prodigious size; but to the latest period of their existence, especially in Windsor forest, we often see an aged thorn at the foot of a venerable oak, forming the most picturesque and interesting group— like the fond but decrepid nurse, still clinging to her foster child, though it no longer needs her assistance.

Avenues.—It seems to have been as much the fashion of the present century (*originally written in* 1794) to destroy

* It is difficult to lay down rules for any system of planting, which may ultimately be useful to this purpose ; time, neglect, and accident, will often produce unexpected beauties. The gardener, or nurseryman, makes his holes at equal distances, and, generally, in straight rows ; he then fills the holes with plants, and carefully avoids putting two of the same sort near each other ; nor is it very easy to make him ever put two or more trees into the same hole, or within a yard of each other : he considers them as cabbages, or turnips, which will rob each other's growth, unless placed at equal distances ; although, in forests, we most admire those double trees, or thick clusters, whose stems seem to rise from the same root, entangled with the roots of thorns and bushes in every direction.

avenues, as it was in the last to plant them; and while many
people think they sufficiently justify their opinion, in either
case, by saying, ' I like an avenue,' or ' I hate an avenue,' let
us endeavour to analyze this approbation or disgust.

The pleasure which the mind derives from the love of
order, of *unity,* of *antiquity,* and of *continuity,* is, in a
certain degree, gratified by the long perspective view of a
stately avenue; even when it consists of trees in rows so far
apart that their branches do not touch: but where they grow
so near as to imitate the grandeur, the gloomy shade, and
almost the shelter of a Gothic cathedral, we may add the
comfort and *convenience* of such an avenue to all the other
considerations of its *beauty.* A long avenue, terminated by
a large old mansion, is a magnificent object, although it may
not be a proper subject for a picture; but the view *from* such
a mansion is, perhaps, among the greatest objections to an
avenue, because it destroys all variety; since the same
landscape would be seen from every house in the kingdom,
if a view between rows of trees deserves the name of landscape.

If, at the end of a long avenue, be placed an obelisk, a
temple, or any other *eye-trap* (as it is called), it will only
catch or please the eye of ignorance or childhood. The eye of
taste and experience hates compulsion, and turns with disgust
from such puerile means of attracting its notice. One great
mischief of an avenue is, that it divides a park, and cuts it
into two distinct parts, destroying the unity of lawn; for it is
hardly possible to avoid distinguishing the ground on the two
sides of such an avenue into north and south park, or east
and west division of the lawn.

But the greatest objection to an avenue is, that (especially
in uneven ground) it often acts as a curtain drawn across the
most interesting scenery: it is in undrawing this curtain at
proper places that the utility of what has been called breaking
an avenue consists.

If the *fashion in gardening,* like the *fashion in dress,* could
be changed with no other difficulty than that of expense, we
might follow its dictates, without any other consideration; we
might boldly modernise old places, and reduce all improvement to the whim and caprice of the day, and alter them

again on the morrow ; but the change of fashion in gardening destroys the work of ages, when lofty avenues are cut down for no other reason but because they were planted in straight rows, according to the fashion of former times.*

Works of Art.—It is not, therefore, in compliance with the modern fashion for destroying avenues, that I advised the removal of a few tall trees near the house at Longleat; but that the character of greatness, in a work of art like this palace, should not be obliterated by the more powerful agency of nature. Without going back to that taste when this vast pile was surrounded by lines of cut shrubs, and avenues of young trees newly planted, much of its grandeur might be restored, by judiciously removing the encroachments of vege-tation : of this kind are some of the tall, shattered elms remaining, of the avenue near the house, which evidently tend to depress its importance.

Appendages to a Palace.—When the artificial but magni-ficent style of *geometric gardening* of Le Nôtre was changed to the more natural style of landscape gardening, it often happened that too little respect was paid to the costly appendages of English palaces; for, although near the small houses of country gentlemen, the barns, and rick-yards, and kitchen gardens might give way to the shaven lawn in the front of such houses; yet, to place a palace in the middle of a grass field, was one of those excesses of innovation to which all kinds of reform are ever liable.

---

* Every sacrifice of large trees must be made with caution ; at the same time, there may be situations in which trees are not to be respected for their size ; on the contrary, it is that which makes them objectionable. We find that all trees grow more luxuriantly in valleys than on the hills ; and thus it is possible that very uneven ground may be reduced to a level surface, if we judge of it by the tops of the trees. The hills at Longleat have been boldly planted, and, at the same period, many fast-growing trees were planted in the valleys; these latter are become, in many places, too tall for their situation. There are some limes, and planes, and lofty elms near the water, in situations where maples and crabs, thorns and alders, or even oaks and chestnuts, would be far more appropriate.

There is no error more common than to suppose that the planter may not live to see his future woods, unless they consist of firs, and larches, and Lombardy poplars, and other fast-growing trees ; but every day's experience evinces that man outlives the beauty of his trees, where plantations do not consist of oak. On the contrary, tall mutilated planes, or woods of naked-stemmed firs, remind him that groups of oak and groves of chestnut might have been planted with greater advantage.

2 x

Example: Longleat.—The first object of improvement at Longleat, within the department of ART, should be to restore its architectural importance, to increase its greatness, by spreading its influence; but this requires some caution. The stables and the offices should form parts of one great whole; but if they be too much extended, or too rich in design, they will counteract this effect.

Appendages attached.—A palace must not be a solitary object; it requires to be supported and surrounded by subordinate buildings, which, like the attendants on royalty, form part of its state; but a building of greater length than the house, becomes a rival rather than an humble attendant: and there is some danger in making stables and meaner offices dispute with the house in richness of ornament. It will be sufficient if the gates, or some elevated turrets of such buildings, present the same character and date, without exactly copying the detail of those costly ornaments in which the palace abounds.

Detached.—This remark is applicable to all such large buildings as may be necessary near the house; but in the small buildings at a distance, the same richness of ornament may prevail, where it is not inconsistent with the respective uses of such buildings. For this reason, I recommend the entrance of this place to be marked by magnificent gates, rather than by humble cottages, however picturesque. The farm-house, and the poultry-house, and pheasantry, half buried in wood, may preserve their humble and appropriate character; but if any building be made conspicuous, it should be ornamented in proportion to its situation and uses. Thus, a keeper's lodge, or a huntsman's kennel in the valleys, may be useful, without affecting to be ornamental; but, when it occupies an elevated station, it should make a part of the scenery worthy of the general character of the place.

Separate Establishment.—One of the greatest errors in modern gardening has been that of placing a large house, not only on a naked lawn, but in the centre of it: to accomplish this, in some places towns have been removed, and villages destroyed, that the modern park might surround it in every direction. There are many comforts and agremens which, by this practice, must be banished to an inconvenient distance, such as the gardens, the pheasantry or menageries, the dairy-farm, paddocks, &c., and where, as at Longleat, each of these is on a very large scale, they become so many separate detached establishments; and, provided the lines of communication be well managed, they become so many separate objects of interest in the place.*

Dairy-farm Buildings.—The dairy-farm is as much a part of the place as the deer park, and, in many respects, more picturesque; consisting of such varied and pleasing inclosures, and so enriched by groups of trees, that it would not be improved by the removal of any hedges: its character is strictly preserved by the style of the buildings; an old farm-house, a labourer's cottage, a hay-stack, or a thatched hovel, are far more appropriate than the pseudo Gothic dairy, or the French painted trellis in a useful dairy-farm; but, in a park, something more is expected.

Park building.—The park is an appendage of magnificence rather than of utility, and its decorations, therefore, should partake of the character of the palace; they should appear to belong to its state and ornament; they should rather consist of covered seats, a pavilion, or a prospect room, than objects of mere use, as a hay-barn, or a cottage; because the latter may be found in any grass field, but the former denote a superior degree of importance.

It has been a practice, of late, to erect a lofty tower, column, or obelisk, on the summit of the highest hill in the park; but such practice tends to lessen the apparent greatness of a place; for, as we can seldom lose sight of so conspicuous a landmark, we are, in a manner, tethered to the same object.†

Conclusion of the Inquiry.—After tracing the various past changes of taste in gardening and architecture, I cannot

* Here the distant kitchen garden is connected with the house by a pleasure-ground, so perfect in its kind that it only requires to be brought in closer contact with the house.

† This would not be the case with the building proposed for an eminence in Longleat Park, because this spot is everywhere surrounded by more lofty hills, and, therefore, it would only be seen occasionally along the several valleys, and would, from every point of view, become a pleasing embellishment, and not an obtrusive feature of the place.

suppress my opinion that we are on the eve of some great future change in both those arts, in consequence of our having lately become acquainted with scenery and buildings in the interior provinces of India. The beautiful designs published by Daniell, Hodges, and other artists, have produced a new source of beauty, of elegance, and grace, which may justly vie with the best specimens of Grecian or Gothic architecture: and, although the misapplication of these novel forms will, probably, introduce much bad taste in the future architecture of this country, yet we may reasonably expect that some advantage will be taken of such beautiful forms as have never before been adopted in Europe. When a partiality for such forms is patronised and supported by the highest rank, and the most acknowledged taste, it becomes the duty of the professor to raise the importance, by increasing the variety of his art. It is, therefore, with peculiar satisfaction that my opinion has lately been required in some great works of this style, which are in too early a stage of progress to be referred to in this volume, although an Inquiry into the past Changes in the general Taste of a country may properly conclude with such notice concerning the future changes probably to be expected.

PART II.

SCIENTIFIC DISCUSSIONS.

Of Situations and Characters.—All rational improvement of grounds is necessarily founded on a due attention to the CHARACTER and SITUATION of the place to be improved: the *former* teaches what is advisable, the *latter* what is possible to be done. Nothing can be more distinct than these two objects, yet they must be jointly taken into consideration, because one is often influenced by the other.

The *situation* of a place always depends on Nature, which can only be assisted, but cannot be entirely changed, or greatly controlled by ART: but the *character* of a place is wholly dependant on ART: thus the house, the buildings, the gardens, the roads, the bridges, and every circumstance which marks the habitation of man, must be artificial; and although, in the works of art, we may imitate the forms and graces of nature, yet, to make them truly natural, always leads to absurdity.

Natural Situations changed by Art.—Where the ground near a mansion is evidently unnatural, it is necessary to begin the inquiry by endeavouring to discover to what extent art has interfered: three cases, nearly similar in this respect, have occurred at *Welbeck,* at *Woburn Abbey,* and at *Kidbrook.* The ground near each of these houses consists of a plain, which has been formed by levelling and filling in the cavities produced by the junction of two brooks, although scarcely any traces remain of their original courses. It has been remarked that, in many parts of America and the West Indies, the destruction of woods has rendered the brooks and rivers almost dry; and, doubtless, the same cause has operated in this country, as may be observed in the vicinity of former great forests. In Leland's "Itinerary," Welbeck is described as standing at the conflux of two streams, one of which is now become so small as to be carried through an arch under ground. The same thing is done at Woburn Abbey, and also at Kidbrook. It is now, perhaps, equally impossible and unadvisable to restore the ground to its natural shape ; but an inquiry into such original shape of ground facilitates the operations of any change in the surface.

Levelling Ground.—In the "Observations, &c.," many examples are given of changing the surface, or, as it is technically called, "moving ground;" to these I may add, that one of the greatest difficulties I have experienced in practice proceeds from that fondness for levelling, so prevalent in all Brown's workmen : every hillock is by them lowered, and every hollow filled, to produce a level surface ; when, on the contrary, with far less expense, the surface may be increased in apparent extent by raising the hills and sinking the hollows. Such operations must, of course, be confined to subjects of small extent, and it is in these that they produce great beauty and variety.*

Example from Cadogan-square.—In disposing of the area opposite Sloane-street, a new mode of treatment for a square was adopted. Instead of raising the surface to the level of the street, as had been usually the custom, by bringing earth from a distance, I recommended a valley to be formed through its whole length, with other lesser valleys flowing into it, and the hills to be raised by the ground so taken from the valleys. Although, in compliance with the general custom and use of a square, the walk on two sides is carried straight, yet the other walks are made to take such curves as the supposed natural shapes of the ground might warrant; and thus the appearance of nature is, in some degree, preserved in this evidently artificial subject. I cannot omit to mention, that, in the plan, a brook was proposed to pass through the valley, which might have been supplied with the overflowings from the Serpentine River ; but this was omitted in the execution.

Russell-square.—The different character and situation of *Russell-square* may furnish another example. The ground of this area had all been brought to one level plain at too great an expense to admit of its being altered; and the great size of this square is, in a manner, lost by this insipid shape.

* I may refer to examples of this mode of levelling ground at Bulstrode, where two small dells in the flower-garden are united into one valley ; and at Wilton Park, in that neighbourhood, where a small valley has been formed between the house and the orangery with great effect.

The Statue.—*Equestrian Statues* have usually been placed in the centre of public squares, but, in one of such large dimensions, no common sized object could be sufficiently distinguished; it was, therefore, very judiciously determined (by a committee) to place the fine statue of the late Duke of Bedford, now preparing by the ingenious Mr. Westmacot, on one side of the square facing Bloomsbury, and forming an appropriate perspective, as seen through the vista of the streets crossing the two squares.

This pedestrian statue, supported by a group of four figures, on a lofty pedestal, will be of sufficient magnitude for the breadth of the vista; though it would have been lost in the middle of so large a square. Much of the effect of this splendid ornament will depend on its back-ground; for, although the white pedestal may be relieved by the shrubs immediately behind it, the bronze figures should be seen opposed to the sky. This is a circumstance which I hope will be attended to in the future pruning of those trees in the grove behind it.

Details and Intention.—As this square is a subject easily referred to, and as, for the first few years of its growth, it will be liable to some criticism, because few are in the habit of anticipating the future effects of plantation, the intention of the plan is here inserted.

To screen the broad gravel-walk from the street, a compact hedge is intended to be kept clipt to about six feet high; this, composed of hornbeam and privet, will become almost as impervious as a hedge of laurels, or other evergreens, which will not succeed in a London atmosphere. Within the gravel walk is a broad margin of grass, on which the children may be kept always in sight from the windows of the houses immediately opposite; and, for this reason (founded on the particular wishes of some mothers), the lawn is less clothed with plantation than it might have been on the principle of beauty only. This circular lawn, or zone of open space, surrounds the central area, in which have been consulted the future effect of shade, and a greater degree of privacy or seclusion.

The outline of this area is formed by a walk under two rows of lime-trees, regularly planted at equal distances, not in a perfect circle, but finishing towards the statue in two straight lines directed to the angle of the pedestal. It is possible that some fanciful advocates for natural gardening will object to this disposition of the trees as too formal; and they will be further shocked at my expressing a wish that the arch formed by these trees over the walk should be cut and trimmed so as to become a perfect artificial shade, forming a cloister-like walk, composed of trees. For this purpose the suckers or sprays from the stems should be encouraged, to make the interior perfectly secluded. In the due attention to the training and trimming such trees by art, consists the difference between a garden and a park, or forest; and no one will, I trust, contend that a public square should affect to imitate the latter.

The area inclosed within these lime-trees may be more varied; and, as it will consist of four distinct compartments, that nearest the statue is proposed to be shaded by a grove of various trees, scattered with less regularity, while the other three may be enriched with flowers and shrubs, each disposed in a different manner, to indulge the various tastes for regular or irregular gardens; yet always bearing in mind that the trees should not be suffered to rise too high in the line immediately behind the statue.

As, from the great extent of Russell-square, it is advisable to provide some seats for shade or shelter, a *reposoir* is proposed in the centre, with four low seats, covered with slate or canvas, to shelter from rain, and four open seats to be covered with climbing plants, trained on open lattice, to defend from the sun; these seats surround a small court-yard, to be kept locked, in which may be sheds for gardeners' tools, and other useful purposes.

A few years hence, when the present patches of shrubs shall have become thickets,—when the present meagre rows of trees shall have become an umbrageous avenue,—and the children now in their nurses' arms shall have become the parents or grandsires of future generations,—this square may serve to record, that the Art of Landscape Gardening in the beginning of the nineteenth century was not directed by whim or caprice, but founded on a due consideration of utility as well as beauty, without a bigoted adherence to forms and lines, whether straight, or crooked, or serpentine.

Examples of Houses dependant on peculiar Circumstances.
—In those places where the house already exists, and the
character is fixed, the grounds must, in a certain degree, be
accommodated to the style of the house: but where a new
house is to be built, its proper site and character will depend
on various circumstances, of which I shall give two singular
examples.

Within the last forty years, the property and even the
characters of individuals, have undergone more change than in
any period of the English history : we daily see wealth,
acquired by industry, or by fortunate speculations, succeeding
to the hereditary estates of the most ancient families ; and we
see the descendants of these families reduced by the vain
attempt to vie in expense with the successful sons of commerce : this will often account for the increase of novel or
fantastic edifices, and the decrease of those venerable specimens of former grandeur, the baronial castle, or the castellated
mansion.   Few instances occur where the honest pride of
ancestry is blended with the prudence and success of commercial importance; yet, in one of these, I had occasion to
deliver the following opinion :

" The antiquity, the extent, and beauty of ——— Park,*
" together with the command of adjoining property, might
" justify the expenditure of ten times the sum to which I am
" instructed to limit my plans. I shall, therefore, describe what
" *may* be done, and not what *might* be done, to fix the true
" character for this house, since it cannot be a palace, and,
" perhaps, ought not to be a castle: from its situation, it cer-
" tainly ought not to be a villa ; it ought not to be a cottage ;
" and, as a shooting-box, the present rooms in the farm-house
" are sufficient for a bachelor : but this must be the residence
" of a family ; and, being amid the mountains of Wales, at
" some distance from society, we must not only provide for
" the accommodation of its own family, in all its various
" branches, but for the entertainment of other families in the
" neighbourhood, and for the reception of friends and visitors
" from distant parts ; all this cannot be expected in a very
" small house ; and since (without great expense) the ancient

* The name is omitted, at the request of the proprietor.

2 Y

"baronial castle cannot be imitated, we may, perhaps, with
"less difficulty restore that sort of importance which was
"formerly annexed to the old Manor House, where the lord
"of the soil resided among his tenants, not merely for the
"purpose of collecting his rents, but to share the produce of
"his estates with his humble dependants, and where daily
"plenteous hospitality was not sacrificed to the occasional
"ostentatious refinements of luxury and parade.

"It is not meant to condemn the improvements in comfort
"or convenience enjoyed in modern society, or to leave,
"unprovided for, every accommodation suited to the present
"habits of life; but to furnish the means of enjoying them
"without departing from the ancient character of the place,
"by erecting, or restoring, on the same identical spot, and in
"nearly a similar style, the *Grange*, or old Manor House,
"which will not be found incongruous with the surrounding
"scenery, when spread out and connected with all its appen-
"dages on the cavity between the two hills on the summit of
"this beautiful mountain."*

Longnor.—I shall conclude these examples by a remarkable circumstance of another house being restored in the same
style and character on the original site.   At the corner of the
old mansion of the Burtons, at Longnor, is a tomb erected over
the body of an ancestor of the present family, who, having
early become a protestant, died through excess of joy at the
news of Queen Elizabeth's accession to the throne, and was
refused burial in St. Chad's church, at Shrewsbury.   On this
tomb (though now scarcely legible) is the following inscription,
in characters of that date :—

HERE LIETH THE BODY OF

E D W A R D   B U R T O N,   E S Q.

WHO DIED ANNO DOMINI 1558.

Was't for denying Christ, or some notorious fact,
That this man's body Christian burial lack't ?
Oh no, not so ! his faithful, true profession,
Was the chief cause, which then was held transgression.

---

* One of these hills, within a short walk from the house, commands a
view of a rich cultivated valley winding through this mountainous scene.
Such a prospect derives additional interest with the proprietor of an estate,

When Popery here did reign, the see of Rome
Would not admit to any such a tomb
Within her idol-temple walls; but he,
Truly professing Christianity,
Was like Christ Jesus in a garden laid,
Where he shall rest in peace till it be said,
Come, faithful servant, come receive with me
A just reward for thy integrity.

I advised this tomb to be repaired, and the inscription preserved on a brass plate, covered with a Gothic canopy of the same date with the event. This forms an appropriate ornament at the angle of the house, which stands on a bold terrace in the garden, commanding an extensive view of the Severn, and the distant Welsh mountains.*

Since most of our pleasures may be traced to mixed sources, and are always heightened by those of association, I am indebted to a periodical critic for the following remark: " Round-headed trees are more particularly well associated " with the Gothic style of architecture, as they are the only " species of trees, in this country at least, that appear coeval " with antique structures." Perhaps from hence arises part of the disgust at seeing modern Gothic buildings, however well designed, surrounded by firs and Lombardy poplars.

Water.—There is something so fascinating in the appearance of water, that Mr. Brown thought it carried its own excuse, however unnatural its situation; and therefore, in many places under his direction, I have found water on the tops of hills, which I have been obliged to remove into lower ground, because the deception was not sufficiently complete to satisfy the mind as well as the eye.

who must naturally feel the satisfaction of looking upon hills and dales, and villages and farms, which he may call his own; a satisfaction which, however the vanity of property and the pride of possessions may be ridiculed, may innocently be gratified, when the proprietor has humanity to reflect how far his influence and benevolence may be extended over the prospect he admires.

* It was deemed necessary to take down the old house entirely, and I hope it is rebuilt with little variation from the original character: but as in this case I was not consulted as the architect, and have never seen the present house, I can only speak of its situation, and not of its character.

On high Ground.—A common observer supposes that water is usually found, and, therefore, most natural, in the lowest ground; but a moment's consideration will evince the error of this supposition. Places abounding in lakes and pools are generally the highest in their respective countries; and without such a provision of nature the world could not be supplied with rivers, which take their source in the highest mountains, and, after innumerable checks to retard and expand their waters, they gradually descend towards the sea. If nature be the model for art in the composition of landscape, we must imitate her process, as well as her effects. Water, by its own power of gravitation, seeks the lowest ground, and runs along the valleys.* If in its course the water meets with any obstruction, it spreads itself into a lake, or meer, proportionate to the magnitude of the obstruction: and thus we often see in the most picturesque countries a series of pools, connected by channels of the rivers which supply them. From certain points of view, these pools, though on different levels, will take the appearance of one continued lake, or river, only broken by islands or promontories, covered with brushwood; and from hence was taken Mr. Brown's frequent attempt of uniting two pools, which could not be brought together in reality, but which become apparently united by an effect of perspective, not always attended to in gardening.

Objects in Motion.—A scene, however beautiful in itself, will soon lose its interest, unless it is enlivened by moving objects; and, from the shape of the ground near most houses, there is another material use in having cattle to feed the lawn in view of the windows. The eye forms a very inaccurate judgment of extent, especially in looking down a hill, unless there be some standard by which it can be measured; bushes and trees are of such various sizes, that it is impossible to use them as a measure of distance; but the size of a horse, a

---

* Indeed I have sometimes fancied, that, as action and re-action are alike, and as cause and effect often change their situations, so valleys are increased in depth by the course of waters perpetually passing along them: thus, if the water only displaces one inch of soil in each year, it will amount to 500 feet in 6000 years; and this is equal to the deepest valleys in the world. In loose soils, the sides of the hills will gradually wash down, and form open valleys; in hard soils they will become narrow valleys: but ravines I suppose to be the effect of sudden convulsions from fire, or steam, and not made by any gradual abrasion of the surface.

sheep, or a cow, varies so little, and is so familiar to us, that we immediately judge of their distance from their apparent diminution, according to the distance at which they are placed; and as they occasionally change their situation, they break that surface over which the eye passes, without observing it, to the first object it meets to rest upon.

Strange Absurdities.—In the affected rage for following nature, as it is called, persons of acknowledged good sense and good taste, have been misled into the strangest absurdities. Thus, forgetting that a road is an artificial work of convenience, and not a natural production, it has, at one time, been displayed as the most ostentatious feature through the centre of a park, in the serpentine line described by the track of sheep; and, at another, concealed between two hedges, or in a deep chasm between two banks, lest it should be discovered: and such, alas! is the blindness of system, that, in a place where several roads are brought together (like the streets at the Seven Dials), within two hundred yards of the hall door a direction post is placed, as necessary to point out the way to the house.*

Of Bridges, as Roads.—A road is as much an artificial work as a house or a bridge: indeed, a bridge is only a road across such a chasm as cannot be passed without one.   There are, indeed, two uses of a bridge; the first to pass over, the second to pass under: the first is always necessary, the second only occasionally so, as where the water under it is navigable: yet, self-evident as this fact may appear, bridges are often raised so high as to make the passage over them difficult and dangerous, when no passage under them is required; and, perhaps, a form of bridge, adapted to the purposes of passing over, which may unite strength with grace, or use with beauty, is a desideratum in architecture; for this purpose I have suggested, for several places, what may more properly be called a *viaduct* than a bridge, of which no idea can be given by description only.

* This example of *practical taste* is taken from the approach to the picturesque mansion of the Author of the " Inquiry into the Principles of Taste."

Form of Roads.—The width of a road must depend on its uses: if much frequented, there should be always room for two carriages to pass on the gravel: if little frequented, the gravel may be narrower, but there must be more room left on each side; yet we often see the broadest verges of grass to the broadest roads, where, in strict propriety, the breadth should be in an inverse ratio.

If the gravel be wider than the traffic upon it requires, so much more labour will be necessary to preserve it neatly: yet it can never be right to put gravel in recesses that no horse or carriage can possibly reach. If a corner projects too far into the road, the driver will certainly go over it, unless prevented by some obstacle; yet it never can be right to endanger the safety by unnecessary obstacles.*

Park Entrance.—The courts, or garden-gates, through which old mansions were approached, prevented the intrusion of improper persons, who were stopped by the porter of the gate: but since it has become a fashion to remove these, and to place the house, a naked, solitary, and isolated object, in the middle of a large park, or grass field, it is become necessary to remove the porter to the entrance of the park; and this is the origin of all that bad taste so often displayed in the entrance of parks.

Ridiculous Park Lodges.—In some places it is a triumphal arch, like a large hole in a wall; in another it is a wooden gate between two lofty piers, attached to a rough park pale; but the most common expedient is a pair of small square boxes on each side of the gate, making, together, one comfortless, smoky house of two rooms, separated by a gate into the park. It is the gate, and not the habitation of the man who keeps the key, which requires to be marked with importance; and if distinguished by architectural embellishments, they should partake of the style of the house, and announce its character: where (as at Stonelands) the entrance is the most

* However obvious and self-evident this may appear when pointed out, yet such is the slowness in the progress of improvement, that a witty author observes, "although *spoons* have been in use two thousand years, yet it is only within our own memory that the handles have been turned the right way." In like manner, although streets have existed in London from time immemorial, yet it is within everybody's memory that the corners were first begun to be *rounded off*.

obvious in point of convenience, and is rather to shew the beauties of situation than the character of the place, a woodman's cottage near the gate is quite sufficient: and if such a cottage is built in the style and date of the old cottages, on the borders of a forest, it will less betray the innovation of modern improvement. It is not by a pointed arch to the door, or a sham Gothic window, that such style is to be imitated, but by a nice observance of the costume, forms, and construction of such buildings as actually existed in the days of Queen Elizabeth, from which the smallest deviation will betray the attempt to deceive: the deception, if complete, is allowable, since it is the " business of art to deceive;" but the spruce or clumsy effort, that is sure to betray, is also sure to be ridiculous—" The attempt, and not the deed, confounds."

PART III.

LITERARY AND MISCELLANEOUS REMARKS.

Fashion is not to be controlled.—If taste in the fine arts be under the influence of fashion, it may, perhaps, be supposed that fashion may be influenced by the professors of the fine arts; but this has seldom been the case, except in some very extraordinary discovery of novelty. Fashion is neither to be directly opposed nor imperiously guided, either by the theory of authors, or the practice of professors. I have occasionally ventured to deliver my opinion freely in theory, but in my practice I have often feared to give offence, by opposing the taste of others, since it is equally dangerous to doubt a man's taste as his understanding; especially as those who possess least of either are generally the most jealous of the little they possess.

In addition to these difficulties, I have had to contend with the opposition of stewards, the presumption or ignorance of gardeners, and the jealousy of architects and builders; yet my *practice* has been supported by the first characters in the kingdom; but my *theory* has been confounded with that of Brown and his followers, although by my writings I thought the difference had been fully explained.

The elegant and gentlemanlike manner in which Mr. Price has examined my opinions, and explained his own, left no room for further controversy; and it might reasonably have been supposed the subject had been dropped: but I find myself again personally (though not by name) called upon to defend the Art of Landscape Gardening from the attacks of a late work, published under the title of "An analytical Inquiry "into the Principles of Taste, by R. P. Knight, Esq.," Author of the "Landscape," a poem, and other ingenious works; it is full of allusions to landscape gardening, without taking any notice of those opinions delivered to the world in my two works on that subject; and which, from their scarcity and costliness of the plates, will probably be less read than the volume which now calls for my notice.

Answer to Mr. Knight's Inquiry.—In perusing these works,

the candid reader will perhaps discover that there is no real difference between us; but, in contending with an adversary of such nice discernment, such deep investigation, and such ingenious powers of expression, it is difficult to say how far we are actually of the same opinion. I thought I could discover a shade of difference between the opinions of Mr. Knight and Mr. Price, although the world confounded them as joint and equal adversaries to' the art of modern gardening. We are now told that in both his volumes "his friend (Mr. Price) " equally mistakes ideas for things, and the effect of internal " sympathy for those of external circumstances, and thence " grounds the best practical lessons of taste upon false prin- " ciples and false philosophy." Under such severity of criti- cism both Mr. Price and I may console ourselves in our mistakes from the following remark: " When Montesquieu " and Burke thus differ upon a subject of common sense and " feeling, which each had made the particular object of his " investigation, who shall hope to escape error in any theo- " retical inquiry ?"

Whatever trifling differences may still exist in our theories, it is no small satisfaction to me to discover that many of my opinions have been confirmed, and many of my thoughts repeated, although new clothed, or disguised in other words, by Mr. Knight, especially those on the subject of Gothic architecture,* on the absurdity of concealing the offices to a

---

* OBSERVATIONS, p. 304.

" Whether we take our models from a *Grecian Temple*, or from a *Gothic Abbey*, from a *Castle*, or from a *College*, if the building does not look like a house, and the residence of a nobleman, it will be out of cha- racter. It may, perhaps, be objected, that we must exactly follow the models of the style or date we profess to imitate, or else we make a pasticcio, or confusion of discordant parts. Shall we imitate the thing, and for- get its application ?—No ; let us rather, &c.......Let us, in short, never forget that we are building a *house*, whether we imitate the bold irregular outline of an ancient Castle, the elegant forms and tracery of a

INQUIRY, p. 179.

" *Grecian Temples, Gothic Abbeys,* and feudal *Castles*, were all well adapted to their respective uses, cir- cumstances, and situations: the dis- tribution of the parts subservient to the purposes of the whole, and the ornaments and decorations suited to the character of the parts; and to the manners, habits, and employments of the persons who were to occupy them : but the house of an English nobleman of the eighteenth or nine- teenth century is neither a Grecian Temple, a Gothic Abbey, nor a feu- dal Castle; and if the style of distri- bution or decoration of either be employed in it, such changes and modifications should be admitted, as

house,* and on the use of terraces, and particularly on the
neatness near a house, in which he very strongly expresses my
sentiments in these words : " Immediately adjoining the
" dwellings of opulence and luxury, everything should assume
" its character, and not only be, but appear to be, dressed
" and cultivated. In such situations, neat gravel walks, mown
" turf, and flowering plants and shrubs trained and distributed
" by art, are perfectly in character."†

This, I apprehend, is the result of an experiment made by
the author near his own mansion, where large fragments of
stone were irregularly thrown amongst briers and weeds, to
imitate the foreground of a picture. Can anything more
strongly prove, that a landscape in nature and a landscape in

Gothic Abbey, or the harmony of
proportions and symmetrical beauty
of a Grecian Temple."

* OBSERVATIONS, p. 271.
After describing six different forms
of houses and offices, at different
dates, of which the fifth and sixth
had wings, " the seventh and last
invented, consists of a compact square
house, with three fronts ; and to the
back of it are attached offices, forming
a very long range of buildings, courts,
walls, &c. all supposed to be hid by
plantation. Such is the horror of seeing
the offices, that, in one instance, I was
desired by the architect to plant trees
on the earth which had been brought
and laid on the copper roof with
which the kitchen offices had been
covered for that purpose !"

† OBSERVATIONS, pp. 202, 240, 277,
279, 284.
" Various examples are given of
terraces in the front of houses, as
forming a basement for the house to
stand upon, which at once gives it
importance, and supplies it with accompaniments....these, it may be
supposed, were the source of that
prophetic remark concerning another
revolution in taste at no great distance."

may adapt it to existing circumstances ; otherwise the scale of its
exactitude becomes that of its incongruity, and the deviation from principle proportioned to the fidelity of
imitation."
INQUIRY, p. 214.
" The practice, which was so prevalent in the beginning of this century, of placing the mansion-house
between two correspondent wings,
in which were contained the offices,
has, of late, fallen into disuse ; and
one still more adverse to composition
has succeeded ; namely, that of entirely hiding offices behind masses of
plantation, and leaving the wretched,
square, solitary mansion-house to exhibit its pert bald front, &c. &c.....
(The offices) are often concealed in
recesses, or behind mounds, the improver generally picking out the most
retired, intricate, and beautiful spot
that can be found near the house to
bury them in."
INQUIRY, p. 215.
" The author recommends ' the
hanging terraces of the Italian gardens,'... as they not only enrich the
foreground, but serve as a basement
for the house to stand upon, which
at once gives it importance, and supplies it with accompaniments. Such
decorations are, indeed, now rather
old-fashioned ; but another revolution
in taste, which is probably at no great
distance, will make them new again."

a picture, are very different things: and that LANDSCAPE
GARDENING is not PICTURE GARDENING ?   This I may
fairly give as my answer to page 214, which I cannot but
suppose directed to me.   " Why this art has been called
" Landscape Gardening, perhaps he who gave it the title may
" explain.   I can see no reason, unless it be the efficacy
" which it has shewn in destroying landscapes, in which,
" indeed, it seems to be infallible; not one* complete painter's
" composition being, I believe, to be found in any of the
" numerous, and many of them beautiful and picturesque
" spots, which it has visited in different parts of the island."

Difference between Painting and Gardening.—The greatest
objection to landscape gardening seems to arise from not
making the proper distinction between *painting* and *gardening*.   The difference betwixt a scene in nature, and a
picture on canvas, arises from the following considerations :—

First, The spot from whence the view is taken, is in a fixed
state to the painter; but the gardener surveys his scenery
while in motion ; and from different windows in the same
front he sees objects in different situations; therefore, to give
an accurate portrait of the gardener's improvement, would
require pictures from each separate window, and even a different drawing at the most trifling change of situation, either
in the approach, the walks, or the drives about each place.

Secondly, The quantity of view, or *field of vision*, in nature,
is much greater than any picture will admit.

* To avoid the imputation of vanity, I could wish that the following
fact were stated by any other person than myself.   In the course of my
practice,
    3000 different sketches, or views, are now extant in private MSS.; from
          these I published fifteen plates in my first work, consisting of
          250 copies; therefore, of these,
    3750 impressions are in circulation.
          Also thirty-five plates were published in my second work,
          which, in the two editions, amounted to
  26,250 impressions are in circulation.
          To these may be added, that, during the last eighteen years,
          I have given thirteen designs to an annual work, making 234
          views, from each of which, I am informed, 7000 impressions have
          been made, and, of course,
1,638,000 impressions are in circulation.
    When this number is compared with the above assertion, that *not one*
landscape has escaped the fatal effects of the art I profess to cultivate and
defend, it must prove, from the numerous purchasers and admirers of these
things, that " de gustibus non est disputandum."

Thirdly, The view from an eminence down a steep hill is not to be represented in painting, although it is often one of the most pleasing circumstances of natural landscape.

Fourthly, The light which the painter may bring from any point of the compass must, in real scenery, depend on the time of day. It must also be remembered that the light of a picture can only be made strong by contrast of shade; while, in nature, every object may be strongly illumined, without destroying the composition, or disturbing the keeping. And,

Lastly, The foreground, which, by framing the view, is absolutely necessary to the picture, is often totally deficient, or seldom such as a painter chooses to represent; since the neat gravel-walk, or close-mown lawn, would ill supply the place, in painting, of a rotten tree, a bunch of docks, or a broken road, passing under a steep bank, covered with briers, nettles, and ragged thorns.

Planting a Down.—There is no part of landscape gardening more difficult to reconcile to any principles of landscape painting than the form of plantations to clothe a naked down. If the ground could be spared, perhaps the best mode would be to plant the whole, and afterwards cut it into shape: it might then be considered as a wood interspersed with lawns; and this must be far more pleasing to the eye than a lawn patched with wood, or, rather, dotted with clumps, for it is impossible to consider them as woods, or groups of trees, while so young as to require fences. The effect of light and shade is not from the trees, but from the lines of posts and rails, or the situation of boxes and cradles with which they are surrounded; and these being works of art, they must appear artificial, whether the lines be straight or curved. Although much has been said and written about the sweeping lines of wood following the natural shapes of the ground, the affectation of such lines is often more offensive than a straight line, which is always the shortest, generally the easiest to disguise, and very often appears curved, and even crooked, from crossing uneven ground.* The sweeping lines of art,

---

* The strongest example of this fact may be taken from a view of large tracts of open country recently inclosed, where the lines of hedges are often drawn on the map, by the commissioners, at right angles, and the fields exactly square: but from the occasional inequality of surface, they generally appear diversified, and each square field takes a different shape in appearance, although, on the map, they may be exactly similar.

when applied to nature, become ridiculous, because they are liable to be compared with works of art, and not of nature.

I have often wished it were possible by any art to produce the outline of Stokenchurch Hill, as seen in the road from Oxford to London; but this is a forest partially cleared of wood by time and accident: in vain will any new place assume the same degree of respectability; it is as impossible to produce the same effect by new plantations, as to produce immediately the far-spreading beech or majestic oak, now become venerable by the lapse of centuries. Every man who possesses land and money may, in a few years, have young plantations and covers for game of many acres in extent; but no cost can produce immediate forest scenery, or purchase the effect of such hedge-row trees as are too frequently overlooked and buried among firs, and larches, and faggot wood, to accomplish the exact monotonous serpentine of a modern belt.

DESIGNS

FOR

THE   PAVILLON

AT

BRIGHTON:

HUMBLY INSCRIBED TO HIS ROYAL HIGHNESS

THE PRINCE OF WALES.

BY H. REPTON, ESQ.

WITH THE ASSISTANCE OF HIS SONS,

JOHN ADEY REPTON, F. S. A. AND G. S. REPTON,

ARCHITECTS.

[*Originally published in* 1808.]

HIS ROYAL HIGHNESS

THE PRINCE OF WALES,

&c., &c., &c.

THE approbation YOUR ROYAL HIGHNESS was pleased to express of the general outline of an opinion I had the honour to deliver concerning the GARDENS of the PAVILLON, induces me to hope that this WORK will meet with the same gracious reception, as it contains the reasons on which that opinion was founded.

I have the honour to be,

With the most profound respect,

YOUR ROYAL HIGHNESS's

Most faithful and most obedient humble Servant,

H. REPTON.

Harestreet, near Romford, Essex;
    February, 1806.

PREFATORY OBSERVATIONS.

In a small Work published in 1806,* it was mentioned (page 41), [see p. 340 of the present volume], " That we were " on the eve of some great change in landscape gardening " and architecture, in consequence of our having lately " become better acquainted with scenery and buildings in the " interior provinces of India :" it was also mentioned, that " my opinion had recently been required in some great works " of this style, THEN in too early a stage of progress to be " referred to." This was in allusion to my having, at that time, completed the original MS., of which the following work is an exact copy.

As many parts of this volume may appear to recommend a degree of novelty, to which I have frequently objected in former publications, it will, perhaps, subject me to some severity of criticism. I must, therefore, plead for candid and indulgent hearing, while I explain the origin of the following work, and endeavour to justify its intentions.

At a time when the wealth of individuals has been increasing in this country, beyond the example of all former periods, it would not be an uninteresting subject of inquiry, to consider how far the more general diffusion of GOOD TASTE has kept pace with the increased wealth of individuals ; or, rather, the effect which that increased wealth has produced on the taste of the country generally.† But in the following

---

* " An Inquiry into the Changes of Taste in Landscape Gardening and Architecture, &c." [See this work, reprinted in the present volume, with the exception of what had appeared in " Hints and Sketches," in p. 321 to p. 357.]

† Amongst the most obvious effects of sudden wealth in the country, is the change of property from the hereditary lords of the soil, to the more wealthy sons of successful commerce, who do not always feel the same re-

pages I shall confine my observations to the united arts of
landscape gardening and architecture.

The natural effect on the human mind of acquired wealth,
is either an ostentatious display of its importance to others,
or a close application of it to selfish and private enjoyment;
and, very frequently, both in the same individual. And this
effect may be traced in the modern practice of what is called
improving both houses and palaces. In the former, if the
inside display of magnificence or comfort be accomplished,
the external architecture is little attended to; while, in
gardening, the perfection of improvement seems to consist in
the extent of ground appropriated to the private enjoyment
of the possessor and his friends. It has frequently been observed, " that England would, in time, become the garden of
" Europe, by the continual increase in the number and extent
" of its improved places:" but the improvement of individual
places has rather injured than benefited the traveller, because
all view is totally excluded from the highways by the lofty
fences and thick belt with which the improver shuts himself
up within his improvement. This arises from the seclusion
which is, perhaps, in some cases, necessary; but which, in
the course of long practice, I have generally observed to be
carried too far; and has introduced the fashion, that in all
places, whether of five acres or of five thousand, the first step
is to inclose with a wall, or pale; and the next, to cover that
boundary with a belt, or plantation.* This gratifies the desire

spect for the antiquity or dignity of venerable mansions: and, although
some may have sufficient taste to preserve the original character of such
places, yet, in general, the display of recent expense in the *newness* of improvement is too prevalent. Hence we have continually to regret the
mutilation of the old *halls* and *manor houses*, where the large bay windows, the lofty open chimneys, and picturesque gables of·Queen Elizabeth's
time, give place to the modern sashes and flat roofs, with all the garish
frippery of trellis, and canvas, and sharp-pointed pea-green Gothic porches,
or porticos of Grecian columns reduced to the size of bedposts.

* This remark will be more striking, when exemplified by a comparison
between a new place and an old one. In the former, a brick wall, or
close paling, is put so near the road as to leave no margin of waste land,
while the old hedgerow thorns and pollard trees are taken away, to make
room for young plantations of firs, and larch, and Lombardy poplars. How

of seclusion and private enjoyment, while that of displaying great possessions has introduced the fashion of considering the importance of a place by its extent, rather than by its variety; and describing it rather by its number of acres, than by its beauties !!

This same false principle of mistaking greatness of dimensions for greatness of character, has, of late, extended itself to the arts of every kind : the statuary surprises by the immense blocks of marble which fill the Abbey, and St. Paul's ! —the painter by an expanse of canvas too large for any private houses!—the jeweller, by large masses of amber and aqua marina, which, by their size, outweigh, though they cannot outshine, the diamond and the ruby!—while, in architecture, the first question concerning a house is, WHAT ARE THE DIMENSIONS OF THE ROOMS ? Indeed everything is swelled out in the same proportion. Thus we continually see, in modern houses, windows too large to be glazed ; doors too large to be opened; furniture too large to be moved; and even beds too lofty to be reached without a ladder! !!

Having long regretted the prevalence of this mistaken fashion, I was rejoiced to receive his Royal Highness's commands to deliver my opinion concerning a place which was deemed by everybody too small to admit of any improvements; and, indeed, such it actually was, according to the modern system, which required UNCONFINED EXTENT WITHIN ITSELF, AND ABSOLUTE EXCLUSION FROM ALL WITHOUT.

On my arrival at Brighton, I found the same system already begun, by the preparation for a belt of shrubs close to

different from the ancient manorial domains! where the public road has a broad margin of herbage, enriched with thorns and spreading timber, under whose twisted branches the rough and knotty pale admits a view into the park, where romantic and decaying oaks denote the old proprietor's taste and preference for picturesque objects, rather than for the intrinsic value of his timber : while, on the contrary, the new possessor, who has, perhaps, lately paid dearly for the timber, is too often anxious to realize the value of his purchase, by converting to profit every tree that has ceased to grow, and is, therefore, deemed ripe for the axe.

the garden wall: and, in conformity to another fashion of modern gardening, there was to have been a coach-road, to enter by a pair of lodges, and to proceed to the house through a serpentine line of approach, as it is called. The principle on which this plan was suggested, arose from confounding the character of a garden * with that of a park ; and it is hardly possible to give a more striking example of the absurdity of applying a general system to every situation. It is not, therefore, to be wondered at, that the acknowledged good taste of his Royal Highness should see the necessity of having recourse to new expedients ; what these are, will appear in the following pages : but I shall candidly acknowledge, that, for many of them, I am indebted to the elegance and facility of the Prince's own invention, joined to a rapidity of conception and correctness of taste which I had never before witnessed.

It was evident, in the present instance, that every attempt to increase the apparent extent of ground on these principles must have betrayed its real confinement ; while, on the contrary, I trust it will appear, that, if there were a thousand acres attached to the Pavillon, such a garden as is here described, would not reasonably occupy more than five or six.

Although it may at first appear that the following observations are more especially applicable to the garden of a palace, under peculiar circumstances of confinement, yet they may be extended to every other place, from the ornamented cottage to the most superb mansion ; since every residence of elegance or affluence requires its garden scenery ; the

---

* This error is so common, that there are few places in which the character of a garden is preserved near the house ; and, therefore, a detached place, called the *flower garden*, has been set apart, occasionally, at such an inconvenient distance, that it is seldom visited. Among those few in which the garden scenery has been admitted to form part of the landscape from the windows, I can only mention, Wilderness, Earl Camden ; Bromley Hill, the Right Hon. Charles Long (now Colonel Long) ; St. Leonard's Hill, General Harcourt ; Longleat, Marquis of Bath ; and Ashridge, Earl Bridgewater. Out of some hundred places, these are all I can recollect where the views from the windows consist rather of *garden* than of *park* scenery.

beauty and propriety of which belong to art rather than
to nature.   In forest scenery, we trace the sketches of SAL-
VATOR and of RIDINGER; in park scenery, we may realize
the landscapes of CLAUDE and POUSSIN : but, in garden
scenery, we delight in the rich embellishments, the blended
graces of WATTEAU, where nature is dressed, but not disfigured, by art; and where the artificial decorations of architecture and sculpture are softened down by natural accompaniments of vegetation.   In the park and forest, let the
painter be indulged with the most picturesque objects for his
pencil to imitate; let the sportsman be gratified with rough
coverts and impenetrable thickets ; let the active mind be
soothed with all the beauty of landscape, and the contem-
plative mind roused by all the sublimity of prospect that
nature can produce; but we must also provide artificial scenes,
less wild, though not less interesting, for

——————— " Retired Leisure,
" That in *trim* gardens takes his pleasure."—MILTON.

For these reasons, I cannot too strongly recommend a due
attention to the following circumstances, which will be deemed
innovations in the modern system, by those who contend that
landscape forms the basis of landscape gardening, *viz.* First,
To reduce the size of the pleasure ground, as it is called,
within such limits that it may be kept with the utmost artificial neatness.   Secondly, Not to aim even at the appearance
of extent in garden scenery, without marking its artificial
boundary, or separature, from the natural landscape.   Thirdly,
When the dressed grounds form part of the view from the
windows, especially those of the principal rooms, let it be
artificial in its keeping and in its embellishments; let it rather
appear to be the rich frame of the landscape than a part of
the picture.   Fourthly, Whether the dressed garden be seen
from the windows, or in a detached situation, let it be near
the house, and, if possible, connected with it by a sheltered,
if not a covered way.   And lastly, As the winter of England

extends from November to May, it is highly desirable to provide a garden for those months, and thereby artificially to prolong our summers beyond the natural limits of our precarious climate.

In the summer, every field is a garden; but, in the winter, our open gardens are bleak, unsheltered, dreary fields. Where the walks are extended to the lengths which too commonly prevail, we find that no one uses them except the nursemaid and children, who are compelled to do so; or the unfortunate visitor, who is not less compelled to walk round the place on the first day of his visit, and who ever afterwards makes his escape into the neighbouring lanes or inclosures to enjoy the country; while in the artificial garden, richly clothed with flowers, and decorated with seats and works of art, we saunter, or repose ourselves, without regretting the want of extent any more than while we are in the saloon, the library, or the gallery of the mansion.

The luxury of a winter garden has of late been, in some degree, supplied by adding large conservatories to the apartments of a house; but this is not, in all cases, practicable, nor in some advisable; yet, in most situations, it is possible to obtain a covered line of connexion with the green-house, and other appendages of a winter garden, at a little distance from the house.*

If, by the various expedients suggested, I have succeeded in lengthening the summers, by shortening the walks; or if I have increased the comforts or pleasures of a garden, by diminishing what is too often miscalled the pleasure garden, I shall not have exercised my profession in vain; since I hope

* The covered walk and corridor at *Woburn Abbey* is the most extensive of the kind in this country. It is a shelter from rain at all seasons, and furnishes a line of connexion with the conservatory, flower-house, tennis-court, stables, riding-house, &c. But this is not covered with glass. Among those on a small scale, I may mention the flower passage at Mr. Manning's villa, at Totteridge; the corridor at Earl Sefton's, at Stoke Farm; and the winter walk at the Hon. J. B. Simpson's, at Babworth, Nottinghamshire: all which add great comfort to the interior, while they contribute, by their exterior, to ornament the garden scenery.

it will tend to curtail the waste of many thousand acres which may be more profitably employed.

I shall now proceed to explain the reasons for recommending, in the present instance, a departure from the styles of architecture hitherto used in this country. It happened that, a little before my first visit to Brighton, I had been consulted by the proprietor of Sezincot, in Gloucestershire, where he wished to introduce the gardening and architecture which he had seen in India.* I confess the subject was then entirely new to me: but, from his long residence in the interior of that country, and from the good taste and accuracy with which he had observed and pointed out to me the various forms of ancient Hindû architecture, a new field opened itself; and, as I became more acquainted with them, through the accurate sketches and drawings made on the spot by my ingenious friend Mr. T. Daniell, I was pleased at having discovered new sources of beauty and variety, which might gratify that thirst for novelty, so dangerous to good taste in any system long established; because it is much safer to depart entirely from any given style, than to admit changes and modifications in its proportions, that tend to destroy its character. Thus, when we are told that "a pediment is old fashioned, and a Doric column too thick and clumsy," the corruption of Grecian architecture may be anticipated. And since the rage for Gothic has lately prevailed, the sudden erection of spruce Gothic villas threatens to vitiate the pure style of those venerable remains of ancient English grandeur, which are more often badly imitated in new buildings, than preserved or restored in the old. It is not, therefore, with a view to super-

* Although I gave my opinion concerning the adoption of this new style, and even assisted in the selecting some of the forms from Mr. T. Daniell's collection, yet the architectural department at Sezincot, of course, devolved to the brother of the proprietor, who has displayed as much correctness as could be expected in a first attempt of a new style, of which he could have no knowledge but from drawings, but who has sufficiently exemplified, in various parts of his building, that the detail of Hindû architecture is as beautiful in reality as it appears in the drawings, and does not shrink from a comparison with the pure Gothic in richness of effect.

sede the known styles, that I am become an advocate for a
new one, but to preserve their long-established proportions,
pure and unmixed by fanciful innovations.

Immediately after I had reconciled my mind to the adoption of this new style at Sezincot, I received the Prince's
commands to visit Brighton, and there saw, in some degree
realized, the new forms which I had admired in drawings.   I
found in the gardens of the Pavillon a stupendous and magnificent building, which, by its lightness, its elegance, its boldness
of construction, and the symmetry of its proportions, does
credit both to the genius of the artist, and the good taste of
his royal employer.   Although the outline of the dome
resembles rather a Turkish mosque than the buildings of
Hindûstan, yet its general character is distinct from either
Grecian or Gothic, and must both please and surprise every
one not bigoted to the forms of either.

When, therefore, I was commanded to deliver my opinion
concerning the style of architecture best adapted to the
additions and garden front for the Pavillon, I could not
hesitate in agreeing that neither the Grecian nor the Gothic
style could be made to assimilate with what had so much the
character of an eastern building. I considered all the different
styles of different countries, from a conviction of the danger of
attempting to invent anything entirely new.   The Turkish
was objectionable, as being a corruption of the Grecian; the
Moorish, as a bad model of the Gothic; the Egyptian was
too cumbrous for the character of a villa; the Chinese too
light and trifling for the outside, however it may be applied
to the interior; and the specimens from Ava were still more
trifling and extravagant.   Thus, if any known style were to
be adopted, no alternative remained but to combine, from the
architecture of Hindûstan, such forms as might be rendered
applicable to the purpose.   After various experiments, the
original MS. and drawings of this present work had the
honour to receive his Royal Highness's most flattering appro-

bation, with gracious permission to lay this fac simile before the public.

However fruitless the attempt to avert the cavils of criticism, I must not conclude these prefatory observations without endeavouring to anticipate some of the objections that I suppose will be urged against this novel application of the most ancient style of ornamented architecture existing in the world.* These objections may, perhaps, be classed under the following heads :—

I. The difference in the climate from whence this style is taken.

II. The brevity of remarks for so important a subject.

III. The want of positive data and accurate measurement.

IV. The want of space for its introduction at Brighton.

V. The costliness of its ornaments and decorations.

The first objection will obviously arise from the difference between the climates of India and of England : but this would apply with equal force against the adoption of architecture from parts of Greece and Italy,† which are hotter than those mountainous tracts of Hindûstan, where the climate differs less from that of England than in the southern provinces near the sea-coasts.

In answer to the second objection, I shall observe, that this work was not intended as a detailed treatise on Hindûstan architecture, but as an essay describing the reasons for recommending that particular style for a particular spot, where the confinement of the place, the character of the

---

* Some of the forms here introduced are taken from the ornaments of the subterraneous and excavated remains, which being worked in the hardest grey granite, were found by Mr. Daniell to be as fresh as if just finished from the chisel of the sculptor ; although they are of a date beyond all record, and are mentioned as being found in the same state at the time when Alexander the Great conquered India.

† The Grecian style was introduced without any attention to the difference of climate ; and so rare is the combination of fashion with good taste, or the union of genius with common sense, that even to the present day we see lofty porticos to shade the north side of houses, where the sun never shines ; and balustrades on the tops of houses, where no one can ever walk, and where the slanting roof marks the absurdity.

garden, and other circumstances, justify its adoption; and it is now before the public to judge how far the beauties and advantages of the same style may deserve to be extended to other places. I may also observe, that, as there was no occasion to discuss more at length the inapplicability of Grecian or Gothic forms, when both had been previously rejected, it became my duty to compress the subject into the narrowest possible compass.

It has frequently been remarked, that a spirit of party and prejudice is so natural to man, that it extends from religion and politics to the arts and sciences of a country. Thus, in philosophy, in poetry, and in all the liberal arts, a difference of opinion is supported or condemned with all the zeal of party bigotry. The admirers of Grecian architecture, and those who have studied the ruins of ancient Italy, from the time of INIGO JONES and Sir CHRISTOPHER WREN to the artists of the present day, speak with contempt of all other styles, and reproachfully call them GOTHIC; while those who have directed their attention to the variety and beauty of forms among the old English remains, glory in changing the term GOTHIC, to ANCIENT ENGLISH ARCHITECTURE, as a style doing honour to their country. Whatever is written in praise of one style, will be condemned by the partisans of the other. What, then, must an author expect, who dares to become an advocate of a style totally different from either, especially where his opinions appear under the high sanction with which the following pages have been honoured?

So far, therefore, from regretting the brevity of this work, it may be feared that I have said too much on a subject which few can understand, and in which my own knowledge must have been derived from the representations and drawings of others, and not from an actual view of the existing models. This naturally leads to the third objection, *viz.* That there are no certain data for the style recommended; that in our knowledge of Grecian forms, we have the most minute

admeasurements of the detail of ancient buildings; that in the
Gothic forms, we can have recourse to a thousand examples in
the remains of various dates; but in that of Hindûstan we
have few or no details, and those from drawings made by
artists who considered the subject as painters, and not as
architects.* This objection I shall answer, by observing, that,
although the Grecian proportions are nearly reduced to fixed
rules, yet such occasional deviations may be discovered in
every fragment remaining, that no two writers on the subject
exactly agree; therefore, in the application of Grecian forms,
both in modern Italy and in this country, the correct eye is
continually offended by false proportions to suit modern purposes of habitation; but in India, the same forms are applied
to buildings of very different sizes; and, therefore, in adapting
the Hindû architecture to the purposes of European houses,
we have only to satisfy the eye of the painter with pleasing
forms of beauty, and the eye of the mathematician with the
safety of its construction; while that infinite variety of proportions, which this new style admits, may be adapted to
every possible purpose, and every kind of material, unfettered
by the restraint which so painfully operates in the Grecian or
Gothic proportions.†

* Although the works of Mr. Thomas Daniell, hitherto published, relate
to the general forms and picturesque effect of Hindû buildings, yet he has
measured many of them with such accuracy on the spot, and has collected
such ample materials for the detail of this style, that the architects who
have access to them can be at no loss for the minutiæ. These he means to
lay before the public; and after the unreserved manner in which he has
permitted me to avail myself of his sketches, it would be unpardonable in
me to do anything which might interfere with his future views respecting
the detail of Hindû architecture, of which my knowledge is chiefly derived
from his liberal communications.

† A very trifling departure from the relative proportions in any of the
Grecian orders will be detected immediately by a correct and classic observer; but the laboured littleness with which some of our artists have
imitated in stone, or wood, or plaster, what they found in marble, without
any consideration of the difference in the materials, is a melancholy proof of
the distinction between genius and science : the one will occasionally dare
" to snatch a grace beyond the reach of art," while the other will only
attend to the feet and inches; and, after having studied amidst the finest
remains of the ancient world, is as contracted in his ideas as the most
ignorant " Ripley with a rule."

The fourth objection, respecting the want of space, will, in some degree, be answered, by considering the expedients proposed; but this objection arises chiefly from the absurd idea,* that every house requires to be insulated, and surrounded on all sides by its own territory. If we consider the Pavillon as a palace in a large town, we shall find it connected with its garden to the west, open to the parade towards the north, and to the Steyn towards the south-east, and only contiguous to the town by its offices towards the south. Since this degree of local freedom is only interrupted by one or two adjoining houses, it is not too much to suppose such a reasonable degree of accommodation as may remove every objection, and give the Pavillon all the space that a palace in a town can require.

The fifth objection which I propose to answer, is founded on the costliness of the ornaments that appear, at first sight, to belong to this style. On a more minute investigation, it will be found that the Hindû enrichments are much more simple than they appear, and far less costly than either those of the Grecian or Gothic styles.†

I trust it will not be contended, that all external ornaments and enrichments in architecture are to be abolished; and that the palaces of our princes are to resemble the villas of wealthy individuals, who study only INTERNAL comfort and

---

* I call this a most absurd idea, because I have so often witnessed the demolition of whole villages, that the mansion may stand in the middle of its park, or lawn, to give it imaginary importance; while, on the contrary, many of our ancient palaces derive true consequence and dignity from being contiguous to the town, or village, over which their influence is supposed to extend. The same mistaken partiality for insulation would, perhaps, extend to the demolition of Windsor, for being too near the Castle; and of Pall Mall, for being too near St. James's Palace.

† If we compare the workmanship, whether in stone or in composition, of Grecian mouldings, even when not carved, with the plain fillets and chamfers of the Hindûstan; or the laboured detail of pinnacles and crockets, and perforated battlements of Gothic work, with the ornaments of Hindûstan, which are chiefly *turned by a lathe*, with very little carving, the difference in economy will be found greatly in favour of the latter, notwithstanding the general appearance of richness and magnificence in the outline.

magnificence, neglecting all rules of architecture in the out-side of their houses. This effect, of combined ostentation and economy is exemplified in the vicinity of every wealthy town, where large rooms, with sumptuous furniture, are " BOXED UP," under the direction of carpenters, builders, and surveyors, who may be ingenious artisans, but who have no science as architects. We are, therefore, often led to regret, that much bad taste is propagated by the fanciful mixture of false GRECIAN with pseudo Gothic forms.*

Every individual claims the right of indulging his own taste, in what relates to himself; but, in the public edifices of a country, the honour of the country should be considered. If we were to judge from the public buildings of the metro-polis, or from the unfurnished state of its churches and theatres, we might suppose that there were no funds for their completion, or no artists competent to the task of adding ornament to utility. But a very different cause must be acknowledged: so soon as such buildings are in a state to receive the admission fees of their audiences, their purpose is completed. They may be considered as manufactories or warehouses for carrying on a species of traffic, and the external appearance is neglected as useless.

In this commercial country, wealth is more generally dif-fused than good taste; and private gratification more prevalent than national dignity. While, therefore, the security of private property is the chief motive for the only public buildings now erecting in the country, which are prisons and workhouses, we cannot wonder that our royal mansions should have more the appearance of workhouses and prisons, than of palaces worthy the residence of royalty!

[* Though the mixture of Grecian with Gothic, in villa architecture, is not so common as it was in the days of Mr. Repton, yet it is still occasion-ally to be met with. Near Hampstead, between West End and Child's Hill, a villa is now (August, 1839,) building, consisting of a centre in the Roman manner, with Grecian architraves to, and with pediments over, the windows, while at the extremities of the two wings are two tall octagon towers, finished with Gothic battlements.—J. C. L.]

[Fig. 124 Fanciful composition, exhibiting various objects more or less connected with, or allusive to,
ornamental gardening. With respect to the latter part of the sentiment inscribed on the
rock, see our Note in p. 77.]

OF THE SITUATION, CHARACTER, AND CIRCUMSTANCES.

THE PAVILLON, originally erected on a small scale, with very little adjoining territory, is now become surrounded by houses on every side; and what was only a small fishing-town, is now become equal to some cities in extent and population. Such must ever be the influence of a royal residence, which cannot long exist in solitude. The situation of the Pavillon is, therefore, that of a palace surrounded by other houses, to which great extent of garden is neither possible nor desirable : yet the ground on which the Pavillon is built (including its offices and gardens), occupies more space than generally belongs to houses built in towns, and includes as much ground as is necessary for a garden so situated. This supposes the proper distinction to be made between garden and park scenery, which have, of late, been confounded: the park may imitate nature in its wilder forms, but the garden must still be an artificial object. The park, by its formal clumps, its sweeping plantations, and meandering gravel roads, has, of late, become an overgrown and slovenly garden; while the garden, by its naked lawn, and its invisible boundary, has become a mere grass field, without interest or animation. The magnificent terraces of former times have been sloped, to unite with the adjoining pasture; while shrubs and flowers, and all the gay accompaniments of a garden, are banished from the windows of the palace, that it may appear to stand in the middle of a lawn, less cheerful than a cottage on a naked common. This defect in modern gardening is to be attributed to the misapplication of the sunk fence, which gives freedom in appearance, but, in reality, confinement. Fortunately, the sunk fence cannot be applied to the gardens of the Pavillon; we cannot blend the surface of the grass with adjoining streets and parades; we cannot give great ideal extent by concealing the actual boundary; we cannot lay open the foreground of the scene to admit distant views of sea or land, while impeded by intervening houses; and, therefore, both the character and situation of the Pavillon render these common rules of landscape gardening totally inapplicable.

GENERAL OUTLINE OF THE PLAN.

Since, therefore, the real extent of this garden cannot be increased by uniting it with surrounding objects, the imagination can only be deceived by such variations in the surface of the ground, and such a position of intervening embellishments, as may retard the eye in its too rapid progress, and amuse by the richness, the variety, and the intricacy of the scene. This will produce greatness of character, without greatness of dimension; and will delight by its beauty, where it cannot surprise by its extent: such is the general outline of the plan as it relates to nature. As a work of art, the garden of the Pavillon is further to be considered.

It has been beautifully observed by Lord Bacon, " That " in the royal ordering of gardens, there should be a garden " for every month in the year;" but, in my humble endeavours to gratify the royal commands, it would be my pride to make a garden which should not be affected by any variations of season, or soil, or weather, or situation; and thus form a perpetual garden, enriched with the production of every climate.

" Hic Ver assiduum, atque alienis mensibus Æstas."

[Here blooms perpetual spring, and summer shines
In months not hers.]

REMARKS ON THE GENERAL PLAN.

To accomplish the great object of a perpetual garden, it will be necessary to provide for a regular succession of plants; and the means of removing and transplanting. It will also require certain space for various other uses;* yet, as the present area cannot be increased, we can only obtain such USEFUL space by contracting the limits of that which is merely ORNAMENTAL. The parts so intended to be thrown out are distinguished in the plan, fig. 125, by the letters w and z; and the ornamental limits, by the letters k, k, k, k, k. This boundary is supposed to be disguised by various expedients;

* Such as the stowage of frames, glasses, coals, wood, mould, garden-pots, and all the unsightly appendages of a working garden.

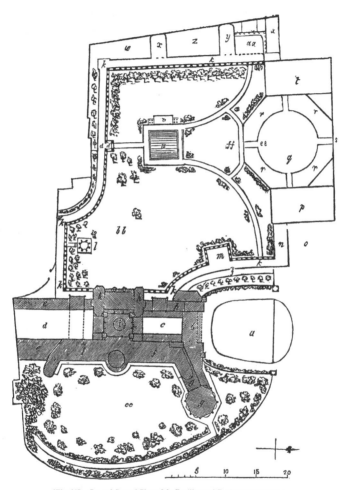

[Fig. 125. General Ground Plan of the Pavilion and Gardens at Brighton.]

a Outer court, entering from the Parade, or North Steyn	l Aviary	y Greenhouse
b Entrance-hall	m Orangery and chiosk	z Yard for mould, pots, barrows, and work sheds, &c. &c.
c Inner court	n Hothouse	a a Pheasantry
d Office-courts, &c.	o Back-yard	b b Garden lawn
e, e Offices, &c.	p Tennis-court	c c Chinese garden
f, f The present suite of Chinese rooms	q The great dome	d d Porch, in front of a foot entrance from the town
g, g Proposed private apartment	s Stable-courts, &c.	e e Point from which the view, fig. 128, in p. 381, is taken
h, h, h Proposed new buildings	t The riding-house	f f Point from which the view, fig. 129, in p. 382, is taken
i Music room	v Orchestra	
Passage for servants to stables, &c.	w Yard for bark, manure, &c. for the hothouses and greenhouses	
k, k, k Corridor	x, x Hothouses	
	r Stables	
	u The pool	

where the aspect will admit any sunshine (although not always to the south), a conservatory, or a green-house, may be most advisable, because they will draw off the attention from the interior of the garden to the interior of the conservatory, a circumstance which constitutes the most interesting part of the garden, in summer, by its exotic productions; and, in winter, by the permanency of its vegetation: each of these, from the diversity of their plants, the studied contrivance in their arrangements, and the contrasted forms and character of their embellishments, will arrest the attention, and increase the imaginary extent of the area.

These different stations may be connected with each other, and with the house, by corridors, or flower passages; in some places, under cover; in others, occasionally covered with glass in winter, which, in summer, may be taken away, leaving only such standards of wood or cast-iron as may serve to trail climbers and creeping plants.

SHAPE OF GROUND.

THERE is so little inequality of surface, that the ground may be almost described as perfectly flat, except that the stables are placed rather higher than the general level: this alone would render it necessary to form a small valley, or hollow, betwixt the stables and the house, to prevent the latter from being oppressed by the former.* Every valley in nature has a fall in some one direction, generally serving as a drain for the surface water; and wherever this is interrupted, either by natural or artificial obstructions, water is formed into a lake in large valleys, and into a pool in small ones.

The most natural shape for the surface will be a valley trending from north to south. The great dome has been placed at the north end of this valley, and an artificial obstruction from the adjoining town has stopped the valley towards the south: this forms a dell, or hollow basin, which

* Supposing the two objects to be kept distinct: but, as I should rather wish to consider them so connected as to form one magnificent whole, under the name of the PAVILLON, I cannot treat them as distinct objects, but as different parts of the same scenery.

ought, if possible, to have a pool of water, however small it may be; and, as it is evidently caused by the interference of art, its form should be artificial; any attempt to make it natural, would look like affectation.

[Fig. 126. View of part of the Stable-front of the Pavillon at Brighton, previous to the removal of certain trees, which obscured the stable-dome.]

In the drawings [our figs. 126 and 127], the slide [our fig. 126] represents the trees which have been removed,

[Fig. 127. View of the Stable-front of the Pavillon at Brighton, as seen from the lawn.]

not only because they stood in a line through the middle of the ground, and hid the dome, but also because the ground on which they stood has been lowered to form the valley.

It may, perhaps, be deemed too great a refinement in taste, to say that a pool is absolutely necessary in this place, because no Indian building is ever seen without; it is not, therefore, to preserve the character of such scenery that the pool is advisable, but, rather, for its utility in supplying the garden with water attempered by the air, and for its beauty in reflecting the surrounding objects.

The pool is proposed to be square, rather than round, for the following reasons:

First, That a small square pool will appear larger than a round one of the same dimensions, because the eye is checked in its progress, and the angles being seen perspectively, it varies its shape with the position of the spectator; while the round pool is always seen in the same point of view.

Secondly, That the inverted picture, formed by the reflection of its margin, is larger and more varied.

And, lastly, that such pools in India are generally of this shape.

ARTIFICIAL CHARACTER.

THE magnificent building, which, by its situation and magnitude, must form the leading feature of the place, ought, therefore, to extend its influence over the scenery: at present, its character is contrasted with all the surrounding objects of art, and its great dimensions withdraw the eye from all the surrounding objects of nature: hence it becomes separated from, or, rather, contrasted to, the scenery; and being thus, in a manner, isolated and detached, we are apt to suppose it too large for its situation.

If the same character be extended throughout the gardens, and the whole scene be enriched by buildings of the same style, this large dome will cease to be unconnected; it will, in a manner, blend with them, although it will always form the leading feature of the scenery.* There might be some reason for objecting to a multiplicity of buildings, if they were all

---

* Another objection to this building, as a separate object, arises from its uses. We are in the habit of supposing that the house should be a more lofty object than its stables, or offices, and are apt to annex dignity to loftiness. For this reason, at Chatsworth, at Hardwick, and some other places,

merely introduced as ornaments, like a public garden crowded with seats; but if each object has a separate use, and each contributes to the comfort as well as the magnificence of the scene, it is hardly possible to make it too rich.

THE GARDEN ENTRANCES.

THE central view of the great dome is, doubtless, the most striking; and it is, therefore, proposed to make an entrance

[Fig. 128. View of the Garden of the Pavillon at Brighton, shewing the corridor and the pool as they will appear from the stable-yard, at the point *e e*, in fig 125, page 377.]

the principal apartments were at the top of the house; yet we do not object, in St. Paul's, or St. Peter's at Rome, that the choir, or most dignified part of the building, is not placed immediately under the dome.

from the town, at the spot from whence the sketch [our figs. 126 and 127, and *d d* in the ground plan, fig. 125, in p. 377] is taken.   Another entrance to the garden will be very striking from the stable-yard [see fig. 128]; the long perspective through the several arches, requires an appropriate termination for the centre of the vista; this same porch [*d d* in fig. 125], or entrance, forms also the central object from the windows of the Pavillon.   On one side of the pool is also represented the orchestra, or platform, for a band of music, which is an essential part of the state and pleasure of such a garden, and to which some central spot must be appropriated; the cupola on the chapel is not of the same character, and cannot be hid from the garden: the appearance of this orchestra will divide the attention, and lessen its influence, although it is, fortunately, not so correct a specimen of Grecian architecture as to do much injury by its intrusion [see fig. 129].

[Fig. 129.   View from the Lawn in front of the stables, taken from the point *ff*, in fig. 125, p. 377.]

In a garden so surrounded by buildings, it is not to be expected that all can be excluded by plantation only; and as, in some places, architectural ornaments must be called in aid of vegetation, it becomes necessary to determine what style such ornaments should assume, especially as these buildings must have a reference to the style of the mansion, as well as that of the stables: this naturally leads to the following inquiry concerning the various styles of architecture which have been, at different times, introduced into England.

[Fig. 130. Imaginary composition, shewing, in the background, the castellated Gothic style of architecture ; next, the ecclesiastic Gothic ; then, the mixed Gothic ; next, the Grecian, or classical style ; and, lastly, Indian architecture.]

AN INQUIRY

INTO THE

CHANGES IN ARCHITECTURE,

AS IT RELATES TO

Palaces and Houses in England ;

INCLUDING

THE CASTLE AND ABBEY GOTHIC,

THE MIXED STYLE OF GOTHIC,

THE GRECIAN AND MODERN STYLES :

WITH SOME REMARKS ON THE INTRODUCTION OF

INDIAN ARCHITECTURE.

In obedience to the royal commands, " THAT I SHOULD " DELIVER MY OPINION CONCERNING WHAT STYLE OF " ARCHITECTURE WOULD BE MOST SUITABLE FOR THE " PAVILLON," the following Inquiry into the Changes which Architecture has undergone in this Country, will not, I hope, be found irrelevant.

Architecture has been classed under two general characters, Gothic and Grecian :* these have been jointly and

---

* The Grecian style was introduced by Inigo Jones, under the auspices of his royal master, James the First.

separately discussed and explained in volumes without number; yet these discussions have furnished no fixed standard
for determining the question, which style is most applicable
to a palace; for such must always be the residence of
royalty, whether it be large or small, and wherever it be
situated.

Until the reign of Queen Elizabeth, the large buildings in
this country had either been castles for security, or colleges
and religious retreats; many of these had been converted into
palaces, or altered to adapt them to royal residences, by such
changes in their original form, as, at length, introduced that
mixed character, called QUEEN ELIZABETH'S, or HOUSE
GOTHIC; which is, in reality, the only Gothic style that can
be made perfectly characteristic of a palace. This assertion is
confirmed by the numerous attempts to revive the Gothic
style in modern-built houses, which evidently shew how inapplicable are these ancient models for the present purposes
of habitation.

THE GOTHIC STYLE.

THE CASTLE CHARACTER requires massive walls, with very
small windows, if any are allowed to appear externally. The
correct imitation of this, in modern times, must produce the
effect of a prison.

The ABBEY CHARACTER requires lofty and large apertures, almost equally inapplicable to a house, although, in
some few rooms, the excess of light may be subdued by
coloured glass. But in the Abbey Character it is only the
chapel, the collegiate church, the hall, and the library, which
furnish models for a palace; all the subordinate parts were
the mean habitation of monks, or students, built on so small a
scale, and with such low ceilings, that they cannot be imitated
in a modern palace, without such mixture and modification
as tend to destroy the original character; therefore, it is
necessary now (as it was formerly) to adopt the MIXED STYLE
of Queen Elizabeth's Gothic, for modern palaces, if they must
be in any style of what is called GOTHIC.

Yet, a mixed style is generally imperfect: the mind is not easily reconciled to the combination of forms which it has been used to consider distinct, and at variance with each other: it feels an incongruity of character, like an anachronism in the confusion of dates; it is like uniting, in one object, infancy with old age, life with death, or things present with things past.

THE GRECIAN STYLE.

UNDER this character are included all buildings in England, for which models have been furnished from Greece, from Italy, from Syria, and from other countries, unmixed with the Gothic style; for in all these countries some intermixture of style and dates, in what is called the Grecian character, may be discovered: and we are apt to consider, as good specimens, those buildings in which the greatest simplicity prevails, or, in other words, those that are most free from mixture. Simplicity is not less necessary in the Gothic than in the Grecian style; yet it creates great difficulty in its application to both, if no mixture of dates is to be allowed in the respective styles of each. Thus, the English antiquary will discover, and, perhaps, be offended at, the mixture of Saxon, Norman, and the several dates of subsequent buildings called Gothic: but the man of taste will discover beauty in the combination of different forms in one great pile, or he must turn with disgust from every cathedral and abbey in the kingdom. In like manner, the traveller and connoisseur in Grecian antiquities, will not only object to more than one of the five orders in the same buildings, but will detect the intermixture of even the minutest parts in detail; while the man of taste will discover beauty and grace in combination of forms, for which there is not authority in the early, and, therefore, most simple edifices of those countries. It is by such combinations only, that the Grecian style can be made applicable to the purposes of modern habitation.

The best models of pure and simple Grecian architecture, were temples, many without a roof, and all without windows

3 D

or chimneys. Such models might be imitated in our churches, or public edifices; but houses built from such models would become inconvenient, in proportion as this external simplicity is preserved. For this reason, INIGO JONES, and our early architects in the Grecian style, took their models from buildings of later date (chiefly Roman), where the different floors are marked by different orders placed one over another.

As the taste for Grecian architecture became more correct, and, by the works of STUART and others, the more simple original models became better known in England, various attempts have been made to adopt it in modern houses; but a palace, or even a moderate sized residence, cannot be entirely surrounded by a peristyle, like a Grecian temple; and, therefore, the portico alone has been generally adopted.*

THE MODERN STYLE.

THE numerous difficulties in reconciling the internal convenience of a house to the external application of Grecian columns of any order, at length banished columns altogether, and introduced a new style, which is, strictly, of no character. This consists of a plain building, with rows of square windows at equal distances; and if to these be added a Grecian cornice, it is called a GRECIAN BUILDING: if, instead of the cornice, certain notches are cut in the top of the wall, it is called a GOTHIC BUILDING. Thus has the rage for simplicity, the dread of mixing dates, and the difficulty of adding ornament to utility, alike corrupted and exploded both the Grecian and the Gothic style in our modern buildings.

Without a bigoted attachment to EITHER, every one must confess, that there are a thousand beauties and graces in

* The difficulty of adapting any order of columns to the windows of a house, is evident, from the portico being sometimes confined to the ground floor only, sometimes extended through two, or even three, floors, and sometimes raised on a basement of arches, unknown to the Grecian character. A more classic expedient has been devised by the ingenious author of the Antiquities of Grecia Magna, in his designs for Harford and Downham colleges; but such lofty portion of windows, though allowable in a public building, would be inapplicable to the purposes of a private house.

EACH, which deserve our admiration, although they cannot, without violence, be made subservient to modern residence.

In this inquiry, no mention has yet been made of the difference of climate, and the influence it may be supposed to have on the different styles, because grace and beauty of form, in ornament and decorations, may be considered, without always annexing ideas of utility; if they can be blended, it is the perfection of art in every province; and, in the choice and adaptation of new forms to new uses, consists the genius of the artist.

But there is another consideration of greater importance, which relates to the MATERIAL of which the building is constructed.

The EYE will not be pleased with THAT to which the MIND cannot be reconciled: we must be satisfied that the construction is safe, and that the material is equal to its office. The resistance of iron is greater than that of stone; but if iron columns be made to represent stone, they will appear too light and weak. On the contrary, if stone columns be made to resemble metal, they will appear too heavy and massive: and if either of those materials be made to imitate wood, not only the relative strength of each must be considered, but also the PRINCIPLES OF CONSTRUCTION, which are totally different in the Grecian and Gothic styles.*

OF GRECIAN CONSTRUCTION.

ACCORDING to the law of gravitation, all matter at rest keeps its place by its own weight, and is only to be removed by superior force, acting in a different direction. A perpendicular rock, or a solid upright wall, will preserve the same position so long as its substance endures: on this principle of perpendicular pressure, all Grecian architecture is founded [see fig. 131, a]. Hence have arisen the relative proportions and

---

* This remark is every day confirmed by the too slender groins of Gothic arches, to imitate stone, in plaster, or cast-iron, and the too slender columns of Grecian architecture in wood, painted to imitate stone and marble.

intercolumniations in the different orders, from the heaviest
Doric to the most graceful Corinthian, the distances being
regulated by the strength of the parts supporting and supported.

Although it is probable that the first buildings were of

[Fig. 131. Sketch exhibiting the principles of pressure in Grecian, Gothic, and Indian architecture: *a*, Grecian
b, Gothic ; *c*, Indian.]

wood, and that rude trees suggested the proportions of the
Doric order, yet, the origin of Grecian architecture was,
doubtless, derived from one stone laid flat upon another, and
the aperture, or void, between two upright stones, was covered
by a third placed across them : thus, the width of the opening

[Fig. 132. Sketch exhibiting the progress of Grecian architecture, from the columns and beams formed of the
trunks of trees, with the bark on, to the Doric order, with fluted shafts, &c.]

was limited by the length of the cross-stone; consequently,
this mode of structure required large blocks of stone, when
that material was used [see fig. 132].

The difficulty of procuring such large blocks as were
required for this mode of construction, suggested the idea of

producing wide apertures by a different expedient; and this introduced the arch.

OF GOTHIC CONSTRUCTION.

In every arch, whether a segment of a circle, an ellipsis, or in the pointed arches, called Gothic, there is a great lateral pressure. This constitutes the leading principle of construc-

[Fig. 133. Sketch exhibiting the principle of forming abutments for Gothic arches, as generally adopted in ecclesiastical buildings.]

tion in Gothic architecture, which depends on its abuttals [see fig. 131, *b*]. An arch may sometimes abut against a rock, as in bridges; or against a pier of masonry, as in castles, &c.; but, in light Gothic structures, the abuttals consist of buttresses to counteract the lateral pressure; and where such buttresses are not sufficiently heavy, additional weight is used under the various forms of pinnacles, or finials, which have often been mistaken for mere ornaments, of no use in the construction; and these are sometimes placed at a distance when they are connected by what are called flying buttresses, like those at Henry the Seventh's chapel [see fig. 133].

OF INDIAN CONSTRUCTION.

UNDER the name of INDIAN ARCHITECTURE, may be included Hindûstan, Gentoo, Chinese, or Turkish; which latter is a mixture of the other three. But this construction is distinct from the Gothic, in having little or no lateral pressure; and from the Grecian, in having a different mode of applying the perpendicular pressure; for although, at the first sight, we might be led to suppose the arches constructed on a centre, like those of Europe, yet, on a closer examination, they will be found to consist of horizontal strata, supported by the process of what is technically called " CORBELLING OUT," or placing the materials in such a position that the aperture may be larger at the bottom than the top, by each stratum of stone over-hanging the other [see fig. 131, c]. From the specimens discovered in the Indian excavations, there is no doubt but the original idea was taken from those subterraneous caves or grottos.

The people who formed these awful wonders of antiquity, instead of erecting buildings on the surface of the ground, began their operations by cutting away the foundation of a rock, to obtain room below, without endangering the superstructure; and thus, by degrees, the Indian architecture seems to have grown from the rudest excavations of Troglodite savages, to the most beautiful forms discovered in the temples of Salsetta, of Elora, and Elephantis.

When these natural subterraneous vaults were imitated

[Fig. 134. Imaginary sketch, exhibiting the principle of perpendicular pressure in the artifical vaults made in the native rock in India, and also in the arches of buildings in the Indian style, and even in their domes. See the elongated dome in the background of the vignette.]

above ground, in buildings of later date, the same construction prevailed; and, therefore, both in the arches and domes of the Indian style, we observe the same principle of perpendicular pressure [see fig. 134].

APPLICATION OF INDIAN ARCHITECTURE.

HAVING already shewn the difficulty of adapting either the Grecian or Gothic styles to the character of an English palace, this newly discovered style of architecture seems to present a new expedient for the purpose, in the forms made known to this country by the accurate designs of Mr. THOMAS DANIELL, and other artists, which have opened new sources of grace and beauty.

To the materials of wood and stone we have lately added that of cast-iron, unknown in former times, either in Grecian or Gothic architecture, and which is peculiarly adapted to some light parts of the Indian style.

In Grecian architecture, the artist is confined to five (or, rather, only to three) different orders of columns, so restricted in their relative proportions, that they are seldom used externally, with good effect, in modern houses, and are generally found too bulky for internal use. Indian architecture presents an endless variety of forms and proportions of pillars, from the ponderous supports of the cavern, to the light, airy shafts which enrich their corridors, or support their varandahes. This alone would justify the attempt to adapt a style, untried, for the purpose to which other styles have been found inapplicable or inadequate.

It is difficult for an artist at once to divest himself of forms he has long studied: this will account for the confusion of Grecian and Gothic in the works of JOHN OF PADUA, INIGO JONES, and others, about the same date, which occasioned that mixture of style, condemned in after-times for the reasons already assigned. The same thing may be observed in the first introduction of Gothic, mixed with the Saxon and Norman which preceded it: and the same will, doubtless, happen in many instances, during the introductory application of Indian architecture to English uses, while a false taste will

both admire and condemn, without any true standard, the
various forms of novelty.

If I might humbly venture to suggest an opinion on the
subject, I should recommend the use only of such Indian
forms, or proportions, as bear the least resemblance to those
either of the Grecian or Gothic style, with which they are
liable to be compared.   If the pillars resemble Grecian
columns [compare fig. 135 with fig. 136], or if the apertures
resemble Gothic arches, they will offend, by seeming to be
incorrect specimens of well-known forms, and create a mixed
style, as disgusting to the classic observer as the mixture in
Queen Elizabeth's Gothic.   But if, from the best models of
Indian structures, such parts only be selected as cannot be
compared with any known style of English buildings, even
those whom novelty cannot delight, will have little cause to
regret the introduction of new beauties.

On these grounds, therefore, I do not hesitate to answer
the question, concerning which I am commanded to deliver
my opinion, that the Indian character having been already

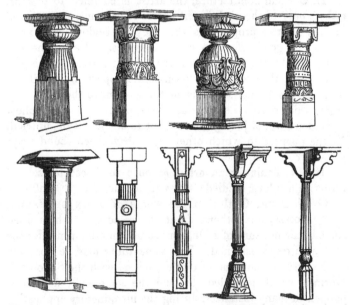

[Fig. 135. From an endless variety of columns used in Hindû architecture, the above few examples are inserted,
    that their relative proportions may be compared or contrasted with those of the orders to which Grecian
    architecture is necessarily confined.]

introduced (in part) by the large edifice at the Pavillon, the
house, and every other building, should partake of the same
character, unmixed either with Grecian or Gothic; and without strictly copying either the mosques, or the mausoleums,
or the SERAIS, or the hill-forts, or the excavations of the east,
the most varied and graceful forms should be selected, with
such combinations, or even occasional deviations and improvement, as the general character and principles of construction
will admit; for which purpose the specimens [see figs. 135 and
136] are submitted for consideration as general hints, rather
than as finished designs [see fig. 137].

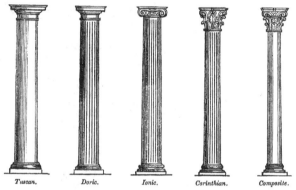

Tuscan.   Doric.   Ionic.   Corinthian.   Composite.

[Fig 136. Specimens of columns of the different orders of Grecian architecture, given with a view of
facilitating the comparison between them and columns from Hindû buildings.]

INTERIOR.

In ancient Gothic mansions, whether castles or abbeys,
converted to domestic purposes, or of the mixed style of
Queen Elizabeth, the rooms, though long and large, consisted of such irregular shapes, or were so broken by the deep
recesses of windows, or enriched by the projection of timber
groins in the ceilings, that the eye was amused and entangled
by a degree of intricacy unknown in modern rooms. The
rage for what is called SIMPLICITY, and the common error of
substituting greatness of dimensions for greatness of character,
have introduced plain walls without the smallest break or projection, and plain ceilings without the smallest enrichments of

3 E

[Fig. 137.   West front of the Pavillon.]

painting or sculpture; while large windows, and large piers, and doors too large for common use, have been made the criterion of GRANDEUR. On the contrary, these only tend to lessen the apparent dimensions of space, because (as in the case of a large naked plain) the eye is immediately led to the boundary, which is the only object that arrests attention. To remedy this defect in modern rooms, it has, of late, become the fashion to cover the ceilings with lustres, and to crowd the floor with tables, and sofas, and musical instruments, which, in some degree, create separate compartments and recesses, although the comfort and enjoyment of them can never be compared with the deep bays, and retired cavities, observed in the galleries of some ancient palaces. The plainness, or simplicity (as it is called), in modern houses, has been extended to every room alike; and often causes, in dining-rooms, an excess of echo and noise, which is intolerable.

In Italian houses of the last century, an ENFILADE was deemed essential to the state of a suite of rooms; but it was always made through small doors, and seldom in the centre of the rooms. The modern fashion of laying two or three rooms into one, by very large folding doors, is magnificent and convenient, where the rooms can be used together: but as great effect of enfilade (north and south) is preserved in the Chinese suite of rooms at the Pavillon, and may be also created in the attached corridors; and as magnificence of extent may be produced by INTRICACY and variety, as well as by CONTINUITY, perhaps the enfilade from east to west may not be so desirable, the distance being comparatively shorter.

DINING ROOM.

IN a dining-room, as the number of guests may be different at different times, some provision should be made for either enlarging it by recesses at the end, or on the side; and these recesses might occasionally be detached from the large room, for a small or select party.

In this sketch [fig. 138] some ornaments are introduced to enrich the ceiling, which, from their novelty, may appear too fanciful; but the difficulty of reconciling the mind to new forms will operate, at first, against every attempt to introduce them.   These ornaments of the ceiling may be subservient to the framing of the roof, and may also supply expedients for

[Fig. 138.   View of the Dining-room in the Pavilion at Brighton.]

ventilating the upper part of the room, which is apt, in dining-rooms especially, to retain the rarefied air and vapour, that cannot descend to the common apertures of doors, windows, or fire-places.

OF ORNAMENTS, &c.

The English language does not admit of a distinction between those ornaments which comprehend utility, and those which are merely ornamental, or, rather, enrichments; thus, columns may be called architectural ornaments, but the sculptured foliage of the capitals are decorations and enrichments.   In the progress of sculpture, we may trace it as an imitative art; from its origin, in the rude mis-shapen blocks of granite in Egypt, to its perfection, in the works of Greece, which are selected or combined forms of beauty, IDEAL FORMS, surpassing those of nature.   We may, afterwards, trace its decline, in the laboured exactness of imitation, as in Chinese figures, where individual nature is so closely copied, that even COLOUR and MOTION are added to complete the resemblance.

Much has been said, of late, concerning the study of nature in all works of art; but, if the most exact imitations of nature were the criterion of perfection, the man who paints a panorama, or even a scene at the theatres, would rank higher than CLAUDE or POUSSIN. In that early stage of painting in England, when the exhibitions were first opened, they were crowded with portraits in coloured wax, artificial flowers and fruits, and boards painted to deceive and surprise by the exactness of their resemblance; but they never excited admiration like the MARBLE of WILTON, the wood carved by GIBBON, or the animated canvas of REYNOLDS. Mr. BURKE observes, that " it is the duty of a true artist to put a " generous deception on the spectators;" but in too close an imitation of nature, he commits an absolute fraud, and becomes ridiculous, by the attempt to perform impossibilities. If it is the mark of a low imagination to aim at the VASTNESS OF NATURE, an endeavour to copy the MINUTIÆ OF NATURE is not less a proof of inexperience and bad taste, since both are equally inimitable.

> " Si la Nature est grande dans les grandes choses,
> " Elle est très grande dans les pétites."

[If Nature is great in great things, she is very great in little ones.]

The model furnishes hints, not portraits; yet such is the love of exact imitation in common minds, that copies are made from copies, without end.

For this reason, houses are built to resemble castles, and abbeys, and Grecian or Roman temples, forgetting their uses, and overlooking the general forms of each, while their minutest detail of enrichment is copied and misapplied. In works of art we can only use the FORMS of nature, not the EXACTNESS. Thus, in furniture, if we introduce the head or the foot of an animal, it may be graceful; but if we cover it with hair, or feathers, it becomes ridiculous. And in the parts taken from the vegetable kingdom, to enrich the ornaments of architecture, imitation goes no farther than the general forms, since we scarcely know the individual plant;

although some writers have mentioned the reed, the acanthus, and the lotus.*

It is a curious circumstance, that the general forms of enrichments may be thus classed: The GOTHIC are derived from the BUD, or GERM [see note † in p. 400]; the GRECIAN from the LEAF; and the INDIAN from the FLOWER; a singular coincidence, which seems to mark, that these three styles are, and ought to be, kept perfectly distinct [see fig. 139].

[Fig. 139. Imaginary sketch, to shew the forms of enrichment in Gothic architecture from the bud; Grecian from the leaf; and Indian from the flower.]

[* In this and the preceding paragraph, Mr. Repton appears to have obtained a glimpse of the Theory of Imitation, so beautifully developed by Quatremère de Quincy. The fundamental principle of this theory is—" To imitate in the fine arts is to produce the resemblance of a thing, but in some other thing, which becomes the image of it." When a thing is imitated, in such a manner as renders the imitation liable to be mistaken for the thing imitated, the object so produced has no claim to be considered as belonging to the fine arts. In general it may be said, that resemblance, by means of an image, renders an object artistical; while similarity, by means of identity, constitutes an object a mere mechanical production. Hence it is, that the close imitation of nature in park scenery, by planting indigenous trees, ferns, thorns, &c., and by breaking the ground, and otherwise introducing or modifying objects, so as to produce a picturesque effect, has but a very subordinate claim to be considered the work of an artist. It has little more claim to this kind of merit, than figures of coloured wax-work dressed so as to imitate life. There are only two modes by which landscape gardening can be brought within the pale of the fine arts. The first is, by the disposition of the indigenous trees and shrubs of a country in a manner decidedly artificial, as in the geometrical style of laying out grounds; and the second is, by employing trees and shrubs foreign to the country in which the landscape is to be produced, as shewn in the " Gardener's Magazine," vol. x. p. 558. The planting these exotic trees, shrubs, and plants, is not less a mechanical operation than planting indigenous ones; but the similarity to nature would not be quite so identical as in the other case : art would, at all times, be recognised in the production, and when the trees were fully grown, and the effect at once picturesque and exotic, the result, we think, might be considered as belonging to the fine arts; or, at all events, as something superior to the mere mechanical art of fac simile imi-

a Outline, shewing the heights of the trees as they appeared in winter, forming three distinct distances. It also shews the relative height of a man, with a rod ten feet long, at different stations.

[Fig. 140. General view from the Pavillon before it was improved agreeably to Mr. Repton's designs.]

b Dome of the stables, which was totally hid by the trees near the Pavillon, some of which have been removed.
c Situation on which stood a great number of young trees,

some of which are removed to the other side of the garden, in order to hide the town.
d The first avenue cut down.
e The farther avenue allowed to remain.

[Fig. 141. The general view from the Pavillon.]

CORRIDOR.

THIS sketch [fig. 142] represents the perspective of the west corridor, as supposed to be seen from the Pavillon; and although, in reality, this conservatory can only be about fifty

[Fig. 142. Perspective view of the west corridor, as supposed to be seen from the Pavillon.]

feet long, the ENFILADE is increased to an indefinite length, by a mirror so placed as to reflect the whole of the north corridor, which goes off at a right angle. This deceptive ornament will not only have a similar effect from the north corridor, but, in the summer, when the glasses are removed, the garden itself will be repeated, and doubled in extent.

In the sketch, a gardener is represented at the angle, to shew the only spot where any moving object can be reflected. In this respect, it differs materially from the mirrors commonly placed at the end of ENFILADES, where the spectator always sees his own image reflected.

tation. In this view of the subject, the modern style of landscape gardening is just as artificial as the ancient style, and this it ought undoubtedly to be, in order to bring it within the pale of the fine arts. See our Introduction to this volume. J. C. L.]

[† Metzger derives them from the flower and leaves. See his work, entitled, "Gesetze der Pflanzen und Mineralienbildung angewendet auf altdeutschen Baustyl, von Metzger." Stuttgart, 1835. 8vo. Plates. The bud, or germ, represented in Mr. Repton's figure, appears to be taken from the flower-stem of the succory or wild endive, Cichorium Intybus. J. C. L.]

[Fig. 143. Design for an Orangery, as it is supposed to appear in the winter season. See *m*, in fig. 125, p. 377.]

[Fig. 144. View of the Orangery changed into the character of a chiosk, by the removal of the glass sashes and part of the frame-work, and the substitution of appropriate furniture, drapery, &c.]

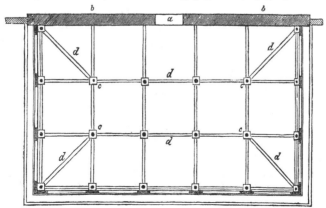

[Fig. 145. Ground-plan of the Orangery designed for the Pavillon.]

Entrance from the corridor. *b* The corridor. *c* Columns which support the roof *d* Plan of part of the roof.

[Fig. 146. View of the Pheasantry at the Pavillon, surmounted by a dove cot.]

The east front of the Pavillon is, at present, so much overlooked by the opposite houses [see fig. 147], that it seems
advisable to inclose this small garden by a thick screen of
plantation, in which openings may afterwards be made, if
necessary: but the only object really worth preserving is the
view to the sea: the annexed sketch [fig. 148] represents that
view, as supposed to be taken from the future private apartments,* the floor of which I should propose to be elevated

[* These private apartments were never built, and the consequence is,
that the sea is not seen at all from any part of the Pavillon; a circumstance
which renders it altogether ridiculous as a marine palace, as Her Majesty
the present Queen is said to have observed when she first saw it. J. C. L.]

(four or five feet), to command a better view of the sea towards
the south, and of the parade towards the north; and also to
prevent its being overlooked.   With this intention, I propose
the wall and the ground to be raised (above the eye) from the
Steyne, which may hereafter furnish a terrace walk, under a

[Fig. 147.   View of the East Front of the Pavillon, as it appeared before the alterations were made.]

double row of trees.   This screen will preclude the necessity
of making much alteration in the east front, which may,

[Fig. 148.   View from the proposed private apartment of the Pavillon, on the supposition that certain
alterations are carried into execution.]

therefore, retain the Chinese character EXTERNALLY, in conformity with the INTERIOR fitting-up of this suite of royal
apartments.

[Fig. 149.   West Front of the Pavillon, towards the garden, before the alterations were made.]

[Fig. 150. West Front of the Pavillon, towards the garden, as proposed to be altered; on the left, the orangery changed into a chiosk, and a part of the stables.]

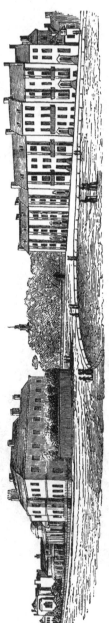

[Fig. 151.   North Front of the Pavillon, towards the parade, before it was altered.]

[Fig. 152.   North Front of the Pavillon, towards the parade, as it would   appear, if the new private apartment were carried into execution.]

FRAGMENTS

ON

THE THEORY AND PRACTICE

OF

LANDSCAPE GARDENING:

INCLUDING

SOME REMARKS ON GRECIAN AND GOTHIC

ARCHITECTURE,

𝕮ollecteꝺ from 𝖁arious 𝔐anuscripts,

IN THE

POSSESSION OF THE DIFFERENT NOBLEMEN AND GENTLEMEN

FOR WHOSE USE THEY WERE ORIGINALLY WRITTEN;

THE WHOLE TENDING TO ESTABLISH FIXED PRINCIPLES IN

THE RESPECTIVE ARTS.

———————

BY H. REPTON, ESQ.

ASSISTED BY HIS SON,

J. ADEY REPTON, F. A. S.

[*Originally published in 1816, in one volume, quarto.*]

TO

THE PRINCE REGENT,

WITH

HIS ROYAL HIGHNESS'S

MOST GRACIOUS PERMISSION,

THIS VOLUME

IS HUMBLY INSCRIBED, BY

HIS ROYAL HIGHNESS'S

MOST FAITHFUL, OBEDIENT,

AND HUMBLE SERVANTS,

H. AND J. A. REPTON.

Harestreet, near Romford,

July, 1816.

PREFACE.

MANY years have elapsed since the production of a former Work, under the title of "Observations on the Theory and Practice of Landscape Gardening;" during which, the Author's attention has been called to such variety of subjects, such interesting scenery, and such novelty of expedients, that a second volume under the former title might have been expected; (the whole of two editions of that Work are entirely sold, and the volume is become very scarce.) The contents of the present volume, which appears under a new title, will be found neither to be a continuation nor a contradiction of the former Observations; but, from the subject's being elucidated by new and more beautiful examples, the Author's former principles in "the Theory and Practice of Landscape Gardening" will be confirmed.

The following fragments having been selected from more than four hundred different reports in MS., an occasional repetition of the same remark will unavoidably, though not frequently, occur; and for this, it is hoped that the variety and beauty of the subjects may compensate, by giving new and more striking examples and elucidations.

The art of landscape gardening (which more peculiarly belongs to this country) is the only art which every one professes to understand, and even to practise, without having studied its rudiments. No man supposes he can paint a landscape, or play on an instrument, without some knowledge of painting and music; but every one thinks himself competent to lay out grounds, and sometimes to plan a house for himself,

3 G

or to criticise on what others propose, without having bestowed a thought on the first principles of *landscape gardening* or *architecture.*

That these two sister arts are, and must be, inseparable, is
obvious from the following consideration. The most beautiful scenes in nature may surprise at first sight, or delight for
a time, but they cannot long be interesting, unless made
habitable; therefore, the whole art of landscape gardening
may properly be defined, *the pleasing combination of art and
nature adapted to the use of man.*

During the last ten years, the art of landscape gardening,
in common with all other arts which depend on peace and
patronage, has felt the influence of war, and war taxes, which
operate both on the means and the inclination to cultivate
the arts of peace; these have languished under the impoverish-
ment of the country, while the sudden acquirement of riches,
by individuals, has diverted *wealth* into new channels; men
are solicitous to *increase* property rather than to *enjoy* it;
they endeavour to improve the *value,* rather than the *beauty,*
of their newly purchased estates. The country gentleman, in
the last century, took more delight in the sports of the field,
than in the profits of the farm; his pleasure was, to enjoy in
peace the venerable home of his ancestors; but the necessity
of living in camps, and the habit of living in lodgings, or
watering-places, has, of late, totally changed his character
and pursuits; and, at the same time, perhaps, tended to
alienate half the ancient landed property of the country.

It is not, therefore, to be wondered at, that the art of
landscape gardening should have slowly and gradually de-
clined. Whether the influence of returning peace may revive
its energies, or whether it is hereafter to be classed among the
" *artes perditæ* " [the lost arts], the Author hopes its memory
may be preserved a little longer in the following pages.

[Emblems of Landscape Gardening and Floriculture.]

FRAGMENT I.

ON RURAL ARCHITECTURE.

NOTWITHSTANDING the numerous volumes on Grecian architecture, from the days of Vitruvius to the present time, to which may be added all that have appeared within the last century on the subject of Gothic antiquities, little or no notice has been taken of the relative effects of the two styles, compared with each other; nor even of those leading principles by which they are to be distinguished, characterized, and appropriated to the scenery of nature. It would seem as if the whole science of Grecian architecture consisted in the five orders of columns, and that of Gothic, in pointed arches and notched battlements.

To explain this subject more clearly, and bring it before the eye more distinctly, I will refer to the following plates [figs. 153 and 154], containing three different characters of

[Fig. 153.]

C   B   A

elevations, supposing each made applicable to a house of moderate size, not exceeding a front of sixty feet, consisting of three stories, with five windows in a line. This is first represented quite plain, as at A; and afterwards with the sur-

face broken by horizontal lines, as at B ; and by vertical or
perpendicular lines, as at C. We may observe, that, without
introducing any order of columns, or any pointed arches, the
eye seems at once to class the former with the Grecian, and
the latter with the Gothic character; and this is the consequence merely of the contrasted horizontal and perpendicular lines.

Let us now proceed one step farther : we must suppose
the same building to be taken from the hands of the mere
joiner and house carpenter, and committed to the architect to
be finished, either in the Grecian, or the Gothic style.

[Fig. 154.]

For the former, recourse is had to the best specimens and
proportions of columns, pilasters, entablatures, pediments, &c.,
represented in books of architecture, or copied from remains
of ancient fragments in Greece, or Italy : but, unfortunately,
these all relate to temples or public edifices, and, consequently,
to make the dwelling habitable in this climate, modern sash-
windows must be added to these sacred forms of remote antiquity.   Thus, some Grecian or Roman temple is surprised to
find itself transported from the banks of the Ilissus, or the
Tiber, to the shores of the Thames, or to the tame margin of
a modern stagnant *sheet of water*.

If the Gothic character be preferred, the architect must
seek for his models among the fragments of his own country :
but again, unfortunately, instead of houses, he can only have
recourse to castles, cathedrals, abbeys, and colleges; many of
which have been so mutilated and disfigured by modern repairs, by converting castles into palaces, and changing convents
into dwelling-houses, that pointed arches and battlements have
become the leading features of modern Gothic buildings. The
detail of parts is studied, but the character of the whole is

overlooked.   No attention is given to that bold and irregular outline, which constitutes the real basis and beauty of the Gothic character; where, instead of one uniform line of roof and front, some parts project, and others recede: but wherever the roof is visible over the battlements, it seems as if it rose to proclaim the triumph of art over science, or carpentry over architecture.   The elevation D, represents one of these spruce villas, surrounded by spruce firs, attended by Lombardy poplars, profusely scattered over the face of the country.   That at F, may be supposed the fragment of some ancient castle, or *manor-house*, repaired and restored to make it habitable; and that at E is something betwixt the two, which will be further noticed.

The remaining part of this subject more peculiarly belongs to the landscape gardener, whose province it is to consider the effect of nature and art combined: let us examine the two different styles in the two landscapes in the next plate.

In the quiet, calm, and beautiful scenery of a tame country, the elegant forms of Grecian art [fig. 155] are, surely,

[Fig. 155. A tame country, for which Grecian architecture is supposed to be most suitable.]

more grateful and appropriate than a ruder and severer style; but, on the contrary, there are some wild and romantic situations, whose rocks, and dashing mountain-streams, or deep umbrageous dells, would seem to harmonize with the proud baronial tower, or mitred abbey, "embosomed high in tufted

trees," as tending to associate the character of the building
with that of its native accompaniment [see fig. 156].

The outline of a building is never so well seen as when

[Fig. 156. A bold rugged country, for which the castellated Gothic is considered best adapted.]

in shadow, and opposed to a brilliant sky; or when it is reflected on the surface of a pool: then the great difference
betwixt the Grecian and Gothic character is more peculiarly
striking.

FRAGMENT II.

RELATING TO SYMMETRY.

The elevations in the first plate will serve to elucidate
some remarks on architecture, not to be expected in treatises
which relate merely to the five orders, and their symmetrical
arrangement.   Such works give a very inadequate idea of that
art which teaches to adapt the habitation of man to rural
scenery, uniting convenience with beauty, and utility with
ornament.   The houses A, B, and C [fig. 153, p. 411], represent that sort of plain front which may be extended to any
length, even till it reaches the dimensions of a barrack or an
hospital.   But in all such fronts, a certain degree of symmetry
is deemed essential; and, therefore, we expect to see the door
in the centre of the building.   This arrangement, in small

houses, tends to destroy interior comfort, by dividing from each other those principal rooms which a family is now supposed to occupy.

If the principal rooms command a south-east aspect (which is, doubtless, the most desirable), the entrance in the centre, with a hall, or vestibule, destroys that uniformity of temperament so obviously useful to the comfort of an English dwelling; and therefore, in at least one half of the houses submitted to my opinion, I have found it necessary to change the hall into a saloon, or the vestibule into an anti-room; making the entrance either in the side, or at the back of the house, and converting the lawn to the south into pleasure-ground, or flower-garden, or a broad terrace, dressed with flowers. This, of course, makes a total change in the arrangement of all those appendages, in which the comfort of houses in the country differs so much from those in a town: in the latter, the offices of every description are under-ground; and the various court-yards, &c., for which there is no space, as in the country, must be provided for in areas and cellars under the street.

If the centre of a building be marked by a portico, or such a visible entrance as invites the stranger to approach it, some impediment, or obstruction, becomes necessary to counteract the habitual respect for symmetry, and prevent our inclination to drive up to a door which is no longer the principal entrance; and this requires a fence, to indicate that it is the garden front, and not the entrance front.

As this is a subject which will be explained farther, I shall, for the present, only mention, that the hint at D and F [fig. 154, p. 412] describes the different styles of fences requisite for Grecian and Gothic mansions.

FRAGMENT III.

ON FENCES NEAR THE HOUSE.

IF there be any part of my practice liable to the accusation of often advising the same thing at different places, it will be true in all that relates to my partiality for a *terrace*, as a

fence near the house.   Twenty years have, at length, by
degrees, accomplished that line of demarcation betwixt art and
nature which I have found so much difficulty in establishing,
viz., a visible and decided fence betwixt the mown pleasure-
ground, and the pastured-lawn; betwixt the garden and the
park; betwixt the ground allotted to the pleasure of man,
and that to the use of cattle.   So many different modes of
producing the same effect may be suggested, that I shall hope
to be useful in describing some of them.

First, where the ground falls from the house in an inclined
plane, the distance of the fence can only be ascertained by
actual experiment on the spot, and, of course, the steeper the
descent, the nearer, or the lower must be the terrace wall.

[Fig. 157.]

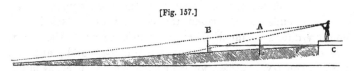

The eye sees the ground over the fence at A [in fig. 157];
but, if carried to B, all view of the ground will be lost [to a
person standing on the floor-line, c].

[Fig. 158.]

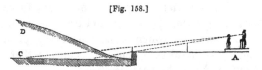

If the ground be flat, as at c [in fig. 158], or rises from
the house, as at D, the fence will admit of being placed much
farther from the house, without obstructing the view of the
lawn.*   The necessity of a fence, to protect the house from
cattle, seems to have been doubted by the followers of Brown,
who generally used the ha! ha! supposing that the fence
ought to be invisible.   On the contrary, it cannot surely be
disputed, that some fence should actually exist between a
garden and a pasture; for if it is invisible, we must either

* In some cases, the ground may be sunk near the wall, which should
be two or three feet high, with an open fence of two or three feet, or more,
if it be a fence against deer.   But in Gothic buildings, the wall may be
much higher, and the fence on the top may be wholly omitted.

suppose cattle to be admitted into a garden, or flowers planted in a field; both equally absurd.

There is no fence more garden-like than that with an open trellis. But, if the house be architecturally Grecian, and the terrace at no great distance, there is no fence so beautiful and proper as an open balustrade, like that of Lord Foley's at Witley Court, Worcestershire. And even if the house be Gothic, an open balustrade may be suited to it; of which an instance will be subjoined under the Fragment concerning Cobham.

In speaking of balustrades, I cannot omit some remarks on the use to be made of them in different situations; such as a defence for a platform, or the parapet of a roof: the latter should be of stone, but the former may, in many cases, be an iron-railing; and in the parapets of bridges, the dimensions ought to relate to those of man, rather than to that of the building.*

A magnificent palace ought not (like many that might be mentioned) to stand in a grass field, exposed to cattle, which are apt to take shelter near the building, and even to enter it, where there is no fence to prevent them; but a ter-race, or balustrade, marks the line of separation. The inside of the inclosure may be decorated with flowers; and we feel a degree of security for them and for ourselves, by knowing that there is a sufficient fence to protect both. This, which I consider a very important part of my own practice, with regard to the fence near a house, will be found elucidated by

---

* It has often occurred to me, in walking along Westminster Bridge, that this has not been sufficiently attended to. The large lofty balustrade is so managed, that the swelling of each heavy baluster exactly ranges with the eye of a foot passenger; and from a carriage, the top of the balustrade almost entirely obstructs the view of the river. Thus, one of the finest rivers in Europe is hid, for the sake of preserving some imaginary proportion in architecture, relating to its form or entablature, but not applicable to its uses as a defence for safety, without impeding the view. If it be urged that we should judge of it from the water, we should consider that this bridge is seen by a hundred persons from the land to one from the water. By the aid of an open upright iron fence, the most interesting view of the river might be obtained, with equal safety to the spectator. I have sometimes seen a drive, or walk, brought to the edge of a precipice, without any ade-quate fence; but good taste, as well as good sense, requires to be satisfied that there is no danger in the beauties we behold. We do not caress the speckled snake, or spotted panther, however we may admire him.

many of the sketches relating to other matter, in the course of these Fragments [see fig. 159].

[Fig. 159   Terrace fence, separating the Italian garden from the park.]

FRAGMENT IV.

CONCERNING COBHAM.

WHETHER we consider its extent, its magnificence, or its comfort, there are few places which can vie with Cobham, in Kent, the seat of the Earl of Darnley; and none which I can mention, where so much has been done, both to the house and grounds, under my direction, for so long a series of years; yet, as the general principles in the improvements originated in the good taste of its noble proprietor, they may be referred to, without incurring the imputation of vanity.

It is now twenty-five years since I first visited Cobham, where a large and splendid palace, of the date of Queen Elizabeth, formed the three sides of a quadrangle, the fourth side being open to the west. The centre building had been altered by INIGO JONES, who had added four pilastres, without any attention to the original style, and without extending his improvements to the two long sides of the quadrangle.

The interior of this mansion, like that of most old houses, however adapted to the customs and manners of the times in which they were built, was cold and comfortless, compared with modern houses. A large hall, anciently used as the dining-room, occupied more than half the centre; and the rest belonged to the buttery and offices, in the manner still preserved in old colleges. The two wings contained rooms, inaccessible, but by passing through one to the other; and the two opposite sides were so disjoined by the central hall, that each was entered by a separate porch.

The great hall at Cobham has been converted into a music-room, of fifty feet by thirty-six, and thirty feet high; and is one of the most splendid and costly in the kingdom. The rest of the central building forms the library, or general living room; which, instead of looking into an entrance-court, as formerly, now looks into a flower-garden, enriched with marble statues and a fountain, forming an appropriate frame, or foreground, to the landscape of the park. The entrance has been removed to the north front, under an archway, or *porte cochère*, over which a walk from the level of the picture gallery (which is up stairs) crosses the road, in the

[Fig 160. Entrance and North front of Cobham Hall, Kent.]

manner described by the annexed sketch [fig. 160], representing the north front, as it has been restored to its original character. In this view is also the bastion, by which the terrace-walk terminates with a view into the park. But no drawing can describe the change made in the comfort of the place, since the improvements were first planned, and which, by the help of the map [fig. 161], may be rendered more intelligible.

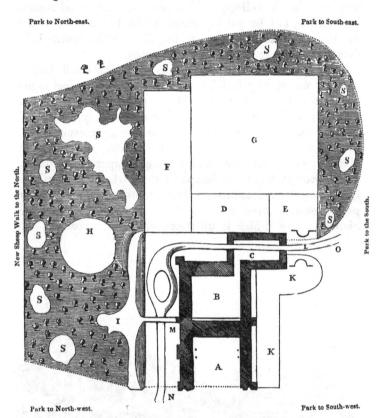

Park to North-east.   Park to South-east.

New Sheep Walk to the North.

Park to the South.

Park to North-west.   Park to South-west.

[Fig. 161. Gardens of Cobham Hall, Kent, a seat of the Earl of Darnley.]

References.

A The entrance-court, now a flower-garden, over which the park is seen from the music-room, library, dining-rooms, hall, &c. By two wings, a quadrangle is formed about 200 feet on each side; that to the south contains the family and private apartments; that to the north, the dining-room, chapel, and state bed, below stairs; and over them the picture-gallery, on the level of the terrace-garden

B The court of offices.
C Stables
D Melon and forcing garden.
E Trellis garden.
F Irregular modern flower-garden.

G Fruit and kitchen-garden.
H Menagerie.
I Regular antique terrace-garden.
K Private garden, (over which the park is seen from the private apartments,) raised on a terrace.
L Subterraneous walk, passing under the back road.
M The present entrance by an arched *porte cochère*, over which the walk passes to the terrace-garden.
N The direction of the principal approach from whence the view is taken.
O The direction of the back-road crossing the Subterraneous walk.
S S Shrubbery and plantations surrounding the whole, and placing it in a garden.

This venerable pile is situated in a valley in the middle of a large park, and was formerly exposed to the cattle on every side, except towards the east, where a large walled garden intervened. The operations were begun by enveloping the whole of the premises in plantations, shrubberies, or gardens; and these, after the growth of twenty-five years, have totally changed the character of the place. The house is no longer a huge pile, standing naked on a vast grazing ground: its walls are enriched with roses and jasmines; its apartments are perfumed with odours from flowers surrounding it on every side; and the animals which enliven the landscape are not admitted as an annoyance. All around is neatness, elegance, and comfort; while the views of the park are improved by the rich foreground, over which they are seen from the terraces in the garden, or the elevated situation of the apartments.

On the whole, Cobham furnishes a striking example of artificial arrangement for convenience, in the grounds immediately adjoining the house, contributing to the natural advantages of its situation and scenery, and enriched by the most luxuriant foliage and verdure. The home views give a perfect idea of what a park ought to be, without affecting to be a forest; for, although its extent of domain might warrant such character, there is a natural amenity in the face of the country, that is more beautiful than romantic, more habitable than wild; and, though in the valleys the view is not enlivened by water, which, in a chalk soil, is not to be expected, yet, from the elevated points of the park, the two most important rivers of England, the Thames, and Medway, form part of the distant prospect.

FRAGMENT V.

ON DATES OF BUILDINGS.

A COTTAGE, or keeper's house, was deemed necessary at Apsley Wood, about three miles from Woburn Abbey. The Duke of Bedford (to whom I am indebted for numerous opportunities of displaying his good taste) one day observed, that out of the numerous cottages called Gothic, which everywhere present themselves near the high roads, he had never seen one which did not betray its modern character and recent date. At the same time, his Grace expressed a desire to have a cottage of the style and date of buildings prior to the

Fig. 162. English cottage built of timber, prevalent from the reign of Henry VI. to Henry VIII., and erected
by the Duke of Bedford at Apsley Wood, near Woburn Abbey, in 1810 and 1811.]

reign of Henry VIII., of which only some imperfect fragments now remain.

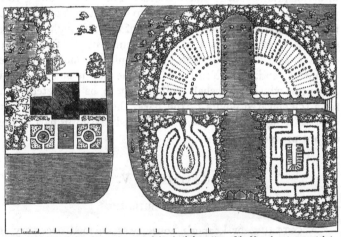

[Fig. 163. General plan of the garden, appended to the timber cottage of the fifteenth century, erected at
Apsley Wood, on the roadside from Newport Pagnel to Woburn.]

Adjoining to this building, an attempt has been made to assimilate a garden to the same character, and the foregoing plates [figs. 162 and 163] will furnish an example of both.

A communication of some curious specimens of timber-houses was made to the Society of Antiquaries, in 1810, which was ordered to be engraved and printed for the Archæologia. But this building does more than any drawing to exemplify many of the parts which have been thus rescued from the effects of time.

To admirers of genuine Gothic forms, the following note may prove acceptable, as shewing the authorities for all the detail of this cottage [fig. 162].*

The hints for this garden [fig. 163] have been suggested by various paintings and engravings of the date of King Henry VIII. and Elizabeth; and even the selection of flowers has been taken from those represented in the nosegays of old

* *Note by J. A. R.* This cottage serves as a specimen of the timber-houses which prevailed in England from about the year 1450 to 1550; that is, from the reign of Henry VI. to that of Henry VIII. As few buildings of this date remain entire, and every year reduces their number, the general plan of this cottage is not copied from any individual specimen, but the parts are taken from the most perfect fragments of the kind, some of which have since been destroyed. The hint of the lower story, being of stone, is taken from a building near Eltham Palace, except that the windows are here executed in oak instead of stone. In some buildings, both of brick and of stone, it is not uncommon to see oak windows used, as at Wolterton Manor House, East Barsham, Norfolk, and at Carhow Priory, near Norwich. Stone and brick corbels, supporting beams, may be found at Lynn Regis and at Ely. The brick-nogging between the timbers is copied from a timber-house in Lynn Regis, built by Walter Conys, in the reign of Henry VI. or Edward IV. The hint of the upright timbers being ornamented with small arches (over the centre building), was taken from a timber-house near Kelvedon, Essex, which has since been destroyed. The gable-board is copied from a house at St. Edmondsbury, and is not uncommon. The form of the pinnacles (of which few specimens now remain, being the parts most exposed to the weather,) is taken from some in brick, or stone: the only one I have ever found carved in oak is at Shrewsbury. The square flag is copied from one at Hornchurch, Essex. The projecting bow is taken from a window in Norwich, but the tracery of it is not uncommon; a specimen in oak is still to be found at Knowle, in Kent. The tracery of the lower window is taken from a timber-house in Coventry, but this, also, is not uncommon. The windows are all taken from an earlier date than the end of the reign of Henry VIII.; that is, before they were divided by a cross-bar, which did not prevail in wood till the reign of Edward VI., Elizabeth, and the early part of the seventeenth century. The design of the porch is a hint from various specimens of open porches, and particularly the cloisters of old alms-houses, or short galleries leading to dwelling-houses, as at Clapton, near Lea Bridge (since destroyed), &c. The design for the door of the cottage is taken from one remaining at Sudbury, in Suffolk. The chimneys are copied from those at Wolterton Manor House, at Barsham, Norfolk, published in the fourth volume of the Vetusta Monumenta. The ornaments painted on the posts and rails are taken from the picture of King Henry VIII. and family, now in the possession of the Society of Antiquaries.

portraits of the same period, preserved in the Picture Gallery of Woburn.* This attention to strict congruity may appear trifling to such as have never considered, that good taste delights in the harmony of the minutest parts to the whole; and this cottage, however small, compared with modern mansions, is a tolerably fair specimen of the style and size of private houses three hundred years ago: for, although the castles and collegiate buildings were large, some of the dwelling-houses of respectable persons did not much exceed this cottage in dimensions or comfort, when one living-room was often deemed sufficient for all the family.

The change in customs, during three or four centuries, makes it very difficult to build such dwelling-houses, as shall contain all the conveniences which modern life requires, and, at the same time preserve the ancient forms we admire as picturesque: yet, the prevailing taste for the Gothic style must often be complied with; and, after all, there is not more absurdity in making a house look like a castle, or convent, than like the portico of a Grecian temple, applied to a square mass, which Mr. Price has not unaptly compared to a clamp of bricks: and so great is the difference of opinion betwixt the admirers of Grecian and those of Gothic architecture, that an artist must adopt either, according to the wishes of the individual by whom he is consulted; happy if he can avoid the mixture of both in the same building; since there are few who possess sufficient taste to distinguish what is perfectly correct, and what is spurious in the two different styles; while those who have most power to indulge their taste, have generally had least leisure to study such minutiæ. To this may, perhaps, be attributed the decline of good taste in a country, with the increase of its wealth from commercial speculation.

By the recent works of professed antiquaries, a spirit of inquiry has been excited respecting the dates of every spe-

[* The plan of this garden, as given in Forbes' Hortus Woburnensis, plate XV., differs from that here given, though not materially. Mr. Forbes has given an extract from the *Red Book* of Woburn Abbey, by which it appears that Mr. Repton recommended the following flowers, as still to be found in very old gardens,—*viz.*, " Rosemary, columbine, double-crowfoot, clove-pinks, marigold, double-daisy, monkshood, southernwood, pansies, white rose, yellow lilies, turk's-cap," &c.—*Hort. Wob.* p. 296.   J. C. L.]

cimen that remains of ancient beauty or grandeur; and the strictest attention to their dates may be highly proper, in repairs or additions to old houses; but, in erecting new buildings, it may reasonably be doubted, whether modern comfort ought to be greatly sacrificed to external correctness in detail; and whether a style may not be tolerated, which gives the most commodious *interior*, and only adopts the general outline and the picturesque effect of old Gothic buildings.

Among the works professedly written on architecture, there is none more effective and useful than that by Sir William Chambers: and it were much to be wished that a similar work on the Gothic style could be referred to; but it has been deemed necessary for artists to study the remains of Greece and Rome in those countries, from whence they generally bring back the greatest contempt for the style they call Gothic. The late much-lamented James Wyatt was the only architect with whom I was acquainted who had studied on the Continent, yet preferred the Gothic forms to the Grecian. As the reason for this preference, he told me, about twenty years ago, that he conceived the climate of England required the weather mouldings, or labels, over doors and windows of the Gothic character, rather than the bolder projections of the Grecian cornices, which he often found it necessary to make more flat than the models from which they were taken, lest the materials should not bear the change of weather to which they were exposed in this country: and this accounts for the occasional want of bold-ness imputed to him in his Grecian designs. In his Gothic buildings, to unite modern comfort with antiquated forms, he introduced a style which is neither Grecian nor Gothic, but which is now become so prevalent, that it may be considered as a distinct species, and must be called *modern Gothic*. The details are often correctly Gothic, but the outline is Grecian, being just the reverse of the houses in the reign of Queen Elizabeth and King James, in which the details are often Grecian, while the general outline is Gothic. In buildings of that date, we observe towers rising boldly above the roof, and long *bower* windows breaking boldly from the surface; but in the modern Gothic all is flat, and the small octagon turrets,

which mark the corners, are neither large enough to contain a
screw staircase, nor small enough for chimneys: yet this style
has its admirers, and, therefore, I have inserted a specimen,
although I conceive it to be in a bad taste, and have placed it
betwixt the Grecian and Gothic fronts of D and F [see fig. 154,
p. 412], not knowing to which it more properly belongs. If a
door, or window, or even a battlement, or turret, of the true
Gothic form, be partially discovered, mixed with foliage, it
stamps on the scene the character of picturesqueness, of which
the subsequent vignette [fig. 164] may serve as an example;

[Fig. 164. Gothic window—exterior.]

and thus the smallest fragment of genuine Gothic often recon-
ciles to the painter its admission into the landscape; even
although the great mass of the building may offend the eye of
the antiquary, or man of correct taste, by its occasional
departure from the true Gothic style.

FRAGMENT VI.

ON CASTLES.

It has been frequently observed, that an artist's fame must
depend on what he has written, or designed, rather than on the
imperfect manner in which his works have been executed.

The annexed sketch is a picturesque attempt to add a house and garden to a romantic situation, near the head of a spring, which spreads its waters through the whole course of a narrow, but richly clothed valley. The old mansion was so deeply placed in the bottom, that the sun could never cheer it during the winter months; I proposed, therefore, that a part of this old building should remain as offices, and a new suit of rooms be built on a higher level: and, although it was deemed more expedient to add to the old house, this airy castle rose in my imagination; I will, therefore, avail myself of this imaginary specimen, to explain certain leading principles, for all of which combined, I can refer to no irregular Gothic buildings, except such as are in ruins; for, although many attempts have recently been made to produce modern Gothic castles, yet, the great principle on which the picturesque effect of all Gothic edifices must depend, has too generally been overlooked, *viz.* irregularity of outline; first, at the top by towers and pinnacles, or chimneys; secondly, in the outline of the faces, or elevations, by projections and recesses; thirdly, in the outline of the apertures, by breaking the horizontal lines with windows of different forms and heights; and, lastly, in the

[Fig. 165.   A design to exemplify irregularity of outline in castle-gothic.]

outline of the base, by the building being placed on ground of different levels. To all these must be added, detached buildings, which tend to spread the locality, and extend the importance of the principal pile, in which some one feature ought to rise boldly above the rest of the irregular mass, while the whole should be broken, but not too much frittered into parts, by smaller towers, or clusters of lofty chimneys. After all, no building can appear truly picturesque, unless, in its outline, the *design* be enriched by vegetation (such as ivy, or other creeping plants); and the *colouring*, by those weather stains, which time alone can throw over the works of art, to blend them with the works of nature, and bring the united composition into pleasing harmony.

The usual manner in which books of architecture have represented the elevations of buildings, has been either *geometrically*, without perspective to denote the projecting and receding parts, or else *perspectively*, as a *bird's-eye view* supposed to be taken from an imaginary spot in the air. However intelligible these may be to professed architects, they are as little comprehended by general observers, as the ground plan of a building by those who are not ashamed to acknowledge they do not understand a plan.*

FRAGMENT VII.

ON UNITY OF CHARACTER.

In a house entirely new, character is at the option of the artist or proprietor: it may be Gothic or Grecian, whichever best accords with the face of the country: but where great part of the original structure is to remain, the additions should, doubtless, partake of the existing character. This we have attempted at Harlestone Park, the seat of Robert Andrew, Esq., near Northampton: and as few places have undergone so much alteration, both in the house and grounds, it may

* The antiquities of this country, and the beauties of Gothic outline, have been, of late, more forcibly elucidated by various picturesque works; of which, that by Hearne and Byrne took the lead; and I cannot omit a tribute to the beautiful combination of correct Gothic remains, and landscape scenery, displayed in the antiquities of England, published by Mr. Britton. These works have already been followed by many other ingenious productions, tending to increase the knowledge of English antiquities, and the study of picturesque effect.

serve as a specimen of the combined arts of landscape gardening and architecture, in adapting the improvement to the
original character of the place.

The house was formerly approached and entered in the

[Fig. 166. View of the house at Harlestone Park, previously to Mr. Repton's improvements.]

south front, which was encumbered by stables and farm-yards:
the road came through the village, and there was a large pool
in front [see fig. 166]: this pool has been changed to an ap-

[Fig. 167. Harlestone House and Park, as altered.]

parent river, and the stables have been removed [see fig. 167].
An ample garden has been placed behind the house; the

centre of the south front has been taken down, and a bow added, with pilasters in the style of the house : the entrance is changed from the south to the north side, and some new rooms to the west have been added. Of the useful and modern appendages to this house, the drawing can give little idea : the more essential part of landscape gardening is apt to be overlooked in the general attention to the picturesque, which has often little affinity with the more important objects of comfort, convenience, and accommodation.*

FRAGMENT VIII.

ON BLENDEN HALL, KENT.

A VILLA BELONGING TO JOHN SMITH, ESQ., M.P.

From the relative situation of this place with respect to the capital, it must be treated as a villa, rather than a constant residence. This distinction is necessary to explain the principle of improvement, because, in the art of landscape gardening, two things are often confounded which require to be kept perfectly distinct, viz. the landscape and the garden. To the former belong the lawns, the woods, the water, and the prospect; these may be improved by imitating nature, but a garden, as I have often repeated, is a work of art. At Blenden Hall, the lawn is beautiful in shape, and its surface enriched with venerable trees, which are sufficiently numerous,

* In speaking of Harlestone Park, I cannot omit mentioning a remarkable fact, connected with its improvement. This park abounded in large oaks, irregularly scattered over its uneven surface; but amongst them were everywhere intermixed many very tall elms, not all planted in avenues, but some in single rows, casting their long shadows over the lawn, oppressing the venerable oaks by their more lofty growth, and spreading shade and gloom over the surface of the park. I could not help observing, that the greatest improvement of which the place seemed capable, might be deemed too bold for me to advise, as it was no less than the removal of almost all the elms, to shew the oaks, and diffuse sunshine over the lawn. A few days after having delivered this opinion, on the 10th of November, 1810, a furious storm of wind tore up by the roots eighty-seven of the largest elms, and only one oak ; producing exactly the effect of improvement which I had anticipated, but had not dared to recommend. This occurrence is recorded on a tablet inscribed—GENIO LOCI. [To the genius of the place].

without the aid of firs and Lombardy poplars; and the boundaries are generally well concealed, or blended with distant woods.

The water, at present, consists of two distinct pools; these may be united in appearance, without altering the levels, which would sacrifice too many good trees, if the lower water were raised; and make the banks too steep, if the upper water were sunk. A bridge, however, may be so constructed as to give continuity to the water, making it resemble a river: and this idea would also be assisted by extending the water to the east, as marked on the ground. With such alterations, the water will become a very important feature in the scenery, which, without it, would require some more distant views beyond the place; but a river is always sufficient in itself to form the leading feature of a natural landscape; and with such interesting objects of lawn, wood, and water, in the home scenery, the distant prospect may be dispensed with.

It has been suggested, that the approach from Eltham ought to be removed to the corner of the premises, in conformity with a commonly received practice in landscape gardening: but I prefer the present entrance for the following reasons: I seldom advise entering at the corner of the premises, and, in this case, the house would present itself almost immediately; a road would cut up the lawn, and oblige us to continue the water, as a river, along the whole valley, which is not otherwise advisable, because there are no rooms in this front of the house to require such waste of lawn and expenditure. Perhaps the fence ought to be kept very low at the corner, to give the public a view into the lawn, which would increase the importance of the place more than by leading a road through it. And, lastly, the cottage is well placed to act as a lodge, and may easily be ornamented for that purpose.

The entrance may serve as an example for a general remark, which will frequently be applicable to other places. The gate, at present, being in the continued line of the paling, there is hardly room left to enter commodiously. If the gate be set back a few yards, the trees, thrown out into the road, will give that degree of importance to the place, which we may suppose belongs to the manorial right; while a pale, enclosing every tree and bush near the road, counteracts this impression.   One other general remark may be useful, however trifling, *viz.*, although the interior fences (to be less visible) may be *dark green,* yet the entrance gate, and its immediately detached fence, should be *white,* a little subdued, to avoid the offensive glare of paper whiteness, yet sufficiently white to prevent accidents, which an invisible gate is apt to occasion after sun-set.

The house having adopted a new character, from its late alterations, I have subjoined a sketch of its south and east fronts [figs. 168 and 169], combined in perspective, which

[Fig. 168.   View of Blenden Hall before it was altered.]

may serve to explain the effect of removing some tall trees, by which it is now oppressed, and deprived of that consequence which its Gothic character has assumed.   This sort of comparative influence of trees, on a building, deserves attention ; and the sketch presents a favourable specimen of that species of architecture which has already been mentioned as Wyatt's Gothic, because introduced by that ingenious architect ; although not strictly in conformity with the abbey, castle, or collegiate characters, or even with that of the old manor-house ; but, since it evidently belongs rather to the Gothic than the Grecian style, it will be advisable to adopt such expedients as best assimilate with buildings of the date of Queen Elizabeth, all which relate to the appendages ; especially as they add, not only to the comfort, but to the picturesque effect of the mansion : among these may be reckoned the fore-court, which extends a degree of neatness a little farther into the lawn, and this, being fenced by a dwarf-wall, should be entered by a gate in the centre.

There is an old building at the south-west corner of the house, which may form the back wall of a conservatory; and a similar wall may serve to form a correspondent wing at the north-west corner ; but, if these were laid open, it would rob the house of its importance, the pleasure-ground of its privacy, and the character of the place would take no benefit from the

[Fig. 169. View of Blenden Hall as proposed to be altered.]

Gothic style, because such walls add greatly to the shelter in winter; and there are many plants, such as jasmine and creepers, requiring the support of a wall, which, so clothed, forms a luxuriant decoration to a garden in summer ; and by ivy, and other evergreens, may partly be extended through the year. This naturally leads to the consideration of the gardens, and their improvement.

Under this head must be included every part of the grounds in which art, rather than nature, is to please the eye, the smell, and the taste.   Each part will require fences, and, perhaps, of various kinds.

First, near the house, a walled terrace, to keep cattle from the windows, and protect a border of flowering plants near the eye.   Secondly, an iron fence may be sufficient to exclude cattle from the pleasure-ground; but, in that part which con-tains fruit, a more substantial guard against man must be provided, and brick walls are the best security.

I will here make some remarks on the occupation of land

3 k

belonging to a villa. It is surprising how tenacious every
gentleman is of grass land, and with what reluctance he increases his garden, or contracts his farm; as if land were only
given to produce hay, or to fatten cattle. He forgets the difference in value betwixt an acre of pasture, and an acre of
fruit-garden; or the quantity of surface required to grow a
load of hay, or a load of currants, cauliflowers, or asparagus,
with the prodigious difference in the value of each. For this
reason, the gardens of a villa should be the principal object of
attention; and at Blenden Hall, the ground betwixt the fruit
trees in the orchard, which produces hay, small in quantity, and
bad in quality, might be turned to more advantage by planting
currant bushes, or sowing garden crops; which, even if sent to
market, will yield five times the value of the feed for cattle.
There is a clipped quickset-hedge, which forms the south boundary of the garden; this is as secure as a wall, and, therefore,
worth preserving. I must also advise retaining the lofty wall
to the west, as the greatest protection against the west winds:
but a screen of trees, or, rather, filberts and fruit trees, should
be planted, to hide the wall from the approach, and to secure
a slip on the outside, and make both sides of this lofty wall
productive. If more walls be required, they may be added as
described on the map, so as to shelter each other from blights;
for it is not necessary that the garden should be a square area
within four walls. A fruit-garden may be so blended with
flowers and vegetables as to be interesting in all seasons; and
the delight of a garden, highly cultivated, and neatly kept, is
amongst the purest pleasures which man can enjoy on earth.

The pleasures of a garden have, of late, been very much
neglected. About the middle of the last century, the introduction of landscape operated to the exclusion of the old
gardens of England, and all straight gravel walks. Glades of
grass and clipped hedges were condemned as formal and old
fashioned; not considering that where the style of the house
preserved its ancient character, the gardens might, with propriety, partake of the same. After this, a taste, or almost a
rage, for farming superseded the delights of a garden; in
many cases, for the mercenary reason, that a sack of potatoes
would sell for more than a basket of roses or lavender. It is
with peculiar satisfaction that I have occasionally observed

some few venerable gardens belonging to parsonage, or old manor-houses, where still may be traced the former grass walks, and box-edged borders, with thick and lofty hedges of holly, quickset, or other topiary plants, which, like the yew or ivy, seem to display a peculiar satisfaction in yielding a fence at once secure, and neat, and opaquely trim.

FRAGMENT IX.

CONCERNING WINDOWS.

THERE is no subject connected with landscape gardening of more importance, or less attended to, than the window through which the landscape is seen. In some ancient houses, the windows were glazed in small lozenges, containing the family arms, or crest: to this Shakspere alludes in King Richard II. " From mine own windows torn my household coat :" of course, the light was so obscured, that no view could be expected ; and, indeed, in some old mansions, the windows are so placed that it is difficult to make the rooms comfortable in the interior, while the exterior character is preserved. The style of the early Gothic of Elizabeth, when not disfigured by an unseemly mixture of bad Grecian, seems better adapted to habitation than the castle, abbey, or collegiate Gothic. But houses of that date generally consist of a large hall, like that of a college, and one or more long narrow galleries, with a number of small parlours, badly disposed, and ill connected. Yet, there is something so venerable and picturesque in many houses of this date, that I have always endeavoured to preserve as much of them as could be adapted to modern uses ; and even, in some cases, advised new houses in that style of architecture. The example selected for reference in the present Fragment, is Barningham Hall, in Norfolk, the seat of J. T. Mott, Esq., where the east front has been preserved most scrupulously; not a window has been altered, except in the glazing ; the improvements and additions have all been made in a new front towards the south, as shewn in the sketch. This house, though it appeared large, did not contain one room that was comfortable, or of a size adapted to our modern style of living. It consisted of two parlours, in each of which two modern sashes had been introduced about sixty years ago. The slide of the annexed sketch [fig. 170], repre-

[Fig 170. Barningham Hall, as it appeared in 1805, previously to the alterations made by Mr. Repton.]

sents the house as it was in 1805, and the drawing [fig. 171] shews the additions since made, according to the plan annexed of the ground floor; to which a library, or living-room, is added over the entrance, and extending forty-five feet by eighteen

[Fig. 171. Barningham Hall, as improved.]

feet wide. The original simple style of Gothic has been pre-served and restored in the vestibule and staircase, which are shewn with the plan.

The greatest improvement in this old mansion has been effected, by adopting, in the south and east fronts, a style of window of the same date, though of a different form from those of the west front. The latter were too small in the apertures to be comfortable, while those added to the south front, glazed with plate-glass, are more cheerful within, and more characteristic on the outside than any modern bow with three sash-windows could have been made.*

In treating the subject of windows, some notice may be taken of the modern improvement, borrowed from the French, of folding glass-doors opening into a garden; by which the

[Fig. 172. Section across the hall, at Farningham Hall, before it was altered.]

* There is a circumstance relative to windows which is seldom attended to, and which has never been mentioned in books of architecture, *viz.*, the situation of the bar, which is too apt to cross the eye, and injure the view, or landscape. This bar ought never to be more than four feet nine inches, nor less than four feet six inches from the floor; so that a person in the middle of the room may be able to see under the bar when sitting, and over it when standing; otherwise, this bar will form an unpleasing line, crossing the sight in the exact range of the horizon, and obliging the spectator to raise or stoop his head. If it can be entirely omitted, the scenery will be improved; but if the bar be preferred, the best position of it may be calculated at four feet six from the floor, and the glass may be continued to any depth below, not more than two feet and a half from the floor; otherwise, persons sitting will not have sufficient sight of the ground, and the view will consist, as in many old houses, of sky and the tops of trees.

effect in a room is like that of a tent, or marquee, and, in summer, delightful.* But as these doors are seldom so constructed as to exclude the cold winds in winter, where they are much exposed, it is found expedient to build up a brick wall breast high, which may be taken away during the summer months.

[Fig. 173. Modern Gothic window, at Barningham Hall.]

In making choice of Gothic bow-windows for *additions* to an old house, and, at the same time, to preserve the correct proportion of the whole, the apertures of the glass ought not to be too large, nor the munions too slender. We should select the best specimens of such remains as may be found in our old manor-houses, without copying the defects.

The plates [figs. 172, 173, and 174] annexed, serve to shew the difference betwixt the same window munions, &c. of an ancient bow-window, and a modern one, as adopted at

---

* This, although more applicable to Grecian than Gothic houses, may be adapted to the latter, by making the munion in the centre to open.

Barningham Hall. The engraving [fig. 172] shews an original window, supposed to be erected in the reign of Queen Elizabeth, with all its faults, in respect to modern comfort, such as the glazing in lozenges, heavy iron casements, and upright iron bars; also the great inconvenience of the sill being too high above the floor; and, lastly, the lofty panels, which tend to depress the height of the room.

The engraving [fig. 173] shews a modern window of the same dimensions and proportions, without copying the de-

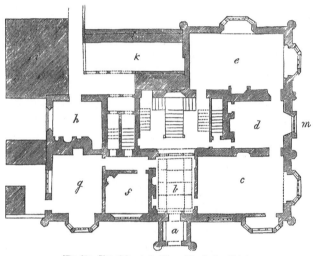

[Fig. 174. Plan of the principal floor at Barningham Hall.]

References:

a Porch	*e* Dining-room	*i* Kitchen
b Entrance-hall	*f* Butler's pantry	*k* Court-yard
c Billiard-room	*g* Servants' hall	*l* Larder, &c.
d Anti-room	*h* Housekeeper's room	*m* South front.

fects; instead of the lozenge-shape, large panes of plate, or crown glass, are inserted. The sills of the windows being lower, admit the landscape. The lofty panels are omitted, and the inside of the room fitted up as best adapted to modern comfort, preserving the character of Gothic architecture externally. There is no impropriety in supposing that the whole bow-window has been preserved, and that the inside only has been altered to suit the taste of the present times.

[Fig. 174 is a ground plan of the principal floor at Barningham Hall.]

FRAGMENT X.

ON GOTHIC OUTLINE.

EXTRACTED FROM THE RED BOOK OF STANAGE PARK.

SITUATION.

BEFORE I speak of the character of Stanage Park, it will be proper to consider its situation with respect to the neighbouring scenery; especially as the opposite opinions of two gentlemen* in its vicinity have produced that controversy, in which I have endeavoured to become a moderator.

When I compare the picturesque scenery of Downton Vale with the meagre efforts of art which are attributed to the school of Brown, I cannot wonder at the enthusiastic abhorrence which the author of " The Landscape" expresses for modern gardening: especially as few parts of the kingdom present more specimens of bad taste than the road from Ludlow to Worcester; in passing over which, I wrote the contents of this small volume. And, while I was writing, surrounded by plantations of firs, and larches, and Lombardy poplars, I saw new red houses, with all the fanciful apertures of Venetian and pseudo-Gothic windows, which disgust the traveller, who looks in vain for the picturesque shapes and harmonious tints of former times.

In the wild scenery of Stanage Park, how discordant would appear the *three window-bows* of these *modern scarlet* sins against good taste ; or, how sprucely would glitter the white-washed villa from the neighbourhood of the capital! Yet some house must be erected, or the park becomes a forest, no longer adapted to the purposes of human habitation. This leads me to the consideration of the character which it ought to assume.

* So many years have now elapsed since the controversy betwixt Mr. Knight and Mr. Price, on the subject of landscape gardening, that it may not be improper to mention, that the former gentleman published, in quarto, a poem, called " The Landscape," and the latter a work in octavo, on " The Picturesque," as distinct from the sublime and beautiful.

CHARACTER.

The antiquity, the extent, and the beauty of this park, together with the command of surrounding property, would justify very great expenditure in preserving the character of the place: but it is the duty of the professor to confine his plans within the limits prescribed by the person who consults him; and being, in the present instance, restricted to one-tenth of that expenditure which a luxuriant fancy might suggest, I shall not indulge myself in describing what might be done; but rather propose such plans as, I trust, may be accomplished within the bounds of good taste, restrained by prudence. I must, therefore, begin, by inquiring what character of house is best adapted to the scenery, since it cannot be a palace. It ought not to be a villa; it ought not to be a cottage; and, for a shooting-box for a single gentleman, the present rooms in the old house are sufficient; but it must be the residence of a family; and, as it is at some distance from society, we must not only provide for the accommodation of its own family, in its various branches, but for the entertainment of other families in the neighbourhood, and for the reception of friends and visitors, who may come from the capital, or other distant parts. All this cannot be expected in a very small mansion. The most judicious mode of combating the difficulty which prudence opposes to magnificence, will be, to follow the example set at Downton, where the inside was first considered, and the outside afterwards made to conform to that, under the idea of a picturesque outline; but, as the character of a castle must depend on its dimensions, I dare not recommend that character, lest it be compared with the massive fragments of the ruins of Ludlow.

THE HOUSE.

After the literary controversy between Mr. Knight and me, I should be sorry to be misunderstood as casting any reflection on the castle character of Downton; for although, perhaps, some may think that its outline was directed by the eye of a painter, rather than that of an antiquary, yet its general effect must gratify the good taste of both; and I

3 L

should have been happy to have shewn my assent to that style, in adopting the castle character for the house at Stanage [see figs. 175 and 176]; but this would exceed my prescribed limitation; and, since we cannot imitate the ancient baronial castle, let us endeavour to restore that sort of importance which formerly belonged to the old manor-house, where the proprietor resided among his tenants, not only to collect the rents, but to share the produce of his estate with his humble dependants; and where plenteous hospitality was not sacrificed to ostentatious refinements of luxury.

I do not mean to condemn the improvements in comfort and convenience enjoyed in modern society; nor to leave unprovided for every accommodation suited to the present habits of life, but to furnish the means of enjoying them at Stanage, without departing from its original character; and this I propose doing, by restoring, as far as possible, the same kind of mansion, on the same identical site, taking for my model the character of the grange, or old manor-farm; which, I trust, will not be deemed incongruous with the surrounding scenery. But before I shew the present state of the site and the effect, I shall describe the internal arrangements of the additions proposed. The three following principles, however they may be at variance with each other, have all been considered in the plan here suggested, viz., 1st, economy; 2nd, convenience; and 3rd, a certain degree of magnificence.* These are placed according to the respective weight each bears in my mind.

1st. Economy dictates compressing within a compact and small extent, and preserving everything which can be retained without alteration, however little it may be worth preserving: but, as I prefer the old site of the house to any other in the park, I see no occasion to take down for the sake of rebuilding.

2nd. Convenience requires a certain number of rooms, of

* I mean a certain degree of magnificence, when compared with a common farm-house. I could, in this case, have used the word picturesqueness; but that bears no relation to its importance, because the meanest objects may sometimes be deemed picturesque; but the external magnificence of a building will often depend on parts intended rather for ornament than use, such as lofty towers in Gothic, and columns in Grecian architecture.

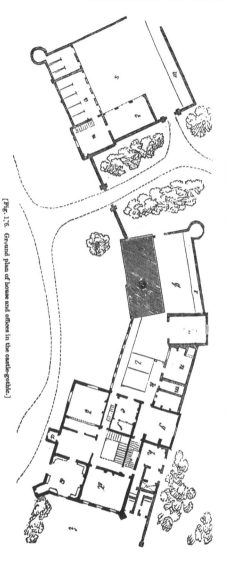

[Fig. 176. Ground plan of house and offices in the castle-gothic.]

[Fig. 175. Example of the outline of a mansion and offices in the castle-gothic.]

References.

a Entrance porch.
b Drawing-room.
c Library.
d Dining-room.
e Housekeeper's room.
f Store closet.
g Kitchen.
h Butler's pantry.
i Flower-garden.
k Passage.
l Pantry.
m Place for cleaning knives, &c.
n Servants' hall.
o Brewhouse and washhouse.
p The present house.
q Office court.
r Sheds for coals, &c.
s Stable court.
t Coach-houses.
u Harness-room.
v Stable.
w Sheds.

certain dimensions; and it will be found that those proposed are ample in size, varied in form, and connected without waste or redundance. The dining-room is detached from the drawing-room, which does away the objection of two immediately adjoining rooms, where conversation in either may be overheard. The drawing-room which, in a house of this date, was called the parlour, may be fitted up with books, musical instruments, and card-tables, to render it the general living-room for the family, according to the modern habits of life, which explode the old absurd fashion of shutting up a large comfortless room, to starve the occasional visitors by damp sofas, and bright steel grates. These two rooms and the study may be fifteen feet high: the approach to them, by the passages and staircase, will be sufficiently ample, without the extravagance of a large and lofty hall, to which much was sacrificed in old houses, because it was used as the dining-room on great occasions.*

In the chamber plan, provision is made for bed-rooms of various sizes, with closets, but not very large dressing-rooms; and the number of bed-rooms, including those in the present old building, and some in the roof over the new part, will be sufficient for all the purposes of convenience or magnificence, as far as the first leading principle of economy will allow. But, should it hereafter be deemed advisable to extend the plan, a provision may be made in the wall to the north (by changing the position of the butler's pantry, &c.) to communicate with a billiard-room, library, or any other rooms which may be required.

The offices contain everything necessary to a house of this kind, connected with each other by passages of communication, with sufficient cellars, &c.

GROUND ABOUT THE HOUSE.

It may, perhaps, be observed by the trim imitators of Brown's defects, that the stables, barns, gardens, and other

---

* After this house was built, an alteration was made, in conformity to my original wishes, that the entrance lobby should be changed to an anti-room, and the real entrance made in the cloisters, or passage, connecting the old and new buildings.

appendages, ought to be removed to a distance. I have, in my former volume, endeavoured to shew the folly of expecting importance in buildings without extent of appendages; and the absurdity of banishing to a distance those objects which are necessary to the comfort of a country residence. There is one point on which I believe my opinion may differ from the theory of the ingenious author of "The Landscape;" at least, so far as I have been told he has endeavoured to reduce it to practice near the house at Downton. I fully agree with him in condemning that bald and insipid custom, introduced by Brown, of surrounding a house by a naked grass-field: but to remedy this by slovenly neglect, or by studied or affected rudeness, seems to be an opposite extreme not less offensive. A house is an artificial object, and the ground immediately contiguous may partake of the same artificial character. In this place, therefore, straight lines of garden walls and walks are advisable, together with such management as may form the greatest possible contrast with that rude character which should everywhere else prevail at Stanage; for, when we cannot assimilate, we must produce effects by contrast, still preserving that character which we attempt to restore: and while I thus draw the bold straight line between art and nature, in defiance of the scoff of ignorance, and the vanity of fashion, I may respect the errors of prejudice, or the excess of enthusiasm.

FRAGMENT XI.

BEAUDESERT.

EXTRACT FROM THE RED BOOK OF BEAUDESERT.

CHARACTER AND SITUATION.

To lay down a rational plan for the improvement of any place, we must begin by considering its character and situation.

The very name of this place should have some influence on the mode of treating it. Great command of territory would of itself suggest, that a character of greatness, insomuch as relates to *art*, should accord with that degree of *vastness* which is the prerogative of *nature* only: at the same time, there is no error so common as the mistaking greatness of dimensions for greatness of character: and this is nowhere better exemplified than in the pictures of the date of Louis XIV., when a great general was represented in a great wig, with a great pair of jack-boots: the truth is, that great and small are only so by comparison. This fact it is the more necessary to establish, because it has been asserted, that everything at Beaudesert must be on a large scale, to correspond with the large scale of its domain; while I contend, that everything should be rather *great* than *large*.

A large massive pile, worthy of the rank and antiquity of its possessors, has been placed on the verge of the forest, a royal demesne and free warren, with paramount rights over the surrounding country: but, though built in a forest, it is evident, from its name and style, that it was not meant to be situated in a desert, the haunt only of wild beasts: on the contrary, it was to be a desert beautified—*un-beau-desert;* rendered habitable with all the elegance, magnificence, and comfort, of which it was capable. We must, therefore, look back to the reign of Henry VIII., when this mansion was presented to the family. It was at that time surrounded, not only by wild scenery of the forest, but by the animals which then made forests terrible; such as stags, boars, and wild cattle: to which might, perhaps, be added the uncivilized human beings, against whom some decided line of defence was absolutely necessary. This was the origin of those court-yards and lofty terrace-walls observable in old pictures of places of this date: so few of these now actually remain in our modern days, that I rejoice to find it the wish of the noble proprietor of this noble pile to restore its pristine character; and, if we are to retain any part of the grandeur of the mansion, we must not surrender its outworks. Although the same motive for defence no longer exists, yet the semblance must be preserved, to mark the limits betwixt the gardens, or pleasure-grounds, which belong to man, and the forest, or desert, which belongs to the wild denizens of the chase.*

---

* In the architectural arrangement of such parts of the following plan as require a knowledge of ancient forms, I am happy to have the advantage and assistance of my ingenious friend, John Shaw, Esq.

THE SITUATION.

It may, perhaps, appear presumptuous in me to assert, that the natural beauties of the situation of Beaudesert are very little known, even to those who are best acquainted with the spot: yet I will venture to assert, that those beauties which are at present hidden, and almost totally lost, far exceed those which are obvious to every eye. The materials by which nature produces the chief beauties of landscape are four in number, *viz.* inequality of ground, rocks, water, and wood: yet, at Beaudesert, it is only the latter which abounds, and to which the other three have all been sacrificed. Inequality of ground is apt to be obliterated by trees, which grow taller in the valleys than on the hills; and, consequently, the surface of a wood, and the surface of the ground on which it grows, are often very different; but, at Beaudesert, this levelling principle of vegetable growth has actually almost effaced the ravines, where tall ashes in the bottom rise above the oaks on the steep acclivities [see fig. 177].

[Fig. 177. Section shewing the bad effects of planting the hollows with tall growing trees, such as the ash, and the sides of acclivities with slower growing trees of less stature, such as the oak.]

Wherever a natural glen, or ravine, exists, we shall generally find rock or water, or both, under the surface; and we know that they abound in the deep dell immediately in front of the house, although, at present, they are hardly visible, being buried under the surface.

The following description will serve to display the capabilities of this place, and discover the true situation of Beaudesert in such a manner as may induce those who only view the present visible state, to say, "Quàm multa vident pictores in "umbris et in eminentiis, quæ nos non videmus!" [How many things painters see in lights and shadows which we see not!]

To bring these hidden beauties into notice, the process is very simple: remove the tall trees in the bottom, prevent the water from burying itself in the ground, by stopping the current of the different streamlets with ledges of large stones, wherever the natural channel is narrowest, make every drop of water visible on a hard surface, and yet let the whole appear the work of nature, and not of art.

CONCERNING THE TREES, VIEW TO THE NORTH.

As much of the improvement of Beaudesert will depend on the judicious removal of certain large trees, which have outgrown their relative situations, I will defend myself from that clamour which he must expect who dares presume to advise the felling of large trees. After forty years acquaintance with the subject, I now am frequently told, as if unconscious of such truisms, that "a large tree has been a long time growing," and, also, "that, when cut down, it cannot be put up again:" but there are situations, and very many of them at Beaudesert, where one tree conceals a wood, and where the removal of half a dozen will shew a thousand others. In winter, we may see, through their branches, objects totally invisible in summer, when a single tree becomes a screen as impenetrable as a wall. I therefore availed myself of this semi-transparent state of Beaudesert, to shew some effects by sketches which were taken when the trees were leafless, although I have supposed them in their full foliage [these sketches have not been engraved].

THE WATER.

It has been said of Beaudesert, that it is on so vast a scale, that nothing less than an arm of the sea can be adequate to the greatness of the place. This remark, however ingenious, is not sufficiently precise; because, as before observed, greatness of dimension does not confer greatness of character; and if a reach of the "smug and silver Trent" were visible in the grounds of Beaudesert, we should not object to it because it was not equal to the Straits of Menai.*

* On the banks of these Straits, in the Isle of Anglesea, stands Plas Newyd, another seat belonging to the noble proprietor of Beaudesert, and, therefore, often compared with it, although their characters are quite opposite.

There can be no doubt, that any glitter of water, visible in the bottom, would be the most desirable feature that could be suggested, to increase the beauty of the scenery: and it fortunately happens, that a number of streamlets all unite near the water-meadow, which by a dam would form a lake of such shape, that it would appear quite large enough for its situation; because its size would be indefinite, the natural shape of the ground favouring the concealment of its terminations. This river-like pool, or meer, would be a cheerful object from every point of view, and an appropriate boundary to the park towards the east, since water is always supposed to be the most natural line of separation betwixt the lawn and arable land. On the banks of this water, a spot might be found for such a garden as would accord with the character of the house; and, by its situation with respect to the house and the water, the most delightful walks of communication might be made, to increase the comfort of the place, while they displayed its interesting features to advantage.

The following sketches [figs. 178 and 179] shew the manner in which the house has been opened by the removal of tall

[Fig. 178. View of Beaudesert before the woods were opened to shew the house.]

trees in that rocky dell through which two streams of water now flow, invisibly, because buried in a deep channel amongst the brushwood. This water may easily be brought into sight, and the meadow may be converted into a lake, from whence

3 M

this sketch [fig. 179] is supposed to be taken. It also represents a part of that new kitchen-garden and fruit-garden
already mentioned.

[Fig. 179. View of Beaudesert as it will appear when the woods are opened and other improvements
effected.]

Having been permitted to make these extracts from my
report of Beaudesert, I will add two fragments from the same
MS., one respecting planting, and the other relating to ancient
interiors and uniformity of character.

CONCERNING PLANTING.

Some additional planting may be advisable at Beaudesert;
although I will confess that I have never seen a place in which
it is less absolutely necessary. In the vast range of chase
and forest attached to the place, a wood of fifty acres would
appear a clump, if the whole of its outline could be discerned
from any elevated station: and to fringe the summit of the
hill with that meagre deformity called a belt, would disgrace
the character of this wild scenery, especially if such belt were
composed of spiral spruce firs and larches, according to the
modern fashion of making plantations. It has always appeared to me, that the miserable consideration of trade has
introduced these quick-growing trees, to make a speedy
return of profit; but, if the improvement of such places as
Beaudesert is to be computed by the rule of pounds, shillings,
and pence, it would certainly be better to cut down all the

trees, kill the deer, and plough up the park. Very different is my notion of the principle of improvement; and, therefore, instead of the conic-shaped trees, which so ill accord with an English forest, and belong rather to Norway, or the Highlands of Scotland, let the staple of our plantations be oak and Spanish chestnut; let the copse be hornbeam and hazel; and let the trees used as nurses be birch: but, above all, let there be at least five or six thorns and hollies for every tree that is planted; these will grow up with the trees, perhaps choke and destroy some, but they will rear many, and in a few years will become an impenetrable thicket, as a cover for game, and a harbour for deer, when the temporary fences will be no longer necessary.

OF THE INTERIOR.

The internal arrangement of houses has undergone great change since this mansion was first erected. Formerly, the great hall at Beaudesert was always used as a dining-room; and there cannot be a better room for that purpose: indeed, it would be a violation of all archæological good taste to dine in any other. I would, therefore, consider the new room to the east, as a music or drawing-room, to be used *en suite* with that venerable gallery of magnificence and comfort, now transformed into the library; and the great hall will then be the rendezvous for breakfasts and dinners, having immediate access to the pleasure-ground on the same level; and, as it will be advisable to exclude all view to the south, by means of stained glass, it is more necessary to attend to the improvement of the view towards the west: this has already been begun, by cutting down some very tall trees, which not only prevented the eye from looking up the rocky ravine, or dell, but also excluded the light of the sky, which now sheds its cheerfulness on many valuable portraits of those worthies who, in remote times, graced the banquets of this room with their presence, and now add dignity to their noble decendants, by still holding a place in the mansion of an ancient family.

VIEW TO THE SOUTH.

This view is confined to the private front of the mansion, and is appropriated to the family apartments.

To those who despise everything they call old fashioned, and who, supposing the perfection of gardening to consist in waving lines, are apt to shrink from everything straight or formal, this part of the plan for Beaudesert may give offence; but the venerable dignity of this place is not to be measured by the scale of a villa, or the spruce modern seat of sudden affluence, *be-belted* and *be-clumped* in the newest style of the modern taste of landscape gardening—no—rather let us go back to former times, when the lofty terraces of the *privy garden* gave protection and seclusion to the noble persons who, with

> " retired leisure,
> In trim gardens took their pleasure."

To this may be added, the modern luxury of hot-houses and conservatories, with all the *agrèmens*, which excite interest in the cultivation of every species of flower that can delight the eye by its colours, or gratify the smell by its perfumes: and it is no small satisfaction to have discovered, from old labourers on the premises, that, in the line of the terrace, and other parts of this artificial and architectural garden, we are restoring the place to what they remember it in the beginning of the last century.

FRAGMENT XII.

CONCERNING COLOURS.

ADDRESSED TO WILLIAM WILBERFORCE, ESQ., M.P.

Sir,

MANY years have elapsed since you first called my attention to that Theory of Colours which your learned friend, Dr. Milner, permitted me to publish in his own words in 1803. During this interval, frequent opportunities have occurred to confirm the truth of his remarks.

Dr. Milner properly observes, that there are only *three*

primary colours—red, blue, and yellow; although Sir Isaac Newton also mentions orange, green, indigo, and violet: but these are compounds of the other three; and, whether in the rainbow or the prism, they appear to melt gradually into each other; and I have always failed in every experiment to fix the precise limits of each tint; but, in the course of some late investigations, I have accidentally discovered certain appearances, from which I have derived some facility in colouring landscape.

Having placed a piece of dark cloth on a wall opposite to the light, I fastened a sheet of white paper in the middle of it; this I looked at through a prism held across my eyes a little above them, when I observed the paper took the curvature of a rainbow: from the upper edge proceeded shades of red-brown, terminating in yellow, and from the lower edge, shades

[Fig. 180.]

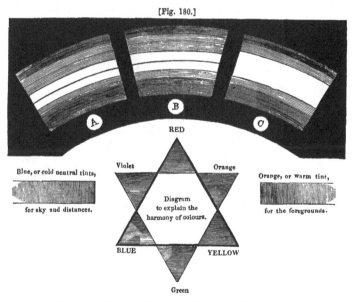

of grey, terminating in blue, as I have shewn at c, in the annexed sketch [fig. 180]. This I observed under various lights, and remarked that, before sun-rise, no red appeared in the brown, and no violet in the grey shades; but they were exactly the same tints which I use in shading and finishing my sketches. I always use for my foregrounds, a warm brown, and for my distances, a cold grey, or neutral tint, and afterwards glaze over the whole with colour; and these two tints exactly coincide with those shewn at c, where a broad mass of white appeared betwixt them. By gently turning the prism, I found that I could contract this white space, as shewn at B; and as the sun began to illumine the horizon, I perceived red in the upper edge, and violet in the lower; but no green appeared, till, by gently turning the prism a little more, I brought the blue and yellow into contact, when the rainbow was completed in all its natural colours, as represented at A. I then considered how I could avail myself of this discovery to facilitate the mechanical process of printing and colouring many thousand plates, to make a fac-simile of my sketches: this I shall endeavour to explain by the annexed two land-scapes [fig. 181], which are supposed to represent the same

[Fig. 181. Morning, after the sun is risen.   Morning twilight, before sun-rise.]
Relative proportions: Red, 45; orange, 27; yellow, 48; green, 60; blue, 60, indigo, 40; violet, 80—360 parts, according to Sir Isaac Newton's scale of quantity.

scene before and after sun-rise; that is, before and after the natural process of colouring takes place. The twilight scene is finished with the two neutral tints, and a harmonizing wash thrown over the whole; the other is also finished with the same two neutral tints, and afterwards washed or glazed over with the colours, in the order described in the diagram, begin-ning with red, then orange and yellow; in the foreground, with blue and violet, for the sky and distances. By this process, in the middle of the picture, green will prevail, partaking more of yellow near to, and more of blue as it recedes from, the eye ; still bearing in mind that all objects will partake of their natural colour: thus, a red cloak must be red, and a green tree must be green, till its distance, or the intervening mass of vapour, called aerial perspective, takes away colour, and blends it with the neutral tint.

As the plates in my former work employed a great number of women and children in colouring them, I expect to render the process much more easy in this present work, by the following instructions given to the printer and colourer:—

" The plates to be printed in a blueish-grey ink (this is " the neutral tint, for the light and shade of the landscape); " the colourer to wash in the sky, with blue or violet, &c., " according to each sketch; also *going over* the distances with " the same colour; then wash the foregrounds and middle " distances with red, orange, or yellow, copying the draw- " ings ; and, when dry, wash over with blue, to produce the " greens in the middle distances: this being done as a dead " colouring, a few touches with the hand of the master, and a " harmonizing tint to soften the whole, will produce all the " effect expected from a coloured print."

FRAGMENT XIII.

CONCERNING INTERIORS.

WHATEVER the style of the exterior may be, the interior of a house should be adapted to the uses of the inhabitants ; and, whether the house be Grecian or Gothic, large or small, it will require nearly the same rooms for the present habits of life, *viz.* a dining-room, and two others, one of which may be called a drawing-room, and the other the book-room, if small; or the library, if large: to these is sometimes added, a breakfast-room ; but of late, especially since the central hall, or vestibule, has been in some degree given up, these rooms have been opened into each other, *en suite*, by large folding doors; and the effect of this enfilade, or vista, through a modern house, is occasionally increased, by a conservatory at one end, and repeated by a large mirror at the opposite end.*

The position of looking-glasses, with respect to the light and cheerfulness of rooms, was not well understood in England during the last century, although, on the Continent, the effect of large mirrors had long been studied in certain palaces: great advantage was, in some cases, taken, by placing them obliquely, and in others, by placing them opposite: thus new scenes and unexpected effects were often introduced.

A circumstance occurring by accident has led me to avail myself, in many cases, of a similar expedient. Having directed a conservatory to be built along a south wall, in a house near Bristol, I was surprised to find that its whole length appeared from the end of the passage, in a very different position to that I had proposed: but, on examination, I found that a large looking-glass, intended for the saloon (which was not quite finished to receive it), had been accidentally placed in the green-house, at an angle of forty-five degrees, shewing the conservatory in this manner: and I have since made occasional use of mirrors so placed, to introduce views of scenery which could not otherwise be visible from a particular point of view. But, of all the improvements in modern luxury, whether belonging to the architect's or landscape gardener's department, none is more delightful than the connexion of living-rooms with a green-house, or conservatory; although they should always be separated by a small lobby, to prevent the damp and smell of earth; and when a continued covered way extends from the house to objects at some distance, like that at Woburn, it produces a degree of comfort, delight, and beauty, which, in every garden, ought, more or less, to be provided: since there are many days in the year when a walk, covered over head, and open on the side to the shrubbery, may

---

* Where there are two or more rooms, with an anti-room, it is always better not to let the dining-room open *en suite*. These rooms may be called libraries, saloons, music-rooms, or breakfast rooms: but, in fact, they form one large space, when laid together, which is more properly denominated the living-room, since the useless drawing-room is no longer retained, except by those who venerate the cedar parlour of former days.

be considered as one of the greatest improvements in modern gardens.

It would be endless to describe the various methods of attaching such covered passages to a house : and, without a plan, as well as drawings, it would be impossible to render the subject intelligible.

To make a house in the country perfect, we must consider in what it principally differs from a house in London : and it is to the inconsistency, I might almost say the folly, and vanity, of transplanting city costumes and fashions into the retirement of rural life, that much bad taste may be ascribed. Some remarks on this subject may not be misplaced, where the comfort, convenience, and elegance of habitation, are to be considered.

First. In magnificent town-houses we expect a suite of rooms, opening by folding doors, for the reception of those large parties for assemblies, when the proprietors are driven out of house and home, to make room for more visitors than their rooms can contain. A provision in the country for such an overflow of society can seldom be required; and, except for an *occasional ball*, the relative dimensions of the rooms should bear some proportion to the size of the dining-room, and the number of spare bed-rooms. The most recent modern costume is, to use the library as the general living-room; and that sort of state-room, formerly called the best parlour, and, of late years, the drawing-room, is now generally found a melancholy apartment, when entirely shut up, and only opened to give the visitors a formal cold reception : but, if such a room opens into one adjoining, and the two are fitted up with the same carpet, curtains, &c., they then become, in some degree, one room; and the comfort of that which has books, or musical instruments, is extended in its space to that which has only sofas, chairs, and card-tables ; and thus the living-room is increased in dimensions, when required, with a power of keeping a certain portion detached, and not always used for common purposes.

Secondly. In town-houses, the principal rooms for company are generally on the first floor, and, consequently, the staircase leading to them requires a correspondent degree of importance; but, in the country, it is generally more desirable

to have the principal rooms on the ground floor, and, consequently, the staircase leading only to the bed-rooms does
not require to be displayed with the same degree of importance as that of a town-house.

Thirdly. We often hear objections to a cross light in a
room, from those who take up their opinion on hearsay, without thinking for themselves; or from having observed in a
town-house, at the corner of a street, the troublesome effect
from noise, and from the thorough light rendering the room
too public to the street, and to the opposite houses. This,
however, in the country, is totally different, where an additional window gives new landscapes, and new aspects. If
the room be lighted from the end by two windows, it will
leave the opposite end of the room dark and dull; while a
window, or glazed aperture, at A [in fig. 182], opening into a
verandah, or green-house, will give cheerfulness, unattended
by the objections mentioned in a town-house, and, consequently, due advantage should be taken of it.

Fourthly. The position of the windows and doors very
much contributes to the comfort and cheerfulness of a room,
and requires very different management in town and country.

[Fig. 182.]

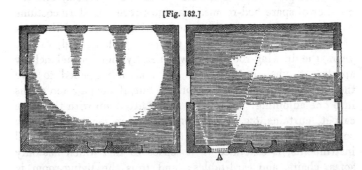

A

A room, longer than it is wide, will be best lighted by three
sashes in the long side opposite to the fire-place, because the
light is more generally spread over the whole area, and a
looking-glass over the chimney will increase the light, and
double the landscape, in the country; but, in a town-house,
where such rooms are more frequently used by candle-light,
the looking-glass may be placed betwixt the piers, as the light

from the lustres and girandoles will be increased by mirrors so placed; and, if the windows be at the end of a long room, which is often necessary in town-houses, the light of candles will be more central, by being reflected from a mirror betwixt two such windows; but, in the country, the daylight is to be studied, and this will be found very defective in a long room, lighted by two windows at the farther end; because the central pier will extend its increasing shadow till it casts a gloom over the dark end of the room. In such cases, the cross light will be found a most enlivening remedy to the dulness of a room, or, I might rather say, to one so darkened by a central pier, which, if it contains a looking-glass, will increase the gloom, by reflecting the dark end of the room.

Fifthly. The favourite proportion for a room, is asserted to be, the breadth to the length, as two to three, or nearly in that proportion: hence, a room twenty feet wide, is to be thirty feet long; and twenty-four wide, to be thirty-six long; and so in proportion, till it reaches any width and length. But when the dimensions are contracted, we must recollect that a certain space betwixt the door and the fire-place ought to be preserved; and, therefore, I have found it expedient, in small houses, to give more space, by placing the chimney at a proper distance, and forming a new centre to the room, as explained in this diagram [fig. 183].

[Fig. 183.]

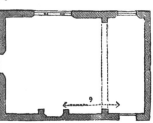

Since the chief object of this collection of Fragments is to bring before the eye those changes, or improvements, which words alone will not sufficiently explain, two sketches [figs. 184 and 185] are introduced, to shew the contrast betwixt the ancient cedar parlour, and the modern living-room: but

as no drawing can describe those comforts enjoyed in the
latter, or the silent gloom of the former, perhaps the annexed
lines may be allowed to come in aid of the attempt to
delineate both.

[Fig. 184. View of the ancient cedar parlour.]

A MODERN LIVING-ROOM.

No more the *Cedar Parlour's* formal gloom
With dulness chills, 'tis now the *Living-Room;*
Where guests, to whim, or taste, or fancy true,
Scatter'd in groups, their different plans pursue.
Here politicians eagerly relate
The last day's news, or the last night's debate;
And there a lover's conquer'd by check-mate.
Here, books of poetry, and books of prints,
Furnish aspiring artists with new hints;
Flow'rs, landscapes, figures, cram'd in one portfolio,
There blend discordant tints, to form an olio.
While discords twanging from the half-tun'd harp,
Make dulness cheerful, changing flat to sharp.
Here, 'midst exotic plants, the curious maid,
Of Greek and Latin seems no more afraid;

There lounging beaux and belles enjoy their folly;
Nor less enjoying learned melancholy,
Silent 'midst crowds, the doctor here looks big,
Wrapp'd in his own importance, and his wig.

[Fig. 185. Modern living-room.]

FRAGMENT XIV.

WINGERWORTH.

EXTRACT FROM THE RED BOOK OF WINGERWORTH, IN DERBYSHIRE,

A SEAT OF SIR WINDSOR HUNLOCK, BART.

CHARACTER AND SITUATION.

THE elevated situation of the house, on one of these broad
hills, peculiar to the most picturesque county in England [see
fig. 186], would alone stamp the character of importance on
the place, in whatever style the house might have been built;
for where we see a large pile of building on the summit of the
hill, we are naturally led to compare its relative importance

with the scenery to which it belongs. And here we shall be surprised, on approaching the mansion, to find it so much larger, richer, and more dignified, than it appears from a distance: the reason is, that the mansion is one square mass,

[Fig. 186. Wingerworth House, before the grounds were altered.]

almost a cube; and every building which partakes of this form, however great its proportions, always appears less than it really is, because the eye is not attracted either by its length, depth,

[Fig. 187. Wingerworth, as proposed to be improved.]

or height, each being nearly equal: and it is only from a subordinate building placed near it, that we form any idea of its real magnitude.

The house at Wingerworth is one of those magnificent piles which were copied from the modern palaces of France and Italy, before our more fastidious architects had discovered the remains of ancient Greece, and applied the peristyle and the portico of a Grecian temple, without any windows, to a dwelling-house in England, requiring more than a hundred such useful apertures.   But the true admirer of pure Grecian architecture is apt to forget the difference betwixt the Hypæthral temple without a roof, and the English mansion, not habitable without doors, and windows, and chimneys.

It is with a combined view to utility and magnificence, that we must look at Wingerworth Hall; and, however it may be necessary to alter its interior, in compliance with the change in modern habits of life, I should regret any alteration in the stately appearance of its exterior; on the contrary, it will be found that what I shall suggest will increase, rather than diminish, its imposing character as a palace [see fig. 187].

VIEW FROM THE HOUSE.

If there were any rooms to the north, and if it were desirable to open the view in that direction, by removing the stables, &c., it would be purchasing a landscape at the expense of all comfort, by opening to the north winds.   It is, therefore, obvious, that the stables can nowhere be better placed; and, fortunately, there are no rooms to the north to require such a sacrifice.

The view to the east is, doubtless, the leading object from this house, and to this, great attention should be given.   At present it is defective in two particulars; first, the ground falls in an inclined plane, and though the lawn is very deep, yet it is so fore-shortened to the eye, that very little of it is visible, and that part of it near the eye, is dirty, and inappropriate as a view from such apartments.   This will be remedied by the terrace and dressed ground proposed.   The distance, consisting of a rich valley, though bounded by the palaces of Bolsover and Hardwick, in the horizon, wants marked and appropriated features.   The smoke and the flame of a foundry attract our notice; but the eye would be more powerfully fixed by the expanse of water, which might be spread over the bottom, and

by removing some trees, to do away the traces of those two
rows that were at one time thought ornamental.   The effect
of such widely separated rows of trees was not like that of the
ancient avenue, whose dark and solemn grandeur amply compensated for its artificial ranks : but this seems to have been
a specimen of the power which art might exert over nature,
by compelling trees to form lines, and take possession of a
country far beyond the limits of the park, or lawn, belonging
to the house ; and, of course, such puerile attempts at mock
importance are not worthy to be retained.

THE WATER.

It very rarely happens, that an object of beauty or taste
can also be made an object of profit; but, at Wingerworth, the
same surface covered by water, may be more profitable than
the richest pasture, because it may be so managed as to admit
of being occasionally drawn down two or three feet to supply
canals, and other circumstances of advantage, in this populous
and commercial part of the kingdom; exclusive of the increased supply of fish, where such food is in constant requisi-
tion. For this reason, I do not hesitate in recommending the
piece of water already mentioned, which forms so striking a
feature in the view from the house, and of which the effect
will not be less striking, when viewed from the ground near
its shores.

The sketch will give some idea of this change in the
scenery, although its appearance, in reality, will be far more
striking than any representation of it by the pencil.

THE APPROACHES.

There is no part of the art of landscape gardening in which
so much absurdity has been displayed by the followers of
Brown, as in the line of road which should lead to the house:
and because, before his time, every road was straight along
an avenue to the front, and in the shortest line from the high
road, it has been supposed that an approach is now perfect, in
proportion to its curvature and to its length : but good taste,
which is only plain common sense, aided by observation,
directs us to make the road as easy as possible, consistently

with the shape of the grounds ; and, if one line shews more
beauty and interest than another, to prefer it ; and if it is not
actually the nearest possible, to make it more natural and easy.
For this reason, I marked a line from the south, through the
plantation, to shew the most interesting scenery of Wingerworth, when the lake shall be completed ; and also to ascend
the hill more gradually than by any other line.

ENTRANCE FROM CHESTERFIELD.

The line of this approach is not only too steep, but it is
very naked and uninteresting. It may be made more easy, by
a little more curvature, to ascend the hill, and direct the eye
to some grass-land beyond the road, which will appear a continuation of the park. The following vignettes [figs. 188 and

[Fig. 188.   Entrance-gate to Wingerworth, before the lodge was built.]

189], will shew the effect of the alteration in the road, and of
placing a lodge across the present line of road.

CONCLUSION.

It would be impossible to enumerate all the points in
which new beauties might be elicited from the natural situation and circumstances of Wingerworth; it seems to have

3 o

been unfortunately treated, in all that regards its pleasure-ground walks, as if it had been a villa at Clapham, or a flat scene in Lincolnshire; but I will not advise the alteration of what has been so recently finished; I will rather turn my

[Fig. 189. Entrance gate and lodge at Wingerworth, with the line of road altered, &c.

attention to the general effect of the whole, and the improvement of its great features, leaving the lesser errors to outgrow and correct themselves.

The too hasty removal of hedges and masses of trees, in compliance with the modern fashion of mistaking extent for beauty, has made it difficult to give the ground, so cleared, the appearance of an ancient park; and we must rather look forward to the future effect of those large masses, which have been more judiciously planted, than to the mistaken assemblage of dots and clumps, with which modern gardening is apt to disfigure an open lawn.

FRAGMENT XV.

ON PLANTING SINGLE TREES.

TO THE RIGHT HON. LORD ERSKINE, &c. &c.

MY LORD,

   IN answer to your Lordship's query, I will begin by stating it in your own words, because it is probable you have kept no copy of them, and I have no recollection of having made the remark you record, although I fully confirm it.

Your letter says, " I have followed your advice in the shel-
" ter given to my cottage, without sacrificing my prospect;
" and you said very truly, that when a man is annoyed with
" sun, wind, or dust, he puts his hand *near* his face, and does
" not depend on distant shelter."

I then recommended you to plant only beech, and now you ask if there may be added a few cedars of Lebanon, pineaster, and silver firs. This I must answer by the help of a sketch, to explain what is so obvious, when explained, that I consider it only as a proof that the most enlightened minds will sometimes hesitate on subjects which they have not studied with the eyes of a *painter* and *landscape gardener :* the former sees things as *they are,* the latter as *they will be.* Indeed, I have frequently observed, that, in planting a tree, few persons consider the future growth or shape of different kinds. Thus, the beech and the ash will admit of a view under their branches, or will admit of lower branches being cut; while the fir tribe and conic-shaped trees will not.

The annexed landscape [fig. 190] is composed of those

[Fig. 190   View shewing the effect of single trees.]

materials, which may rather be called tame and beautiful, than romantic or picturesque. It consists of a river quietly winding through a valley, a tower on the summit of a wooded promontory, and a cottage at the foot of the hill; a distant village spire, and more distant hills, mark the course of the valley: to all this is added a foreground, consisting of two large trees to the left, and three small ones to the right.

The former can never be supposed to grow much larger, but the latter may, in time, fill the whole space now occupied by the dark cloud over them; and, in so doing, they will neither injure the landscape, nor hide any of its leading features. Let me now direct the attention to the two small fir-trees in the foreground, which appear so out of character with the scene, and so misplaced, that they offend even as they are here represented; but we must remember, that in a few years these trees will grow so high as to out-top the tower on the hill; and also spread out their side branches till they meet, to the total exclusion of all the valley, and all that we admire in the landscape.

Thus, the hand which should only shade, will then be placed before the eyes; and the landscape, as well as the sun, wind, and dust, would be better excluded by a Venetian blind at once.

I have seldom known trees planted singly, or dotted on a lawn, with permanent good effect; because, few who have planted such trees, have courage to take them away after they have begun to grow: for this reason, I have always asserted that it is better to cut down a large tree, which actually impedes the view, than to plant a small one, which will very soon do the same thing.

FRAGMENT XVI.

CONCERNING VILLAS.

It has often been hinted to me, when called on for my opinion concerning places of small extent, that I can hardly be expected to give to them the same attention as to those of many hundred acres. My answer has generally been, that, on the contrary, they often require more attention than larger places. They may be compared to the miniature, with respect to the portrait large as life: the former requires to be more highly finished, but the likeness is the chief object; and this likeness in the picture, may be compared, in landscape gardening, to that pecular identity which adapts the place to the wants and wishes of the proprietor, and the character exclusively belonging to each. To pursue the simile one step further: if the nobleman will be painted as a mail coachman, or the plain country gentleman in the dress he wore at a masquerade, we shall look for the likeness in vain: so if the park be ploughed and sown with corn, or a field of twenty acres affect to be a park, the art of landscape gardening becomes useless: it does not profess to improve the *value* of land, but its *beauty :* it does not profess to gratify vanity, by displaying great extent, but to extend comfort, as far as it is feasible; and, if possible, to inculcate the great secret of true happiness —*" not to wish for more."* It is not by adding field to field, or by taking away hedges, or by removing roads to a distance, that the character of a villa is to be improved: it is by availing ourselves of every circumstance of interest or beauty within our reach, and by hiding such objects as cannot be viewed with pleasure: for I have often found, in places of the largest extent, that their principal views are annoyed by some patch of alien property, like Naboth's vineyard; some

> " Angulus ille,
> ......Qui nunc denormat agellum."

> [Oh! might I have that angle yonder,
> Which disproportions now my field.]

It seldom falls to the lot of the improver to be called upon for his opinion on places of great extent, and of vast range of unblended and uninterrupted property, like Longleate or Woburn: while, in the neighbourhood of every city or manufacturing town, new places, as villas, are daily springing up; and these, with a few acres, require all the conveniences, comforts, and appendages, of larger and more sumptuous, if not more expensive places. And as these have, of late, had the greatest claim on my attention, and may, perhaps, be found more generally useful to those who wish to enjoy the scenery of the country, without removing too far from active life, I shall produce some examples of places of this description, and make such extracts from the Reports as may become interesting to all who make new purchases, and create new scenes in the neighbourhood of the metropolis, or of any large town.

REPORT CONCERNING A VILLA AT STREATHAM,

BELONGING TO THE EARL OF COVENTRY.

MY LORD,

    I CANNOT but rejoice in the honour your Lordship has done me, in requiring my opinion concerning a villa, which, when compared with Croom or Spring Park, may be deemed inconsiderable by those who value a place by its size or extent, and not by its real importance, as it regards beauty, convenience, and utility. I must, therefore, request leave to deliver my opinion concerning Streatham at some length, as it will give me an opportunity of explaining my reasons for treating the subject very differently from those followers of Brown, who copied his manner, without attending to his proportions, or motives, and adopted the same expedients for two acres, which he thought advisable for two hundred. Mr. Brown's attention had generally been called to places of great extent, in many of which he had introduced that practice distinguished by the name of a belt of plantation, and a drive within that belt. This, when the surface was varied by hill and dale, became a convenient mode of connecting the most striking spots, and the most interesting scenes, at a distance from the mansion, and from each other. But when the same expedient is used round a small field, with no inequality of ground, and particularly with a public road bounding the premises, it is impossible to conceive a plan more objectionable in its consequences; for as the essential characteristic of a *villa* near the metropolis consists in its *seclusion* and *privacy*, the walk, which is only separated from the highway by a park paling, and a few laurels, is not more private, though far less cheerful, than the path of the highway itself. To this may be added, that such a belt, when viewed from the house, must confine the landscape, by the pale to hide the road; then by

the shrubs to hide the pale; and, lastly, by the fence to protect the shrubs; which, all together, act as a boundary more
decided and offensive than the common hedge betwixt one
field and another.

The art of landscape gardening is, in no instance, more
obliged to Mr. Brown, than for his occasionally judicious
introduction of the ha! ha! or sunk fence, by which he united
in appearance two surfaces necessary to be kept separate.
But this has been, in many places, absurdly copied to an
extent that gives more actual confinement than any visible
fence whatever. At Streatham, the view towards the south
consists of a small field, bounded by the narrow belt, and
beyond it is the common of Streatham, which is, in parts,
adorned by groups of trees, and in others disfigured by a
redundance of obtrusive houses. The common, in itself, is a
cheerful object, and, from its distance, not offensive, even
when covered with people who enjoy its verdure. Yet if the
whole of the view in front were open to the common, it might
render the house and ground near to it too public; and, for
this reason, I suppose, some shrubs have been placed near the
windows; but I consider that the defect might be more
effectually remedied, by such a mass of planting as would
direct the eye to the richest part of the common only; then,
by raising a bank to hide the paling in such opening, the grass
of the common and of the lawn would appear united, and form
one unconfined range of turf, seen point blank from the prin-
pal windows; while the oblique view might be extended to
the greatest depth of lawn, and to some fine trees, which are
now all hid by an intervening kitchen-garden, not half large
enough for the use of such a house.

This naturally leads me to explain the principle of improvement which I have the honour to suggest. The value of
land near the capital is very great; but we are apt to treat it
in the same manner as if it were a farm in the country, and
estimate its produce by the ACRE, when, in fact, it ought to
be estimated by the FOOT. An acre of land of the same
quality, which may be worth two pounds in Worcestershire,
may be worth five pounds at Streatham, for cattle; but, if
appropriated to the use of man, it may be worth twenty
pounds as a garden. It is, therefore, no waste of property,

to recommend such a garden establishment, at Streatham, as may make it amply worth the attention of the most experienced gardener to supply the daily consumption of a town-house, and save the distant conveyance or extravagant purchase of fruit and choice vegetables; especially as such an arrangement will add to the beauty and interest of the grounds, while it increases their value.

The house at Streatham, though surrounded by forty acres of grass-land, is not a farm, but a villa in a garden; for I never have admitted the word *ferme orné* into my ideas of taste, any more than a butcher's shop, or a pigsty, adorned with pea-green and gilding. A garden is of different value in different seasons, and should be adapted to each. In SUMMER, when every field in the country is a garden, we seldom enjoy that within our own paling, except in its produce; but near London, where the views from public roads are all injured by the pales and belts of private property, the interior becomes more valuable, and the pleasure of gathering summer fruit should be consulted in the arrangement of the gardens. In WINTER the garden is only preferable to a field by a broad gravel-walk, from which the snow is swept, except we add to its luxury the comfort of such glass as may set the winter at defiance; and the advantage of such forcing-houses for vines and flowers will be doubly felt in the neighbourhood of the capital.

In SPRING, the garden begins to excite interest with the first blossoms of the crocus and snowdrop: and, though its delights are seldom enjoyed in the more magnificent country residences of the nobility, yet the garden of a villa should be profusely supplied with all the fragrance and the beauty of blossom belonging to " *il gioventu del anno.*" [The youth of the year.]

Lastly, the garden in AUTUMN to its flowers adds its fruits; these, by judicious management, may be made a source of great luxury and delight: and we may observe, that it is chiefly in spring and autumn that gravel-walks are more essentially useful, when the heavy dews on the lawn render grass walks almost inaccessible.

It happens at Streatham, that a long range of offices, stables, and farm buildings, fronts the south, and seems to call for the expedient by which it may be best hid, *viz.*, a continued covered way, extending a vista from the green-house annexed to the drawing-room ; houses of every kind for grapes, peaches, strawberries, vines, &c., &c., to any extent, may here be added, without darkening the windows, which may be lighted under the glass, and a low screen of flowering shrubs in summer, will sufficiently hide this long range of winter comfort, without intercepting the rays of the sun.

REPORT

CONCERNING A VILLA NEAR A COMMON, IN THE NORTH OF ENGLAND.

THE scenery and situations which I have been called in to improve, are not more diversified than are the characters of those to whom they belong; and were I to relate the difficulties attending my endeavours to meet the wishes of all my friends, this volume might, perhaps, be deemed a libel. Some, who never meant to follow my advice, or even to do anything to their places, would be offended if no notice were taken of them: while others, who have literally realized all I suggested, would be equally offended if their places were named ; but, in these times, when the landed property of the country is changing with every generation, I shall sometimes venture to allude to subjects where great improvement has been proposed, and, in many cases, actually carried into effect, without incurring the resentment of my friends from mentioning them by name.

Among the most painful circumstances attending the professor's life, is the time and contrivance wasted to produce plans, although highly approved, yet, from vanity, from indecision, or from the fickleness of human nature, not unfrequently thrown aside, and plans more expensive, but less useful or ornamental, adopted in their stead.

About fifteen years after I had given plans for making a house and grounds perfectly comfortable, at an expense estimated at four thousand pounds, I had the mortification to find, that fifteen thousand had been expended; and the proprietor

told me that he would gladly add five thousand more to make
it as I had originally proposed.

...... " Video meliora, proboque,
Deteriora sequor."

[" I see the right, and I approve it too;
Condemn the wrong, and yet the wrong pursue."]

Tate's Trans.

To give the name of the place, or the proprietor, in such
cases, would be only proclaiming the folly or perverseness of
individuals; but if all examples were suppressed, from this
feeling of delicacy, some of the most interesting specimens
would be lost, and I should be guilty of injustice to the
powers of the art which I have so long professed to cultivate.
For this reason, I will refer to a nameless specimen of improvement in the north of England, where a villa, placed on
the edge of a goose-common, commanded a view of distant
country, enriched with woods and gentlemen's seats; but the
leading feature of the landscape was a row of mean tenements
[see fig. 191], with some of those places of worship too apt to

[Fig. 191. View from a house on the edge of a common in Yorkshire.]

disfigure the neighbourhood of all great manufacturing dis-
tricts.   These white-washed scars, in modern landscape, form
a melancholy contrast to the venerable churches and remnants
of edifices of former times, which are now suffered to moulder
into ruins.

The annexed sketch [fig. 192] represents the advantage
taken of an act of parliament to enclose a common, where the
water, which stood in several small pools, was collected into an

apparent river; and the road [see fig. 191], with all the unsightly objects, is now become a line of plantation, forming a pleasing foreground to the richly-wooded distance.

[Fig. 192. The common in Yorkshire enclosed, and the view improved.]

REPORT CONCERNING A VILLA IN ESSEX,

VERY SINGULARLY SITUATED;

Consisting of four or five acres of Garden, in the centre of Epping Forest, abounding in Deer, and immediately surrounded by a Rabbit Warren of twenty acres.

IN delivering my opinion concerning the improvement of this place, I must state the peculiar circumstances which render it very different from any other.

The house was much out of repair. It had long been the Rein-deer Tavern: occupied, not as an inn by the side of some great road, but as a house in a sequestered part of the forest, with summer access by green lanes, or broad grass glades; and appropriated to the Sunday visits of those who made holiday, fancying they enjoyed solitude in a forest, amidst the crowd of "*felicity hunters*," who came here to forget the cares of London. It was not uncommon to see fifty horses in the yards and stables, and twice as many guests filling the large rooms; but these visits were confined to the summer months, and, on this account, the cool views towards the north were preferred to the sunshine of the south. Now, it must have occurred to all who attend to situation in the country, that, in this climate, the first object of cheerfulness and comfort for a permanent residence is a south or south-east aspect. If a family ever reside here during the winter, they will, perhaps, discover, that the window on the staircase to the south admits more cheerfulness than any other in the house : and this may, perhaps, suggest the idea of opening windows to the south in the great room, although they may only look into a conservatory, or green-house, which may be also a grapery; but the difference between a public-house and a private one operates in so many ways, that I must proceed farther. The public-house required broad glades and free access in all directions; large stables, stable-yards, and out-buildings, far more extensive than are necessary to a private-house; and, consequently, all these may be reduced, and the access to them simplified. There is yet another consideration, which makes this place different from all others: it is not only a spot of four or five acres enclosed from a forest, but it is surrounded by a rabbit-warren, which the late occupier made an object of profit, though with the utmost difficulty could he preserve from these rapacious animals the vegetables in the garden, intended for his scarcely less rapacious guests; and thus the whole is subdivided by unsightly palings, and the place altogether is a scene of slovenliness, with dirty ponds, and numerous puddles, cesspools, and traps for vermin, in every part of the premises; while the surface presents nothing but yawning chasms, or barren mounds of clay, without a blade of grass, which is wholly destroyed by the rabbits. The first thing, therefore, to be done, is to secure the whole by such a fence as shall, at the same time, exclude the deer of the forest, who leap over anything that is less than six feet above ground, and the rabbits, who burrow under anything, but a brick wall, two or three feet beneath the ground. Now, such a fence, quite round the premises, will be expensive; yet, without an effective fence, there can be no enjoyment of the place: I will, therefore, suppose, that much of the present fence may be repaired, and, where a new fence is necessary, it should be constructed according to the annexed section: thus—suppose the dotted line the present surface of the ground; then begin by digging out the earth at A [in fig. 193], five or six feet wide, and throw it up at B, to raise the walk on a terrace; then face the bottom of the bank, about three feet deep, with bricks, and put upon it a paling about three feet high: this will make a fence of six feet against the deer, while a person walking withinside will look over the pale, and enjoy the prospect of the forest.

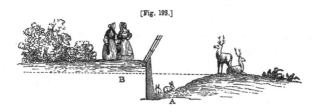

[Fig. 193.]

After effectually inclosing and securing the whole area, the next consideration is, how it shall be occupied: this is described by the map. I first allot the garden for fruit and vegetables, which may be locked up, laying it altogether, and making access to the yard by a road to divide the garden from the melon-ground, &c. This garden may be decorated with neat gravel-walks and beds of flowers and shrubs, with terrace views into the forest; and, including the house and yards, will be about two acres: there will then remain about two acres and a half to the north, which is too much to be all pleasure-ground, and either the whole may be fed with cattle as a lawn, or the part near the house, including the gravel roads, may be fenced with a wire or trellis fence, which will give neatness and comfort, without waste of land.

All rabbit-beds and burrows, within the fence, should, of course, be levelled and destroyed.

***

REPORT CONCERNING EALING PARK.

This is one of the few places which still retain the importance of the last century, in the blended scenery of landscape and gardening: but the trees have outgrown their original intention. Brown, whose work this appears to have been, surrounded the whole place by a narrow belt, or screen, of plantation; and in conformity, doubtless, with the wish of its proprietor, he made a gravel-walk through the whole length of the same, notwithstanding it everywhere runs parallel to a high road, from which it is only separated by

a pale: this was hid while the plantation was young; but now
the trees are grown so naked and open at bottom, that the
proximity of the boundary is everywhere felt; and since it
would be impossible to remedy this defect, without too great
a sacrifice of respectable trees in the belt, we must seek for
new beauties elsewhere, and have resort to different expedients, to shew the situation to advantage. A circular drive
round a place, with views only towards the interior, has little
to excite our admiration, after the first two or three rounds.
It is not sufficient to see the water, and the large group of
trees in the lawn; they are still always out of our reach: we
long to enjoy more of them; we wish not only to see them,
but actually to be on the banks of the water, and under the
shade of the trees: and, like Rasselas, in the happy vale of
Abyssinia, we regret the confinement of this belt, and should
rejoice at emancipation from the magic circle by which we
are restrained. Yet the exercise and pleasure of such a
length of walk is an object not to be hastily relinquished.

There are now but few places where the surrounding
belts, planted in Brown's time, have not been cut down, for
the sake of the timber, or the ground cleared for the sake of
the pasture; but where they exist, especially in a flat country,
the trees have acquired such height as to exclude all distant
view; and, consequently, an air of confinement is produced,
which was not intended in Brown's original belts. Two
instances of this kind have occurred to me in the neighbour-
hoods of Ealing and Acton; where a pleasing offskip, with
wooded distance, and such features as the pagoda and palace
of Kew, were totally hid by the lofty trees which formed the
belt. In one of them, an attempt had been made to break
the continuity; but some few tall trees that were left produced more mischief than all the others before they were
taken away; because, while the belt remained, we might suppose it concealed some unsightly object; or, that nothing
existed beyond it deserving a place in the landscape; but now
we perceive features whose beauty is by no means increased
by being partially concealed.

The sketch [fig. 194] represents the opening made in a
belt, in which the pagoda at Kew forms a striking object:
but the sketch differs considerably from the original; the

spectator, in the former, is supposed to be stationary; but, in the latter, whenever he changes his position, the pagoda is alternately hid by the four trees, which are supposed to be

[Fig. 194. The belt, when it has outgrown its original intention, can only be broken effectually by large and bold openings ; because, by leaving a few trees, the line of belt still remains; especially when marked by a hedge, or permanent fence.]

taken away in the sketch [fig. 195]; and every one must be sensible that the opening requires to be made thus free from

[Fig. 195. Sketch, shewing the effect of opening the belt, which not only admits a view of the distance, but, by throwing light on the water in the foreground, renders the scene much more cheerful.]

impediment and encumbrance: the distance may then be decidedly separated from the foreground. These two sketches also serve to elucidate another remark: the offskip, or distant country, must either be seen *over* or *under* those objects which constitute the foreground; here advantage is taken of both: in the larger opening it is shewn *over* an intervening copse, or mass of brushwood; and in the other, it appears through the stems of tall trees, and *under* their branches. Such openings, if not too frequently repeated, or too artificially made, will improve the landscape, without destroying the continuity of wood and of walk within the same. But, in these sketches, another effect is hinted, by breaking the line of clipped fence, partly by a few thorns planted before it, and partly by suffering some bushes in the hedge to grow taller; this will render the walk more interesting than in its present state, where the same view into the same lawn becomes tiresome and monotonous; and where the house and the water is the axis round which we turn, we feel, in a manner, tethered to a certain point; and it would be a relief to have the attention drawn away to other objects more new, though not so beautiful.

Having classed, under the same head of Small Places, or Villas, several subjects of very different magnitude and importance, one more may be added, to which not an acre belonged; and, therefore, it may serve to shew, that the quantity of acres attached does not make a place large or small; and, also, as yielding a striking example of the difference to be observed betwixt the scenery of a park and that of a garden, blending utility with ornament, and giving privacy to a situation most exposed to the public.

A modern villa, called WHITE LODGE, is situated nearly in the centre of Richmond Park. This has long been granted by the King as a residence to Lord Viscount Sidmouth. When I first visited the spot in 1805, a small quantity of land had recently been allotted from the park; without which, indeed, the house was before hardly habitable; for, although it was surrounded on every side with varied landscape, and the scenery of a forest rather than that of a park, being one of the royal domains, the deer and cattle of the forest had access to the doors and windows, and were only kept from the corridor by a chain, or hurdle, put across the arches. It is obvious, under such circumstances, that there could be no walks, no privacy, no enjoyment of garden luxuries, either for pleasure or for use: and neither fruit nor vegetables could be raised upon the premises.

There were two modes of treating this subject, according to the modern system of landscape gardening. The first was, to enclose the whole area granted, by a belt of trees and shrubs; this would have excluded all view into the park, and reduced the situation to that of any villa on Clapham Common. The other mode was that which I found actually begun, *viz.* to surround the whole with an open or invisible fence, to unite, in appearance, the ground of the park with that of the enclosure, bringing to the same level the surface where it was irregular. This would have completely destroyed every advantage of privacy, of convenience, or of use, in the acquisition from this new grant. I was, therefore, driven to suggest a third expedient, which, in these Fragments, has, or will be, frequently mentioned, *viz.* to adopt a decided artificial character for the garden; boldly reverting to the ancient formal style, which, by some, will be condemned as departing from the imitation of nature : and, by such treatment, is now secured to these premises an ample portion of ground for fruit and vegetables of every kind; yet, these are so enveloped, in screens of shrubbery and garden-flowers, as to be nowhere visible, or offensive. At the same time, by preserving the inequalities in the ground, which were about to be levelled, the walk is made to take advantage of views into the park; and, thus, neither beauty nor utility is banished by the enclosure.

The drawings [figs. 196 and 197], by which this subject is elucidated, will, perhaps, be deemed more picturesque as

[Fig. 196. Lord Sidmouth's house in Richmond Park before the scenery around it was altered.]

a park than as a garden; but it has frequently been observed, that garden scenery seldom presents subjects for a picture. Let us rather consider which of the two is most applicable to the uses of habitation—the neatness and security of a gravel-walk, or the uncleanly, pathless grass of the forest, filled with troublesome animals of every kind, and some, occasionally, dangerous.

The improvement suggested [in fig. 197] has been exe-

[Fig 197. Lord Sidmouth's house in Richmond Park as proposed to be altered.]

cuted in every respect by the present noble inhabitant, with the exception of the treillage ornaments, which may, at any time, be added.*

FRAGMENT XVII.

OF A GARDEN NEAR OPORTO.

To ——— HARRIS, ESQ.

SIR,—When I first received the honour of your letter, desiring me to furnish a plan for the improvement of ground belonging to a villa in Portugal, I doubted whether your

* In a beautiful work, lately published in France, entitled, "Choix des plus celebres Maisons de Plaisance de Rome," by Cha. Percier and P. F. L. Fontaine, the following just distinction is made betwixt the Italian gardens and those of France; to which might be added, the modern English garden also. "Ce n'est jamais, comme on le voit chez nous, un jardin dans le quelle "on a pretendu faire un site, un paysage, mais au contraire, un site dans le "quel on a fait un jardin; c'est l'art qui a paré la nature, et non pas l'art "qui a voulu la creer."

[It is never there, as it is with us, that a garden is pretended to be made into an interesting situation, or a fine landscape; but, on the contrary, a garden is made in a fine situation: it is art which has adorned what nature has supplied, and not art which would create a substitute for nature. J. C. L.]

correspondent from that place could be in earnest, in supposing such an undertaking possible; yet, the novelty of the
experiment induced me to attend to this request: and, notwithstanding all the difficulties of the subject, I have endeavoured to comply with your friend's wishes, " that I should
" furnish a plan for shewing a specimen of English gardening
" in a foreign country."

The difficulties to which I allude are these:—

First. My not having ever seen the spot, or any drawing
of it.

Secondly. My never having been in that country; yet,
from sketches furnished by those who have been there, I had
some general ideas of the face of the country near the spot,
though not of the spot itself.

Thirdly. Having only a map of the surface, without any
sections of the ground, to describe the various levels.   And,

Lastly. The vast difference in climate, soil, seasons, and
costume, between England and Portugal; all which seem to
render the introduction of our English garden into the
grounds near Oporto, almost as difficult as it has been to
introduce the vegetation of a hot country into the cold
regions of the north: yet, this has, by perseverance, been
accomplished; and I am happy in an opportunity of attempting to shew, in this instance, how far difficulties may be
surmounted.

Having, at various times, published my opinion on English, or landscape gardening, I must beg leave to refer your
correspondent to those works, and shall only point out the
peculiarities which seem to render our English style almost
inapplicable to Portugal, or, at least, to call for a different
mode of treatment in the subject under consideration.

The first of the great requisites in English gardening is,
to banish all appearance of confinement, and to give imaginary
extent of freedom, by invisible lines of separation, by a
ha! ha! or sunk fences, &c. If this be difficult in a territory
of two or three hundred acres, how much more so must it be
in a plot of three or four acres, enclosed by walls, and surrounded by neighbouring buildings? All we can hope to effect,
is, to hide this boundary everywhere by plantations of such
varied outline and depth, as to disguise what we cannot
extend or remove.

Secondly. That which peculiarly distinguishes the gardens
of England is the beauty of English verdure: the grass of
the mown lawn, uniting with the grass of the adjoining pastures, and presenting that permanent verdure which is the
natural consequence of our soft and humid climate, but
unknown to the cold regions of the north, or the parching
temperature of the south. This it is impossible to enjoy in
Portugal to any great extent; where it would be as practicable to cover the general surface with the snow of Lapland,
as with the verdure of England—I mean naturally; yet, artificially, it may be effected, on a small scale, by shade and
irrigation; some hint, therefore, will be given for producing
this effect, if only as a specimen of English verdure.

Thirdly. There is another circumstance belonging to English gardens, which may possibly be imitated in any climate;
and that relates to the prevailing lines of walks, &c.

In the old gardens of France and Italy, and I suppose it
is the same in Portugal, the walks are always straight, the
surface generally flat; and all the shapes, whether of land, of
water, of beds, or of parterres and borders, are drawn in
circles, or squares, by straight mathematical rules. These are
evidently works of art, and do not pretend to any resemblance
to nature: indeed, when the ground has been formed into
level terraces, supported by straight walls, perhaps a straight
walk, parallel to the wall, is more rational than the affectation
of a serpentine walk, which can only be justified either by irregularity of surface, or by the variety of views. In a straight
walk, the view forward is always the same; but, in a curving
walk, it varies with every step we take: thus, whether it consist of distant prospects, or of the shrubs and plants near the
eye, the scene is constantly changing, and this characteristic
of an English garden may be imitated in any country. The
annexed plan will fully explain how this may be accomplished
in the present instance; but there is also an expedient suggested to realize the English style in two other circumstances,
viz. the inequality of surface, and its verdure. I am given
to understand, that the broad terrace, or platform, near the
house, is about six or eight feet above the second terrace, and
that water may be brought thither to supply a basin for gold
and silver fish; let us, therefore, suppose the shape of the

ground to be altered according to the annexed plan and sections:* this will make a fall in every direction towards the

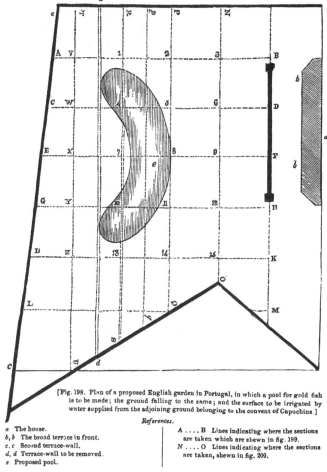

[Fig. 198. Plan of a proposed English garden in Portugal, in which a pool for gold fish is to be made; the ground falling to the same; and the surface to be irrigated by water supplied from the adjoining ground belonging to the convent of Capuchins.]

References.

a   The house.
b, b   The broad terrace in front.
c, c   Second terrace-wall.
d, d   Terrace-wall to be removed.
e   Proposed pool.

A .... B   Lines indicating where the sections are taken which are shewn in fig. 199.
N .... O   Lines indicating where the sections are taken, shewn in fig. 200.

pool; and if the water, by which this pool is to be supplied, be led along a narrow channel of stone, or lead, from every

* In describing this garden of Portugal, the plan for the walks and shrubberies has been omitted; but the map [fig. 198], inserted with the sections of the ground [figs. 199 and 200], may be considered as of a more general nature; serving to explain the process of laying out a surface in such manner, that it may be raised and sunk, according to the stakes by which the ground is divided into squares. This was done, under my direction, at the gardens at the Pavillon at Brighton, and the whole surface altered accordingly.

part of which it may trickle on the grass, it may be spread over the whole surface, in the same manner as water meadows are irrigated, and in the hottest climate the verdure will be preserved; especially, if some trees be planted round the area, to contribute their shade.   By this expedient, I have no doubt, we may produce a specimen of English verdure in the upper ground; and, if it succeeds there, it may also be made to succeed equally in the lowest ground, by the water in the well in the middle of the garden.   Not having seen the

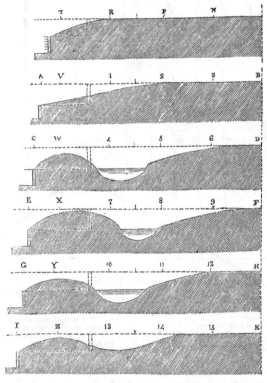

[Fig. 199. Cross sections on the lines shewn in the plan. fig. 198; the letters and figures in the sections corresponding with those on the plan.]

spot, and not knowing what plants or shrubs there may be now growing, or what will best succeed in this situation, I shall give the following general rules for the planting, after the ground is shaped according to the preceding diagrams.

First. If there are any good trees or plants on the spot, too large to be removed, let them not be disturbed, unless very much interfering with the levels of the ground, or line of the walks.

Secondly. The height of the plants must be guided by the objects they are intended to hide, or the views they may obstruct. In those places where good prospects are seen over the walls, they need only be high enough to hide the walls; in others, where houses or other objects require to be concealed, they must be high in proportion.

Thirdly. Let the lowest growing shrubs be nearest to the walks, and some flowers in front; except in those walks which require shade, and there tall plants may be put close to the side of the walks.

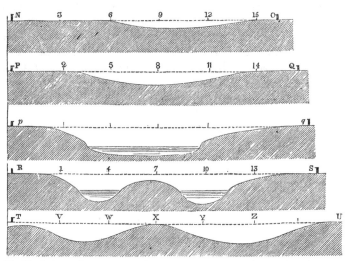

[Fig. 200. Sections on the dotted lines placed lengthways on the plan, fig. 198; the letters and figures on the section corresponding with those on the plan.

N.B. The dotted line from small *p* to small *q* in the plan, and the corresponding section in fig. 200, is only at half the distance between the dotted lines P, Q, and R, S, for the purpose of carrying it through the centre of the pool to shew its depth.]

The water, to irrigate the grass and supply the pool, I understand, is to be furnished from a redundance in the adjoining ground belonging to a convent of Capuchins; and much will depend on the due attention that not a drop shall be lost or wasted. I am happy to add, that this plan was successfully realized, and admired for its effect and novelty in Portugal.

Directions for shaping the Surface with a fall every way towards the Pool.

First, take down the terrace-wall, then put stakes at twenty-five feet apart each way, and mark on each stake the numbers on the plan. Measure the height of the wall to be removed, and divide it into five equal parts, and this will serve as a scale for the depths. The exact height of this wall not having been sent, I suppose it seven feet and a half, making each division one foot and a half; or, if it be only five feet high, then each division will be only one foot.

Dig trenches, crossing each other according to the lines on the plan, and to the depths, corresponding with the sections each way, supposing five feet to the surface of the water; but the pool must be formed and dug out afterwards. The trenches, or pattern lines, being dug two or three feet wide, the whole surface will be easily shaped, by taking away the earth in the intermediate squares.

FRAGMENT XVIII.

UPPARK.

EXTRACT FROM THE REPORT OF UPPARK, THE SEAT OF
SIR HARRY FEATHERSTONE, BART.

SITUATION AND CHARACTER.

BEFORE a rational plan for the improvement of any place can be laid down, it is always necessary to consider its natural situation, and the character which has been given to it by art. The former, at Uppark, is truly magnificent, being on the summit of the south down range of hills ; and when we consider the large masses of wood, the beautiful shapes and verdure of the lawns, with the distant and various views of sea and land, it is difficult to adapt any style of building to such a spot, that may correspond with the great scale of the place.

Of this difficulty the architect seems to have been aware, by the degree of irregularity which was originally adopted in the position of the outbuildings: this is evident, both from the map, and from an old picture on the staircase, wherein the stables and other offices appear to have been placed, not at right angles, but converging from the entrance front.

His reasons for so doing seem to have been well founded. He knew that a correct correspondence of parts in a building tends to diminish its importance; that the Roman style, which was then introduced into England, would not admit of such irregularity; and all that could be done was, to spread out the detached buildings, which produced an appearance of irregularity, when seen from a distance, while the effect of symmetry was preserved in the entrance-court, where the lines converging instead of being parallel, increased the apparent length of perspective.*

THE ENTRANCE.

As the principal object of improvement at Uppark relates to an alteration in the entrance, I shall endeavour to explain the causes which have rendered such alteration necessary.

Before the introduction of Roman architecture into England, all the palaces and large mansions consisted of one or more quadrangles, surrounded by buildings, as at Cowdray, Hampton Court, &c. But at the time when Uppark was built, the fashion of these quadrangles was about to be changed; though, for a long time after, it was continued in the entrance fronts; and, at Uppark, where the entrance was to the east, a *basse cour*, or court, was preserved; and in so lofty and exposed a situation, such an entrance was absolutely necessary to the comfort of the residence; for, however the views might be opened from the other windows of the house, those on the same side with the entrance could command only a view surrounded by walls or buildings.

As the gloomy confined view into a quadrangle, or paved court, forms a great contrast to the cheerful landscape of a park, or more distant prospects, it is no wonder that fashion should open the views in every direction; forgetting that one

* An example of this may be observed in George-street, as viewed from the end of Hanover-square.

3 R

side, at least, of every house must be appropriated to *useful* rather than *ornamental* purposes. In the original plan of Uppark, it is evident, by the little attention given to its appearance, that the architect intended the north front should

[Fig. 201. Uppark, before the new entrance from the park to the pleasure-ground was formed.]

not be seen; but when the buildings, which formed the due importance of the east or entrance front, were taken down, the entrance still continued as before.

[Fig. 202. View of the entrance from the park to the pleasure-ground at Uppark.]

It is difficult to give an adequate idea of the improvement which has been actually executed, without inserting the whole series of sketches and plans by which the report is illustrated, and which consists of nine distinct drawings.

The foregoing sketches [figs. 201 and 202] shew the entrance from the park to the pleasure-ground and flower-garden, which is defended by a wall of flint-work ; and, after passing through the iron-gates, the road continues in the highly-dressed pleasure-ground, till it enters the quadrangle to the north of the house, in which a corridor and portico of Doric columns mark the entrance.

FRAGMENT XIX.

CONCERNING COMBINATION.

I confess myself at some loss for an appropriate title for the subject of this section.   Perhaps it will best be explained by comparing a mansion of the last century with those venerable piles of more ancient date; many of which have been sacrificed to the prim spruceness of that modern fashion which dictates uniformity of style through the whole building, and, consequently, renders it necessary to hide, by planting, all such offices, or appendages, as cannot be made to assimilate with its character.   To this may be added, the prevailing custom of placing a house in the middle of a park, detached from all objects, whether of convenience or magnificence; and thus making a country residence as solitary and unconnected as the prison on Dartmoor.

Let us compare such a scene with the ancient family mansion of two or three centuries ago, and which may, perhaps, have undergone repairs or additions, in different styles, during a long succession of generations.   We shall often observe a combined mass of buildings, irregular in their outline, and, perhaps, even discordant in their style ; but the confused mixture in a mansion, with its offices attached and detached, gives an imposing assemblage, while the church, and even the village, or, at least, some houses of dependants, add to that quantity and variety, without which there can be no real greatness or importance.   It is a mistaken idea, that a place is increased in its grandeur by removing all its subordinate parts to a distance, or out of sight : on the contrary, many of our most venerable palaces are attached, at least on one side, to the neighbouring town; while views into gardens and park scenery are enjoyed from the principal apartments.

Except in the cases of some royal domains, the examples of such venerable edifices are becoming more and more scarce; partly from the prevalence of bad taste, that generally accompanies wealth suddenly acquired, and partly from the propensity which dictates the pulling down and rebuilding, rather than preserving and restoring the ancient specimens of former magnificence. This erroneous practice is further increased by the fashion of detaching the mansion from its natural dependancies, rather than combining it with them. Amongst these, perhaps, there was none deemed more appropriate, in old times, than the church or chapel, or family place of burial and worship. This, so far from casting the gloom which modern times have annexed to such combinations, was formerly considered as an object of pride and pleasure to the living, by combining the associated remembrance of many generations of the same family.

If there be a pleasing association excited by a collection of family pictures, in the hall of an ancient mansion; if we look back, with a degree of pride and satisfaction, on the ancient costume of the stiff portraits of former proprietors, how much more forcibly do the busts, and statues, and recorded deeds of former worthies, arouse the mind to a feeling of respect, when we know that all which can remain is sacredly preserved on the spot where they once enjoyed the regard of their contemporaries! Not having received the permission of the noble proprietor to mention the name of his place, I must take the liberty to explain the subject by the help of a sketch, without giving the name.

Near the ancient seat of a NOBLE FAMILY, the parish church (to which is attached the burial place, used by the family for more than seven centuries) is so close to the site of the mansion, that some may, perhaps, think it too near; but a modern house has been restored on the original site, and is now so intimately connected with the church, the churchyard, and offices belonging to the house, that it is impossible not to be pleased with the combined effect of such a mass of buildings. This I have recommended to be increased, by preserving a picturesque cottage, formerly the parsonage, and adding a cemetery, to form the boundary of the churchyard.

It is remarkable, that, in this neighbourhood, a custom prevails of profusely gilding the tombs and gravestones, which are some of white and some of a black stone; and I suggested the idea of intermixing shrubs and beds of flowers with the gilded and carved ornaments: the novelty of the scene would tend to remove that degree of disgust which ought not to be excited by the emblems of mortality, while we believe in the immortality of man.

FRAGMENT XX.

CONCERNING CONTRASTS.

THE gaudy sketch, which accompanies this fragment, was taken at the moment when a dark and heavy summer's shower was suddenly succeeded by a bright effulgence of light, in a conservatory from which the glass roof had been removed. Although the effect was such as neither this sketch nor any painting can express, it may yet be useful in elucidating the following remarks concerning contrasts. The first contrast here shewn, is, that in the shape of the trees, betwixt the straight, stiff, and upright forms to the right, and those drooping forms to the left; and, though we admire the stately and aspiring character of the hollyhock and larkspur among flowers, with the cedar and cypress among trees, yet, if we turn to the opposite side, we shall confess the justice of Mr. Burke's remark, that a certain degree of weakness is not incompatible with beauty; and that in vegetables, as in the human form, the apparent need of support increases the interest we feel in what is graceful or beautiful.

The sketch first serves to exemplify the contrast * betwixt aspiring and drooping plants, as well as the contrast of colours.

---

* I have occasionally planted, near each other, such trees as the liburnum and the acacia, with weeping birch and willows; adding such flowers and shrubs as gracefully accord, by the pendulous manner of their growth; this makes a source of pleasing variety in our gardens: while, in others, I have collected together all the different species of some beautiful genus: thus, in the thornery at Woburn, are to be found every species of thorn which will bear the climate.

I have, also, endeavoured to delineate, but found it impossible to do justice to, the rainbow, either in its vivid hues, or its transparent effect. I should have wished to give an adequate idea of that harmonious contest, which I witnessed, betwixt the vivid meteor in the sky, and the assemblage of objects seeming to vie with the rainbow in the richness of their colours.

The next contrast I shall mention is that of light and dark, not in shadow and shade, but of a variety in colouring observable in nature, and well worth cultivating in the art of gardening, although difficult to represent in painting. Of this I shall enumerate several kinds.

Firstly. The difference of a leaf with the light *shining full upon it*, which renders it an opaque object, and the same leaf seen transparent, by the light *shining through it.**

Secondly. The contrast produced amidst the more gaudy colouring, by the sort of repose that the eye derives, sometimes from white flowers, as of the jasmine, the passion-flower, and other plants, whose leaves are dark and not glossy : sometimes the same repose is produced by a mass of light foliage, at a little distance, appearing without shape or colour, as in a bed of mignionette.

Thirdly. The contrast in texture ; some plants and flowers appearing as if composed of silk; others, of cloth or velvet; some smooth as satin; others, harsh, rough, and prickly.

Fourthly. The contrast of size; some, like the aloe, the horse-chestnut, or the tulip-tree, bearing their blossoms above the reach of man ; and others, like the diminutive rock-plants, and miniatures of nature, requiring to be raised, or placed on tables, and in flower-pots or baskets. Sometimes plants of the same species assume new dimensions, forming a contrast with their more common measurements; as in the diminutive dwarf Burgundy rose, and the gigantic viola tricolor; which

---

* Having, one day, when at Holwood, pointed this out to Mr. Pitt, as a source of the delight we experience in a sunny day, from an open trellis of vines overhead, or the foliage in the roof of a conservatory, he was so forcibly struck with the remark, that he made several experiments with leaves of different shapes and tints, some of which, from the opaquer ramification of their fibres, or other circumstances of texture, &c. became new objects of delight to a mind like his, capable of resorting to the beauties of nature, as a relief from the severer duties of his arduous situation.

may be used as an example of contrasts in colour, and in relative dimensions.*

The last contrast I shall mention, is that of cloud and sunshine. There is, perhaps, nothing more reviving and delightful, than the sudden effect of a summer's shower, after a long continuance of dry weather : then all nature seems revived : the ground and the plants send forth new and grateful odours ;

[Fig. 203. Effect of sunshine after rain.]

Here Nature's contrasts art attempts in vain ;
Who can describe the joy that follows pain ?
Or paint th' effect of sunshine after rain ?

* In alluding to the contrast from dimensions, I cannot omit some notice of the power of art over nature in this respect.

In China, it is a common practice not only to compress the feet of women, but they have a mode of stinting the growth of trees, by which they can reduce oaks and elms to the size of shrubs in garden-pots, to decorate the decks and cabins of their ships. A curious specimen of this kind of dwarf plant may be seen growing from the roof of a conduit, near the road side, betwixt Hyde Park and Knightsbridge ; where a perfect elm, in miniature, has existed, to my knowledge, nearly half a century, without being now much bigger than a currant-bush. [Trees of this kind are not

The flowers, the birds, the insects, all join to express their pleasure; and even the gold fish in a globe, by their frolic motion, shew that they partake in the general joy; splashing the water, and sometimes leaping out of it, to meet the welcome drops.   An assemblage of contrasts, under such circumstances, I vainly fancied I could fix by my pencil's art [fig. 203]; but a single drop convinced me how feeble is art in her imitation of nature.

> " Si la Nature est grande dans les grands choses,
>   Elle est très grande dans les petites."   ROUSSEAU.

[See page 397.]

Every spray was bespangled by drops, hanging like diamonds, and each changing to all the colours of the rainbow, from whence they appeared to have newly fallen, to cheer the eye, delight the heart, and lift the mind to the contemplation of that source of light and joy from whence alone such beauties and such wonders can proceed.

FRAGMENT XXI.

FROM A REPORT CONCERNING FROME HOUSE, DORSETSHIRE.

A SEAT OF NICHOLAS GOULD, ESQ.

THE character of a place will always be influenced by the style of the house; and I have ever considered, that, without

uncommon on old walls. In 1831, a Scotch pine, not much larger than Mr. Repton's elm, had lived, for many years, on the wall of the corridor, which connects the centre of the house with the wings, at Castle Semple, near Paisley; and, in 1833, there were, on the piers of the court of entrance to Blenheim, from Woodstock, an ash on one pier, and a sycamore on the other.] In England we are apt to err in an opposite extreme, endeavouring rather to increase the size, than diminish it; thus we destroy the original stock (witness the gigantic but tasteless gooseberry of Lancashire), and often injure the flavour, by increasing the size; swelling the pippin to a pearmain, and the nonpareil to a nonsuch. [Opinions of this kind were prevalent twenty years ago; but, in the present day, large fruits, of every kind, are to be found of as exquisite flavour as small ones. In gooseberries, for example, we have the whitesmith, lady Delamere, white eagle; in apples, the new town-pippin, Cornish gilliflower; in pears, the duchesse d'Angoulême, beurré diel, &c.—J. C. L.]

absolute necessity, to destroy an ancient mansion, venerable by age, denotes as bad taste as to erect a modern castle or abbey, where there are no vestiges or pretensions for a building of that description.

About the date of Frome House, there prevailed, in England, a certain character of architecture, holding a middle station betwixt the baronial castle and the yeoman's habitation; it was the *manor-house.** In modern times the habits of life are changed; wealth, from the success of industry or adventure, has frequently become possessed of such ancient mansions; and the rage for novelty has often destroyed all vestiges of ancient *greatness* of *character*, to introduce the reigning fashion for *greatness* of *dimension*. Hence we often see rooms, too large to be warmed, or lighted, or inhabited with comfort; and doors and windows too large to be opened; and sometimes a single house is displayed with a long line of rooms, in which there is not a corner, or recess, to sit in. Our ancestors, when they made large rooms, contrived bays and breaks in its uniform shape; but the modern saloon of (what is called *perfect proportion*) thirty-six by twenty-four feet, must be crammed with tables, and sofas, and instruments, to create some intricacy in this barn-like space.

Having too frequently regretted the demolishing of old mansions, it is a peculiar satisfaction to me to be called upon to improve the ancient mansion of an ancient family: and, while I indulge my fondness for antiquity, I will endeavour to justify the plans I suggest, by answering such objections as may, perhaps, be offered; at the same time that I assign the reasons for my opinion.

The south-east front of Frome House is a specimen of regular, but not enriched, house-gothic; and is in such state of repair as makes it unnecessary to take it down, and unpardonable to replace it by any modern style of building. What remains, however, of the old house, is neither large enough, nor sufficiently convenient for the modern residence of a gentleman's family; being only a single house, without access to any room, except that which was formerly the hall, but now

---

* A most sumptuous specimen of this kind exists in Norfolk (Wolterton manor-house), and has been recently published by the Society of Antiquaries, from drawings by J. Adey Repton.

destroyed, as such, by an intermediate ceiling. It is, therefore, proposed to add so much as will double the present building; and I trust no one can object to this new part corresponding in style with what is left of the old original mansion, by
preserving the character of similar date ; although to the proposal of adding at all to the present house the three following
objections may possibly be urged :—

Firstly. The present site is too low:

Secondly. It is too near the water: and,

Thirdly. It is on the verge of the estate.

To the first I must observe, that our ancestors very judiciously placed their castles on eminences for defence, and their
abbeys and houses in the valleys for shelter : but, in the cham-
paign country of Dorset, it would be absurd to place a new house
on a more elevated part of the property, exposed to every
wind, without a tree to cover it.

Secondly. If the water were a stagnant pool, or one of
those *sheets of water*, as they are called, in imitation of a lake,
it might be objectionable to place a mansion so near its misty
shores : but, where the water is constantly gliding, or in rapid
motion; where a hard, pebbly bottom appears through the lim-
pid stream, and where the banks are not swamps, or bogs, the
current of the stream increases the wholesome current of the
air; and its lively motion constitutes its chief interest. It
should, therefore, be brought close to the windows, in a channel not too deep; as, in such streams, we do not require the
still, sleepy mirror of deep water belonging to a navigable
river, which Milton very beautifully contrasts, as distinct
objects,

"The shallow brook, and river wide."

Thirdly. Where a property is bounded by a natural river,
it is surely advisable to take advantage of so interesting a feature ; especially where islands, or brushwood on the opposite
shore, prevent any nuisance or intrusion from a neighbour;
but, at present, the shores appear more wet and swampy than
they really are, from the willows and aquatic plants, which
have been suffered to grow, in preference to the alders, which
have, not improperly, been called, "the aquatic oak." But
when new channels are dug, and the ground raised on the
island, it will be found capable of bearing all sorts of trees and

shrubs, which will do away the present apparent swampiness of the shores.

There is still one other objection which may, perhaps, be made by those who consider that a house in the country must not only stand in the middle of its own property, but, also, that it should be surrounded by park, or lawn, of great extent. This may be necessary to give the mock importance of space to a modern mansion, but the ancient manor-house generally stood near a public road, and derived its importance from the neighbouring village, or cottages, or pastures, rather than from the destruction of every other dwelling near it, to produce one overgrown grass-field ; or, by " lawning a hundred good acres of wheat," to produce a bald naked park, dotted with starving trees, or belted and clumped with spruce firs, and larches, and Lombardy poplars.

The sketch subjoined [fig. 204], is an accurate portrait of the present view from the south-east front of the house, encumbered by barns, and the water is seen in the furrows along the side of the willow copse. At present, the view is neither land nor water ; but, by digging a channel to connect with the line of the river, the water will become the boundary

[Fig. 204. View from the south-east front of Frome House before it was altered.]

of the dressed ground near the house ; while a pleasing intricacy will be occasioned by the contrasted forms of bridges to connect the several islands ; and the landscape, without being bold, or romantic, will become interesting and picturesque [see fig. 205].

There is a circumstance belonging to the rivulets in Dorsetshire, which requires peculiar treatment. The water of this and the adjoining counties of Hants and Wilts, often consists of small rivers, called *bourns*, some of which are perfectly dry during the summer, and others are so shallow as to be nearly invisible, from the quantity of grass and weeds floating on the surface. Where the water is meant to be ornamental,

[Fig. 205. View from Frome House, as altered.]

it will often be advisable to mow and rake the stream, which requires as much attention as a grass or gravel walk, not to appear slovenly, or overgrown with weeds.

FRAGMENT XXII.

OF ASPECTS AND PROSPECTS.

FROM A REPORT CONCERNING THE SITUATION FOR

WALLWOOD HOUSE,

TO BE ERECTED ON A PROPERTY IN THE PARISH OF LAYTONSTONE, ESSEX, BELONGING TO WILLIAM COTTON, ESQ.

NOTHING is more common than for those who intend to build, to consult many advisers, and to collect different plans, from which they suppose it possible to make one perfect whole. But they might as well expect to make an epic poem, by selecting detached verses from the works of different poets. Others take a plan, and fancy it may be adapted to any situation; but, in reality, the plan must be made not only to *fit the spot*, it ought actually to be made *upon the spot*, that every door and window may be adapted to the aspects and prospects of the situation.

It was a remark of my venerable friend, Mr. Carr, of York, after fourscore years' experience as an architect, that, to " build " a house, we had only to provide all that was wanted, and no " more ; then, to place the best rooms to the best aspects, and " the best views." Simple as this apothegm may appear, it contains more truth in theory, and more difficulty in practice, than all the rules which have ever been laid down in books by architects, or the remarks of all the admirers of rural scenery, with whom I have conversed. The former never think of *aspects*, and the latter think of nothing but *prospects*. I will, therefore, beg leave to enlarge on these two subjects.

I consider the *aspect* as of infinitely more consequence to the comfort and enjoyment of the inhabitant than any *prospect* whatever : and every common observer must be convinced, that, in this climate, a southern aspect is most desirable ; but few are aware of the total difference in the effect of turning the front of the house a few points to the *east* or to the *west* of the south ; because, although the south-east is the best, yet the south-west is the worst of all possible aspects; for this reason, *viz.* all blustering winds and driving rains come from the south-west, and, consequently, the windows are so covered with wet, as to render the landscape hardly visible. My attention was originally drawn to this subject by travelling so much in post carriages, and often remarking the difference betwixt the window to the south-west and that to the south-east, during a shower of rain, or immediately after ; when the sun, shining on the drops, causes an unpleasant glitter, obstructing the prospect, while the view towards the south-east remains perfectly visible.*

---

* At Organ Hall, in Hertfordshire, a seat of William Towgood, Esq., the living-room was towards the south-west, and, during a heavy storm of wind and rain, we accidentally went into the butler's pantry, which looked towards the south-east, where we found the storm abated, and the view from the windows perfectly clear and free from wet; but, on returning into the other room, the storm appeared as violent as ever; and the windows were entirely covered with drops which obstructed all view.

On considering the prevalence of south-west winds, it was determined to reverse the aspects of the house, by changing the uses of the rooms; making a very comfortable house of one which, from its aspect only, was before hardly habitable; since no window, nor hardly any brick walls, will keep out the wet, where a front is exposed to the south-west : for this reason, it has been found necessary, in many places, as at Brighton, &c. to cover the walls with slates, or pendent tiles, and to use double sashes to the windows so situated.

If we had only one front, or one aspect to consider, our difficulty would soon vanish; but the prevailing partiality for variety of prospect seems to require that, in every direction, the views should be retained; and as the opposite walls of the house must be parallel, and the corners at right angles, we must consider the effect on each of the four sides : thus,—

First. The aspect *due north* is apt to be gloomy, because no sunshine ever cheers a room so placed:

Secondly. The aspect *due east* is not much better, because there the sun only shines while we are in bed :

Thirdly. The aspect *due west* is intolerable, from the excess of sun dazzling the eye through the greatest part of the day.

From hence we may conclude, that a square house, placed with its fronts duly opposite to the cardinal points, will have one good and three bad aspects.

Let us now consider the effect of turning the principal front towards the south-east, then the opposite front will be to the north-west; an aspect far better than either due north or due west ; because some sunshine may be preserved, when its beams are less potent than in the west, and the scene will be illuminated by those catching lights so much studied by painters ; especially where, as in the present instance, the landscape consists of large masses of forest trees, and thickets richly hanging down the side of an opposite hill. An aspect open to the north-east would be objectionable, during the cold winds of spring; but, in this instance, it is effectually sheltered by an impervious screen of trees, and large hollies, not drawn across the landscape, but perspectively receding into a deep bay, and forming an admirable defence against the north-east winds ; while the richness and variety of this amphitheatre of evergreens will render the prospect as perfect as the aspect. This warmly sheltered corner will invite the cattle from every other part of the grounds, to enliven the home view near the windows.

It now remains only to mention the side towards the south-west; and, having stated the objection to this aspect, we may consider it fortunate that the prospect, in this direction, is such as requires to be hid rather than displayed ; and, consequently, the detached offices and plantations, to connect the gardens with the house, will defend the latter from the driving storms of the south-west, and give that sheltered and shady connexion betwixt the house, offices, and gardens, which constitutes one of the most delightful *agrèmens* of a country residence.*

While speaking of the three different aspects, I have slightly adverted to their respective views or landscapes; but I will speak further on that towards the south-east, to which all the others may be considered as subordinate, although not sacrificed. [The general result, in regard to aspect, is exhibited in the following diagram, fig. 206.]

It is very common for admirers of landscape or natural scenery, to overlook the difference betwixt a tree and a pole, or betwixt a grove of old trees and a plantation of young ones. We fancy that time will reconcile the difference; but, alas! we grow old as fast as the trees; and, while we dot and clump a few starving saplings on an open lawn, we indulge hopes of seeing trees, when, in fact, we only live to see the clumsy fences by which, for many years, they must be protected. Happy, therefore, is that proprietor of the soil, who becomes possessed of large trees, already growing on the land he purchases; since no price can buy the effect of years, or create a full grown wood: and without that, we may possess a garden, or a shrubbery, but not a landscape. This consideration alone is

[* With great deference to Mr. Repton's opinion on this subject (with which, on the whole, we agree), yet we cannot help stating, that, to a certain extent, we differ from him. In the case of a flat surface, with the views on every side capable of being rendered equally good, then we should have no hesitation in following Mr. Repton's rules; but, in the case of a situation, where there was only a good distant prospect on one side, we would arrange the plan of the house so that the windows of the drawing-room should look to that prospect, whatever might be their aspect. We do this on the principle that a prospect, from the windows, is a feature of the greatest value in a country-house, and one that cannot be created by art; while cold, wind, rain, and excessive sunshine, with all the other evils of bad aspects, may be counteracted by the skill of the architect; and a view, however flat the scene, may always be rendered interesting by the landscape gardener. We cordially agree with Mr. Repton, that the flower garden should always, if practicable, be placed on a warm side of the house, so as to have the morning sun; and, also, that the entrance-front should always be on the worst side of the house, considered with reference to prospect; excepting, indeed, in the case of labourers' cottages, where the entrance-door should generally be placed on the warmest side, in order that the air, which blows in when it is opened, may not be so cold as if it blew from any point north of east and west.—J. C. L.]

sufficient to attach us to the vicinity of that venerable avenue, which it would be a sort of sacrilege to desert, and whose age and beauty will give an immediate degree of importance to the house, which could never be expected in any more open part of the estate.

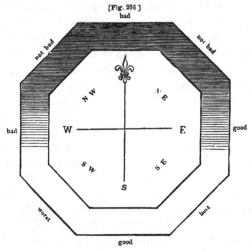

[Fig. 205]

The view towards the south-east, will consist of a glade into the forest, where the distant woods of Wanstead are seen betwixt the stems of large trees in the foreground, producing a purple tone of colouring, so much studied by painters and admirers of picturesque effect. To this may be added, the cheerful moving scene of a public road, not too near to be offensive; for, however some may affect to prize the solitude and seclusion of a forest, shut out from all the busy haunts of men, yet, within six miles of the capital, few places can boast such privacy as Wallwood House commands within its forty acres, surrounded by a forest. Who, then, would regret to see, occasionally, and at a proper distance, the enlivening mixture of man with animal life, and vegetation in its most interesting forms?

From its situation, within so few miles of the metropolis, this place ought to combine all the pleasures of the country, with the conveniences of a town residence.

FRAGMENT XXIII.

OF VARIETY.

In delivering my opinion in a former work, I used an expression respecting the *humility of experience,* which was ridiculed by some critics, but which I now repeat after a lapse of many years. " I do not presume to establish prin-" ciples in taste, but to record my practice, and the motives " which led to it. This I do with all the *humility acquired* " *by experience.*" When I look back on the many hundred places I have visited, and plans I have formed, I can find no two which exactly resembled each other; but where some small similitude might, perhaps, be traced, there ever existed such variety in the circumstances, the wishes, or the characters of the possessors, that it was impossible to class them in such a manner as might lead to general use. Indeed, if we consider that, of the many thousand houses which have been built, no one has ever been exactly copied in the plans for any future house; but, on the contrary, that every plan is either taken from designs which have never been executed, or from the remains of ancient buildings, of whose uses we are almost ignorant; we may ask, *cui bono?* to what good purpose are plans, and designs, and works of art ever published? The only answer that seems plausible, beyond that of amusement to the mind from variety, is, that, by examining and comparing different designs, the best parts of each may be selected; but this is contradicted by every day's experience.

There is no instance in which a good plan has ever been the result of much diversity of opinion: and in no instance, since the effort of Beaumont and Fletcher, has combined genius excelled individual unity of talent or experience.

To the remark, in the preceding Fragment, which recommends " placing the best rooms towards the best views and the " best aspects," I should add, " not placing the entrance on " the same side of the house with the principal apartments:" and thus, after all, it will be found that, nine times in ten, the entrance must be on the north side; and notwithstanding the absurdity of a magnificent portico towards that aspect, where no sunshine can illumine its columns, or require its shade,

3 T

almost all the finest porticos in England are placed to the north; and I have myself, from necessity, been compelled to do so, in many instances, against my better judgment.*

In approaching the close of my active life, it is natural that I should look back on the various objects which have claimed my attention, and called forth my exertions: some of these I can view with delight, and record with exultation; but, alas! in how many have my time, my labour, and my contrivance been employed, without producing fame or profit: the latter was only a secondary consideration, and yet, when that has been withheld, the other has generally suffered in proportion.

By leaving this memorial of some of my works, I shall endeavour to recover a little fame, although I may derive no other emolument: and I insert the annexed sketch of a house [fig. 207], stables, a school-house, a parsonage, and numerous

[Fig. 207. House, stables, &c. alluded to in the following observations.]

plans for buildings, which have been made, approved of, and executed, without our ever being permitted to visit the progress of works on so extensive a scale, and for which we furnished the minutest details, without ever receiving the expected

---

* The late Sir William Chambers asserts, that the entrance to a house, like a nose on a face, was the principal feature, and ought to be the most prominent. Yet in his *own villa*, at *Whitton*, he had five doors in the principal front: that in the centre opened into a shell-work grotto, used as a dairy. Such is the difference betwixt theory and practice, even where the professor may be supposed to have been uncontrolled.

remuneration.   The name both of the place and its proprietor are omitted ; but in the preceding drawing

" Stat nominis umbra."
[Stands the shadow of a name.]

FRAGMENT XXIV.

LONGLEATE, WILTSHIRE,

A SEAT OF THE MARQUIS OF BATH.   EXTRACT FROM THE REPORT OF 1803.

CHARACTER AND SITUATION.

" WHETHER we consider the natural shape of the ground,
forming ample hills and valleys, the great masses of wood
with which these hills have been magnificently clothed, the
extensive range of park and surrounding domain, the vast
command of various distant prospects, or the great style and
magnitude of the house itself, we must acknowledge that the
character of *greatness* makes a strong impression, and is peculiarly appropriated to Longleate.   But as objects are only
great or small by comparison, and it is the duty of the improver to guard against anything that may tend to weaken
this first impression, he must, if possible, increase the apparent *vastness* of the place in all that belongs to *nature*, and
preserve the character of *greatness* in such parts as depend on
the works of art.

" I must here again remark, that there is no error more
common than to substitute greatness of dimensions for greatness of character.   Thus, in landscape, we often see lawns of
great extent, with as little variety or interest as Salisbury
Plain ; and walks and drives of many miles in length, through
shrubberies and plantations, without any change of scenery,
or any diversity of features ; while, in architecture, we occasionally see huge masses without shape or proportion, boast-
ing the ground they cover, or the apartments they contain,
yet with less appearance of a palace than a cotton-mill, or a
manufactory.   I am here led to make a marked distinction
between the improvements relating to *art* and those relating

to *nature*, from the two leading circumstances to which my attention was first called, *viz.* the proper situation for the stables, and the proper management of the water; the latter belongs to landscape gardening, as an art which imitates nature; the former to architecture, as an art that adorns nature, and, indeed, forms the strongest auxiliary to the art of landscape gardening.

" This mansion was built at the period when the Gothic character was giving place to the introduction of Roman and Grecian architecture; and although some would call the house *Grecian*, from its pilasters and entablatures, yet its general appearance is *Gothic*, from the bold square projection of the windows, and the varied outline of the roof, occasioned by the turrets and lofty chimneys, and the open-work enrichments.

" The eight towers in the roof are so placed as to occasion some confusion, or, rather, a certain degree of intricacy in perspective, from whatever point they are viewed; had they been placed at regular distances, the effect of grandeur in this building would have been weakened. To explain this, I must observe that symmetry, or an exact correspondence of parts, assists the eye in viewing and comprehending the whole object at once; but irregularity retards the progress of vision; and, from the difficulty of comprehending *the whole*, its magnitude increases on the imagination. (This subject is further explained in my first printed work, ' Sketches and Hints on Landscape Gardening.') [See p. 25 to p. 116.]

" The hills at Longleate have been boldly planted, and at the same period many fast growing trees were planted in the valleys; these latter were become, in many places, too tall for their situation. There are some lines, and planes, and lofty elms, near the water, in situations where maples, and crabs, and thorns, and alders, or even oaks and chestnuts, would have been far more appropriate : and there are some few tall shattered trees remaining, of the avenue near the house, which tend to depress its importance.*

---

* Since this report was delivered, almost all the objectionable trees have been removed by a spring blight, which destroyed so many planes in every part of England; and the place has been greatly improved in consequence.

" It is a mistaken idea, that the planter may not live to see his future woods, unless they consist of firs and larches, or planes, and other fast growing trees; but every day's experience evinces that man outlives his trees, where plantations do not consist of oak; and that often tall, mutilated planes, or woods of naked stemmed Scotch firs, remind him, that groups of oaks, and groves of chestnut, might have been planted with greater advantage. It is not, therefore, in compliance with the modern fashion of gardening that I advise the removal of a few tall trees, but in conformity to taste, founded on reason, and which dictates that the character of greatness, in a work of art, should not be obliterated by the more powerful agency of nature: and without wishing to go back to that taste which prevailed when this vast pile was surrounded by cut shrubs, and avenues of young trees newly planted, I think some of its grandeur might be restored by judiciously removing some of the encroachments of vegetation.

THE WATER.

" There having been various opinions concerning the management of the water, it may not be improper to state and examine each in the following order.

" The first opinion arises from the natural wish that the water should be in the *lowest ground*, and, therefore, it was proposed to float the valley to the north, making one large lake. To this there are many objections: first, it would not be seen from the house; secondly, if seen, it would not be desirable, being to the north; and, lastly, if it were possible (of which I have some doubts), it could only be accomplished by an enormous dam across the valley: this it would be far more difficult to disguise than the present dam, which only requires to be planted to deceive, and conceal the lower ground; for every piece of water that is made by art to imitate *nature*, must be produced by some degree of deception.

" The second opinion is, that the brook should pass through the valley in a natural channel, instead of being checked by so many different dams, to form so many different pools: the objection to this arises from the supply not being sufficient. Where a rattling, turbulent mountain-stream passes through a rocky valley, like the Derwent at Chatsworth, perhaps Mr. Brown was wrong in checking its noisy course, to produce the glassy surface of a slow moving river; but, as the quantity of water at Longleate might pass through a narrow channel, or watercourse, far beneath the dignity of the place, it ought rather to be carried in a culvert under ground, than be shewn at all in the humble shape and scanty quantity that *nature* has allotted: yet it was a stream of sufficient magnitude for the purposes of *art*, in the ancient style of gardening, when art was boldly avowed; and this stream supplied the fountains, and cascades, and basins, which then constituted the magnificent but artificial scenery near the house. To this may be added, that it supplied the mill, a very important object in old times; and this mill, Leate, gave its name to the place, now called Longleate.

" We next proceed to the third opinion, *viz.* that the water should form an apparent river through the whole valley; this, I believe, was originally the intention of Greenway,* who preceded Brown, and whose fondness for serpentine lines gave the water its present shape, at a prodigious expense. Brown continued the same idea, but reduced the scale of the original design; and though he has, in some degree, produced the effect of a river by various different pools, yet the deceptions are not well disguised, and the part most unfinished is that nearest the house, where the two plans

---

* Or Bridgeman, disciples of Kent, in the first departure from straight lines.

of Greenway and Brown are brought into contact, without being well united or blended together.

" My opinion only goes to the completion of Brown's idea, to imitate nature in the form of a large river, and disguise the art by which this is effected. I will suppose that a large river has always passed through the valley, and, like many large rivers, that it was not originally navigable, but that by art it has been made so to a certain spot, and that near this spot the house was built; under such circumstances a bridge would naturally be placed where rocks present a foundation, and to this bridge, and no further, we may imagine the water navigable.

" The bed of the river being dug so deep as to bring all the water below the bridge to the same level, the house would stand high above the water, instead of appearing on the same level as it does at present.* The shape of the water should be made gradually to swell into the broad river; but as there will then be a disproportion between the channel near the bridge and the broadest part of the river, this might be accounted for by a channel dug near the group of elms, and thus the house would seem to stand on a broad pro-montory, formed by the conflux of two different streams.

" The water above the bridge will not require to be dug any deeper, nor the surface to be much lowered; because the fall at the bridge will fully account for the river being no further navigable than where a ledge of rocks impedes its course; for the largest rivers in the world are interrupted by reefs, called *rapids;* and, therefore, not only at the bridge, but at the junction between the two next pools, this process of nature should be imitated.

" The different levels of the several pools were formerly disguised by plantations; but these, having outgrown their original intention as brushwood, have become trees, between whose stems the deception of the dams is too much betrayed. These screens should be repaired by thorns and alders, so as to produce the general effect of one continued river, as I

* This, in 1814, was completed, and the effect produced exceeds the promise, or any representation made by the drawings. From the ground near the house, there was a fall of only five feet to the surface of the water; that surface has been lowered thirteen feet.

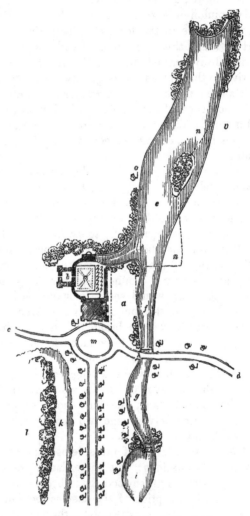

[Fig. 208. Map of part of the grounds at Longleate.]

References.

a Flower-garden in front of the
   house.
b Stables.
c Approach from Bath.
d Approach from Warminster.
e The lake, or large pool, 12 feet
   below the level of the pool, f.
   The pool which has been sunk to
   the level of e.

g The pool, 15 feet higher than f:
   the fall is under the bridge, h.
h The bridge.
i The pool, 10 feet higher than g,
   with a fall and cascade between them.
k Pleasure-ground, leading to the
   kitchen-garden.

l A great wood.
m Oval area, in which the three
   approaches meet.
n, n The old boundary line of the
   pool, e.
o The new boundary line of the
   pool, e.

have represented in all the different views of the water"
[see the plan, fig. 208].

With permission of the Marquis of Bath, the preceding
extracts have been made from the volume of reports concerning the improvements of Longleate. The original MS.
is elucidated by fourteen different drawings, from which is
selected the drawing [fig. 209] representing the south and
east fronts of the house from Prospect Hill, a spot at no
great distance from the Warminster approach. It is difficult
to represent the vast range of country which this hill commands, extending over the whole county of Somerset towards
the Welsh hills, beyond the Bristol Channel. This magnificent park, so far from being kept locked up to exclude mankind
from partaking of its scenery, is always open, and parties are
permitted to bring their refreshments; which circumstance
tends to enliven the scene, to extend a more general knowledge of its beauties to strangers, and to mark the liberality
of its noble proprietor, in thus deigning to participate with
others the good he enjoys.

FRAGMENT XXV.

A PLAN EXPLAINED.

I HAVE frequently observed, that those who perfectly understand a drawing in perspective, have, sometimes, no idea of
a plan, or map, and are not ashamed to confess they do not
understand either. I will, therefore, avail myself of the following letter, addressed to a lady ; and I trust the difficulty
will be removed, by referring to the plate [fig. 210], which
gives an example of both the map of a garden, and the ground
plan of a house, and its appendages.

To LADY * * * *

MADAM,

As you have confessed to me that you never
could understand a plan, I will endeavour to remove your
difficulties, by a reference to the one annexed [fig. 210].
We must begin by supposing that the embroidery at the
corner is a pattern for a flower-garden, the blue patch in the

3 U

[Fig. 209. General view of Longleate from the Prospect Hill; shewing the water as it has been finished, and the surface lowered to raise the house.]

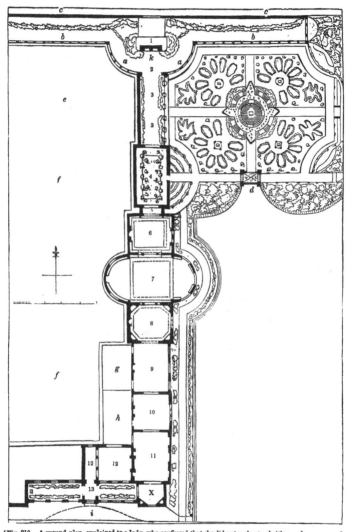

[Fig. 210. A ground plan, explained to a lady, who confessed that she did not understand either a plan or a map.]

References.

1. Aviary, surrounded by the
2. Conservatory and vinery.
3. Flower-passage, glass roof.
4. Orangery, gl ss roof.
5. Lobby, to prevent damp.
6. Tribune, for music-books, &c.
7. Library, or living-room.
8. Tribune, for books of prints.
9. Breakfast-room.
10. Anti-room.
11. Dining-room. old.
12. Hall, lighted from

13. Porch and green-house, visible from dining-room.
aa Arcade of trellis work, covered with grape vines.
bb Range of forcing houses.
cc Flued wall of the hot-houses, which are on one side, while the kitchen-garden is on the other.
d Trellis work in the flower-garden.
e Melon ground.
ff Court, connecting with the stables and other useful courts

such as wood-yards, linen-yards, out-houses, and offices of various descriptions, in the same state in which they were before the old house was altered.
g Billiard-room.
h Stairs.
i Main entrance.
k Large mirror behind the statue of Flora.
x Spot commanding a view of two enfilades. See p. 518.

middle is a fountain, the basin of which is about ten feet wide; the orange lines are gravel walks; the little patches of red and green represent roses, and other flowers, in beds or baskets, standing on the neatest mown grass; and the four circles may be *berceaux*, with hoops to support creepers, or they may be grass plots with vases or statues. I am aware that this will cause some alarm to those who fancy all NA-TURE at variance with ART, and who will exclaim, that it is going back to the old fashioned formal gardening of former days: I answer, by reminding them, that I am not now describing a landscape, but a garden; and " A GARDEN IS A WORK OF ART, USING THE MATERIALS OF NATURE."

Among the infinite variety of flowers which adorn the garden, there are some so minute, that they require being lifted from the ground to meet the eye, and some so formed, that they should be raised even above the eye, to shew their beauties (such as the fuchsia, the American cowslip, and other pendulous plants): to provide for these, we will suppose four beds of fossils, or flints, or rocky fragments, for the reception of that interesting class of plants which requires such a soil and situation; these are described on the plan by purple and yellow. As occasional spray from a fountain might wet the gravel-walk, it should be neatly paved with pebbles round the basin.

On that side of the flower-garden which fronts towards the south, is a house for peaches and strawberries. On the side opposite, and in some degree corresponding, is a row of posts with hoops to train creepers, and an architectural gate communicating with the park, betwixt two projecting lines of shrubbery, which are meant to consist of every kind of thorn, towards the park to the south, and American plants towards the garden to the north. This attention to north and south is very essential, since everything in a flower-garden depends upon its exposure, and, therefore, I must refer you to the compass to ascertain the aspects, of which that to the *north* is cold, sunless, and gloomy; that to the *south* is hot, genial, and cheerful; that to the *east* partakes of both, but requires shelter in spring; and that to the *west* is exposed to more stormy rains and winds than any other; and, therefore, we will suppose the flower-passage, marked No. 3

on the plan, to be defended from the west by a flued wall, and on the side next the flower-garden by glass in spring, but removable in summer; the glass roof may remain constantly. The whole inside of this roof is covered with a wide trellis, to support vines and other climbing plants.

Another sketch [fig. 211] represents this flower-passage terminated by a statue of Flora, which conceals the point of junction in two large looking-glasses placed behind it. In

[Fig 211. Passage, covered with trellis-work and vines, with a flower-border on each side, and behind the statue of Flora a large mirror, as shewn in the plan, fig. 210, at *k*.]

this mirror are repeated all the objects in the passage; nor is the deception discovered, till, on a nearer approach, we find that we can proceed no further in that direction. The passages to the right and left lead to the fruit-houses, or to an aviary at the back of the mirror.

By the help of the plan, let us go back again to the flower-passage, No. 3. No. 4 is the orangery, which is lighted from the roof, and receives only the morning and evening sun, that at noon being impeded by the position of the mansion to the south; but an orangery does not require so much sunshine as a hot-house; and, in the summer, the orange-trees in tubs are placed in front of the house in the two quarters of a circle described on the plan.

The magnificent library, or living-room, consists of three compartments, with a fire-place in each, and a flue near the windows of the bow: the centre is fifty feet by twenty-eight, opening into two recesses, or tribunes, of different shapes, fitted up in very different manners; one being for music-books and instruments, the other for books of prints and portfolios of drawings; and both joined to the large library by a screen of columns, or at pleasure separated from it by drapery and curtains. All this is repeated by a large mirror over the fire-place, which, aided by three apertures for stained glass above the level of the bookcases, prevents this end from being deficient in light, and gives to the whole an air of united cheerfulness, magnificence, and novelty.

On leaving these new rooms, Nos. 6, 7, and 8, we enter the old suit of apartments, Nos. 9, 10, and 11, now opening into each other by large folding doors; and from the spot marked x we have two *enfilades*, one of about three hundred feet, but, in fact, by the mirror of Flora, rendered indefinite; and the other of about seventy feet, along the two green-houses, through the entrance-porch, and terminating either by a statue or fountain, or doubled by another mirror at the end.

On the plan are distinguished, by a light brown wash, the grounds for use and not for ornament, being invisible from the house; and by a green wash, those which are visible from the principal rooms, consisting of landscape and park scenery, where the cattle are prevented from breaking the windows by a dwarf terrace-wall, richly dressed with flowers, which forms the foreground, or frame, of the picture. So magnificent and complicated a plan may, perhaps, appear ideal, but it actually exists, although I have never seen it since I made the plan on the spot.

To explain this, I will relate the following fact. The proprietor called at my door, and took me to the place, to ask my opinion about adding a new room of large dimensions to an old house. I described, by a pencil sketch, the general idea of this annexed plan, with which he was so much pleased, that he desired me, the day following, to explain it to a gentleman, who, I afterwards discovered, was a clerk of the works to an eminent architect. The pencil sketch was all that I was ever permitted to deliver, from which the whole was immediately carried into execution, without having yielded me either emolument or fame, or any other advantage, except the useful lesson—not to leave a pencil sketch in the hands of a clerk of the works. Under such circumstances, I hope I may be excused for claiming my share in a design which I have often heard commended as the sole production of the late proprietor's *exquisite taste*. He certainly made it all his own: but there was not a single idea which I had not furnished.

" Detur suum cuique."
[Let every one have his due.]

FRAGMENT XXVI.

EXTRACT FROM A RECENT REPORT OF A PLACE NEAR THE CAPITAL.

ONE of the most magnificent places in this country, which furnishes examples for the geometric style of gardening, has so recently been submitted to my opinion, that little time has yet been given to the development of plans, to which I shall, therefore, not allude by name; but, in a few years, I trust, there will be no reason for regretting that the following sketches [figs. 212 and 213], and extract from the report, have been allowed to form a part of this volume, intended to record some of the most striking and important of those designs which I have had the honour to deliver.

INTRODUCTION.

CONCERNING THE STYLE AND CHARACTER OF * * * * * *.

If the fashion of gardens could be altered with the same ease as the fashion of dress, or furniture, it would be of less consequence how often it was varied, or by what caprice or whim it was dictated; but the original plan of this place must ever be strongly traced in many parts, though a century has elapsed; and it is impossible to be quite obliterated, in conformity with more modern styles. It is, therefore, an object worthy of consideration, whether the original, or the more recent style, be advisable; and how far both may be admitted, without the incongruous mixture of two things so opposite that they cannot be blended in one rational plan.   I shall call the *ancient style* of gardens that of Versailles, as introduced into this country by Le Nôtre, in the beginning of the last century; and the *modern style*, that called English, as invented by Brown, and practised in England during the latter half of the last century.

THE ANCIENT STYLE.

This consisted in straight lines and geometric figures, and had more reference to art than to nature.   It was distinguished by avenues, or even single straight rows of trees, extended to a great distance, and far beyond the actual limits of the place.   The surface of the ground was cut into slopes, called *amphitheatres*, or raised up into conic shapes, called *mounts;* and even the water was obliged to assume some geometrical outline.   So far from consulting, or following nature, the chief object of art was to display its triumph over nature.   All this had its admirers, and became, at length, so much the fashion, that every garden in the kingdom, whether great or small, was condemned to submit to the same strict rules, till they were brought into ridicule by the admirers of more natural landscape ; as by the satirical allusions of Pope, in this couplet, so often quoted :—

> " Grove nods at grove, each alley has a brother,
> And half the platform just reflects the other."

When every villa had its little symmetrical garden thus laid out, it is not to be wondered that the universal sameness of the design should create disgust in him who could never see anything else, and, at length, that,

> " Tir'd of the scene parterres and fountains yield,
> We find, at last, he better likes a field."

However imposing and magnificent a straight walk may be, between two lines of lofty trees, yet, if every walk about a place be of the same kind, we shall prefer the winding footpath across the forest; and it was, therefore, very natural that the formal and artificial style should have given place to one more free and natural.

THE MODERN STYLE.

This was, in every respect, the reverse of the former. Instead of displaying the means by which art could triumph over nature, it seems to adopt for its motto—"*Artis est celare artem.*" The natural landscape was the chief object to be studied; and while, in the ancient style, every situation, when shut up, became the same; in the modern style, every place open to the country varied with the different surrounding scenery of nature, and, consequently, nature was the model for art to *follow*, but not to *copy :* she was to furnish hints and patterns, but not to be imitated with exact servility. The poet's rule says,

> "To build, to plant, whatever you intend,
> To rear the column, or the arch to bend,
> To swell the terrace, or to sink the grot,
> In all, let nature never be forgot."

From hence it is evident, that the poet no more meant to banish entirely the terrace or grotto of the old style in gardening, than columns or arcades in architecture: but, as mankind always step at once from one extreme to another, so every straight line became curved; and, in contrasting art with nature, it was asserted, that a serpentine line was the true line of beauty, and that nature abhorred a straight line; forgetting that in nature's most sublime works the straight line prevails; as in the apparent horizon of the ocean, and the rays of the sun, which may be broken, but cannot be bent. This favourite meandering and undulating line soon prevailed in everything, whether it was a line of a road, a walk, a canal, or the surface of the ground, or even the fence of a plantation; till, at length, it became as monotonous as the straight line, and every place in the kingdom was alike, whether large or small; from a citizen's villa with two acres, surrounded by a shrubbery and serpentine walk, to the nobleman's park of two thousand acres, surrounded by a belt of plantation and serpentine drive. These, by their uniform sameness, are equally insipid, and have called forth from a modern poet lines not less severe than those of Pope:—

> "Prim gravel walks, through which we winding go,
> In endless serpentines, that nothing shew,
> Till tir'd, I ask, ' Why this eternal round?'
> And the pert gard'ner says, ''tis pleasure-ground.'"

THE LANDSCAPE, BY R. P. KNIGHT, ESQ.

As applied to this place, if all straight lines must be abolished, almost every tree would be sacrificed: and if all the shapes of land or water, which are artificial, must be made natural, the cost would be as much more to undo the work of former times, as the difference between the price of labour and the value of money now, and at the period when these costly works were executed.

VIEW TO THE WEST.

My opinion respecting this view will, I have no doubt, be considered as a dereliction of all the modern notions of taste in landscape gardening. In this view towards the west we can see nothing natural, except the materials which nature has furnished, of land, trees, and water; but all these have been so forcibly brought under the control of art, that they are no longer to be considered as natural objects, any more than the stones and masonry of the house can be considered as natural rocks. The surface of the ground has been shaped to form corresponding and adequate roads of approach, and the trees have been ranged in rows to accompany such roads. The water has been collected into a vast basin by an effort of art, which is avowed in the lofty mound that separates the upper from the lower levels. All these have existed nearly a hundred years; and, whether right or wrong, cannot now be altered; while they afford a magnificent specimen of the ancient style of gardening. The great character of this place must be considered, as it relates to the vicinity of the capital. Those who could treat this splendid palace like the seat of an English country gentleman, at the distance of a hundred miles from the metropolis, would rob it of all its importance, and more than half its interest and beauty. It would be absurd, in this place, to conform to the modern style of placing the house in the centre of its domain, from which everything is banished but the beasts of the forest.   On the contrary, it must be classed with those royal and princely residences, which form the retreats of the great from the court or city : and we do not expect near *a metropolis* anything like perfect seclusion from mankind, either in the palaces of Versailles, Potsdam, or Kensington, any more than in *the metropolis*, as at Carlton House, or St. James's.   To each of these the gardens behind the house may be private, but the entrance-front must be exposed to the public ; and we must not, as in lesser places, consider the entrance of the park as the boundary of the domain : on the contrary, I have always considered the gate which opened immediately into the fore-court, or *bassecour*, as the dressed limit of such palaces.   And if it were possible to exclude from Hyde Park, or Kensington Gardens, the gay assemblage of company which enlivens the scene, we should only produce one dull and cheerless solitude, without a single feature to constitute natural landscape, or to reconcile the mind to artificial rows of trees with their symmetrical formality.

It has been justly observed by a late author on Taste (Lord Kaimes), " that symmetry on a small scale is displeasing; " but, where the object is too large to be comprehended at " once, symmetry assists the eye in developing its parts :" and, therefore, to a very large palace, the richness of a symmetrical parterre is more consonant than a square area of lawn, too small to be fed by flocks and herds, and too large to be considered as a bowling-green.   This, I hope, will be a sufficient excuse for my having advised, or, I should rather say, acceded to the disposing of the area, or garden, in front of this palace.   But there is, also, another reason for it : the principal rooms being raised over a basement story, the interior of this area will be visible from thence ; while the clipped fence, with which such a garden ought to be surrounded, will prevent the public from looking into this private garden, and will exclude even those who actually come into the fore-court, and drive up to the portico.   The contrast betwixt the works of art and of nature will increase the interest of both ; and the foreground may be viewed as a rich carpet spread under the eye, in perfect harmony with the vases, and obelisks, and other works of art, attached to the architectural grandeur of the entrance-front.

THE WATER.

The natural surface about this place is so flat and level a plain, that it must depend for all its interest and beauty on the wood and the water, without which it would be a dreary waste. But these two objects are, at present, unconnected; and the naked banks of the large ponds give rather the appearance of a land-flood than of natural lakes or pools. It will, therefore, be advisable to clothe very amply the heads and banks of the reservoir pool in such manner as may render it unnecessary to alter its form; except that it might be greatly improved by giving it a better connexion with the lake, all view of which is, at present, excluded from the lower level, by the paling along the side of the road, and from the principal floor, by a wood planted in quincunx rows. This wood was originally intended to have been kept low, but it has now outgrown its intention, and not only hides all view of the water, but also the distant prospect of the forest, hanging down to the banks of the lake; and, in the horizon, that view of the metropolis, which, at such a distance, is a more impressive feature, and in perfect harmony with the grandeur of the scene.

The annexed sketches [figs. 212 and 213], may serve to give

[Fig. 212. View from the portico of a villa near London, before it was altered.]

some idea of the magnificent landscape which this villa commands. It is supposed to be taken from the principal front, and shews a part of the alteration proposed in the parterre to

the west, and the improvement of the water towards the south-
west, which is, at present, almost totally hid by the intervening
quincunx of trees [as shewn in fig. 212], which have been
suffered to outgrow the original intention of Le Nôtre, who

[Fig. 213. View from the portico of a villa near London, as proposed to be altered.]

meant it to form a foreground to the picture, which ought to
be seen over it [as shewn in fig. 213].

FRAGMENT XXVII.

GARDENS OF ASHRIDGE.

Of all the subjects on which I have been consulted, few
have excited so much interest in my mind as the plan for
these gardens.   This may partly be attributed to the importance and peculiar circumstances of the place; but, perhaps,
more especially to its being the youngest favourite, the child
of my age and declining powers : when no longer able to un-
dertake the more extensive plans of *landscape*, I was glad to
contract my views within the narrow circle of the *garden*,
independent of its accompaniment of distant scenery.

The large and magnificent palace recently erected in his
best style of Gothic architecture, by James Wyatt, presents
two fronts of more than six hundred feet, of beautiful stone,
by a depth of one hundred and thirty to one hundred and
seventy feet from north to south; and from the richness of its
ornaments, and the quantity of its mass, it must be considered

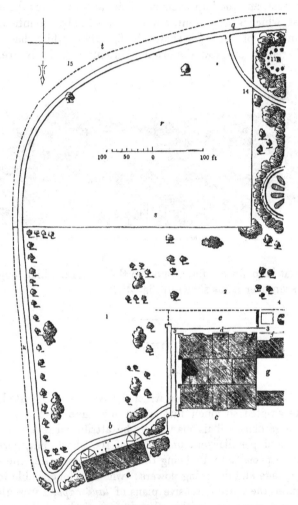

[Fig. 214. Map shewing the arrangement proposed for the gardens at Ashridge.]

References.

a The old house.
b The garden and lawn to the old house.
c Main entrance to the new house.
d Paved terrace.
e Embroidered parterre.
f Conservatory.
g Office-court.
h Chapel.
i Cloister and winter walk.
k Outer court of offices, stables,&c.
l Magnolia conservatory.

m Forcing houses for flowers to supply the conservatory.
n Reserve garden, and place for turning out the green-house plants during summer.
o Heath-house.
p Drinking pool for deer and cattle.
q South terrace-walk, with view to the park.
r A naked flat lawn of about two acres and a half, proposed to be thrown into the park, by

which the deer will be brought nearer to the house, and be more visible. This view being flat and uninteresting, requires this treatment to give animation to the landscape.
s Proposed south terrace-walk and boundary of pleasure-ground.
t The deer park.
u The east mall, with a view to wooded valley in the park.

References continued.

1 Original lawn and pleasure-ground.
2 Addition in the same style.
3 Paved terraces to the house.
4 Broad sanctuary and holy well.
5 Pomarium and winter walk.

6 The monks' garden.
7 Arboretum of exotic trees.
8 Magnolia and American garden.
9 Embroidered parterre.
10 Grotto, and garden for rock plants.

11 Cabinet de verdure.
12 The mount garden, &c.
13 Rosarium and fountain.
14 Connecting interior walks.
15 Open terrace and exterior walks.

as one of the most splended specimens of wealth recently expended under the guidance of taste.

It may, perhaps, be asked by the fastidious antiquary, whether the whole edifice most resembles a castle, an abbey, or a collegiate pile. To which may be given this simple answer :—it is a modern house, on a large scale, where the character of the rich Gothic of Henry VII. has been successfully introduced and imitated. And, knowing the wish of the noble proprietor to direct every part of the improvements, both in the house and grounds, I could not but feel highly gratified on being desired to give my opinion concerning the manner of adapting the ground near the house to the magnificence and importance of the place and its possessor.

The situation of the new house, built over the cellars and foundation of the ancient monastery, has not much beauty of locality to boast; though commanding a very extensive view of park to the south, yet the surface is flat, and without water. Although the park abounds in fine woods and large trees, yet in the view from the windows the landscape is naked and uninteresting.

Under such circumstances, we had only two modes of treating it; either to bear with the nakedness and flatness of the prospect, and enliven it by bringing the deer and cattle near the eye, or else to exclude the landscape altogether, by bringing plantations near the house; and I recommended both these expedients in the manner explained by the map. The boundary fence of the pleasure-ground having been completed before I visited the spot, I have not had sufficient influence to effect its removal; but I was permitted to suggest the plantation of about eight acres, which hides one half of the naked lawn, forming a rich mass of foliage near the eye. It was next to be considered how best to convert the interior of this mass to the purposes of beauty, convenience, and variety, with some degree of novelty in the plan.

Every part of a modern pleasure-ground is alike; and, unless varied by views into the adjoining country, we soon tire of the sameness of gravel walks, in serpentine lines, with broad margins of grass, and flowers, and shrubs, everywhere promiscuously mixed and repeated; and, therefore, I ventured boldly to go back to those ancient trim gardens, which formerly delighted the venerable inhabitants of this curious spot, as appears from the trim box hedges of the monks' garden, and some large yew-trees still growing in rows near the site of the monastery.

I delivered my opinion, elucidated by many drawings, some of which have since been realized, and with some I had hoped to enrich this volume; but, I am informed, the book has been mislaid, and I can, therefore, only describe the general principles of what I had the honour to suggest, by a reference to the map [fig. 214], and a sketch from memory of the rosary [fig. 215] and the conduit, or holy well, for which a Gothic design is given [fig. 216], with a hint of its relative situation, shewing the rosary and entrance to the monks' garden.

No less than fifteen different kinds of gardens were proposed in the map, of which Nos. 1, 2, 3, 14, and 15, belong to the modern style of pleasure-ground, but the others are all different; *viz.* in

No. 4, I proposed a conduit, or holy well, in an inclosure of rich masonry, and decorated by flowers in vases, &c. This is supposed to front the centre of the conservatory.

No. 5, the winter garden, with covered walk open to the south, which is a luxury that no place should be without.

No. 6 is the monks' garden restored.

> " The close clipp'd box, th' embroider'd bed
> In rows and formal order laid,
> And shap'd like graves (for mindful still
> Of their last end, the church doth will,
> E'en in their joys her sons should be
> Pensive in very gaiety)."
>
> HON. MRS. E. ERSKINE.

No. 7, disposed in groups, the various kinds of foreign trees which will bear so sheltered an inclosure.

No. 8, availing ourselves of a very large building, the magnolia, and other American plants, will here find an appropriate situation.

Nos. 9 and 10, are gardens with beds raised to meet the eye, and very unlike any other garden. The grotto is an excavation formed out of an old pool, instead of filling it up; and the whole area of No. 12 has been formed into small hills and valleys, and so surrounded by plantation, that its original flatness is totally disguised. In the rosarium, No. 13, is proposed a fountain, supplied from the holy well, and then led into the grotto, from whence it is finally conducted into the drinking-pool in the park, presenting, from one and the same source, a redundance of water under different appearances.

After almost half a century passed in the parks and gardens
of England, and, during much of that time, having been profes-
sionally consulted on their improvement, I am fully convinced

[Fig. 215. View of the rosary at Ashridge.]

that fashion has frequently misled taste, by confounding the
scenery of art and nature.   And while I have acceded to the
combination of two words, landscape and gardening, yet they
are as distinct objects as the picture and its frame. The scenery
of nature, called landscape, and that of a garden, are as different as their uses; one is to please the eye, the other is for
the comfort and occupation of man: one is wild, and may be
adapted to animals in the wildest state of nature; while the
other is appropriated to man in the highest state of civilization
and refinement. We therefore find, that although painters may
despise gardens as subjects for the pencil, yet poets, philosophers, and statesmen, have always enjoyed and described the
pure delights of garden scenery.
    A garden, as the appendage to a place of such importance
as Ashridge, is no trifling consideration; and it ought well to
be weighed, before we sacrifice one of the most splendid and
costly works of art, to the reigning rage for nature, and all that
is deemed natural.

It will, perhaps, be said, that, where we work with nature's materials, the production should imitate nature: but it might, with equal propriety, be asserted, that a house, being built of rocks and stones, should imitate a cavern.

[Fig. 216. Design for a conduit proposed at Ashridge, with a distant view of the rosary and monks' garden.]

Let us, then, begin by defining what a garden is, and what it ought to be. It is a piece of ground fenced off from cattle, and appropriated to the use and pleasure of man: it is, or ought to be, cultivated and enriched by art, with such products as are not natural to this country, and, consequently, it must be artificial in its treatment, and may, without impropriety, be so in its appearance; yet, there is so much of littleness in art, when compared with nature, that they cannot well be blended: it were, therefore, to be wished, that the exterior of a garden should be made to assimilate with park scenery, or the landscape of nature; the interior may then be laid out with all the variety, contrast, and even whim, that can produce pleasing objects to the eye, however ill adapted as studies for a picture.

If my pencil has given inadequate representations of scenes not yet existing, I may plead, in my excuse, that I am not a painter; and, if I were, my subjects could not be painted; yet they may serve (better than mere words) to realize, and bring before the eyes of others, those ideas which have suggested themselves to my own imagination.

"Segnius irritant animos demissa per aures,
Quam quæ sunt oculis subjecta fidelibus."

["What we hear,
With weaker passion will affect the heart,
Than when the faithful eye beholds the part."
FRANCIS' *Trans.*]

OF ANCIENT GARDENING.

It fortunately happens at Ashridge, that the area proposed to be dedicated to garden and pleasure-ground, is bounded both to the east and to the west by a straight line of lofty trees; these give a character of antiquity and grandeur to the site, and prove it to have existed before serpentine lines were introduced.

I can hardly expect that the sweeping line of wire-fence should be immediately altered; but, as it must very soon perish, it becomes my duty to point out a different line, for the future more durable boundary of the gardens; and this alteration will throw out two or three acres of ground, which must otherwise be kept mown, since no plantations can possibly be made there, without injuring the view of the park. The only use that could be made of these three acres, would be an open cricket-ground, which may either be in the park, or excluded from it, yet appear one surface with the intermediate space of lawn, which I have called the bowling-green. This is an appendage perfectly accordant with the ideal date and character of the building; and would be made still more perfect, by extending the walk from the east terrace, to form the quadrangle complete. These walks may all be considered as part of the original artificial and truly magnificent style of gardening in former times, when the works of art were avowed as artificial, their costliness bespeaking their value.*

OF MODERN · GARDENING.†

When the straight walks and lofty walls of ancient gardening had disgusted by their sameness, prevailing in all places alike, whether great or small, it was naturally to be expected that fashion would run into the opposite extreme, by making everything curved, as the greatest contrast to straight. To the little interest we experience, after the first hundred paces, in a meandering walk betwixt two broad verges of grass, at a great distance from the beds of flowers and shrubs, may be added the mistake of mixing together in such a manner every kind of plant, that no one part of the garden differs from another. Yet there are many pleasure-grounds of this kind, with walks of a tedious length, which I have shuddered to encounter : for this reason I have never advised such walks, except as the connecting lines leading to other objects.

WATER.

The water at Ashridge is, by art, brought from a deep well, dug by the monks, immediately under the chapel; and this must be pumped up into reservoirs. Now, it would be possible to lead pipes from these reservoirs in such a manner that every drop of water used for the gardens should be made visible in different ways, beginning with a conduit in front of the conservatory, and from thence led to supply a *jet d'eau* in the

---

* I cannot here omit mentioning the having been present when Mrs. Siddons objected to the straight braids represented in the celebrated picture in the character of the Tragic Muse, and requested Sir Joshua Reynolds to let the hair flow in more graceful ringlets; but that great master observed, that without straight lines there might be grace or beauty, but there could be no greatness or sublimity; and this same rule applies to gardening, as to painting. It was, therefore, with peculiar satisfaction, that I observed the straight lines of walks near the house, and that mall to the east, in a line with the trees, which Mr. James Wyatt had advised.

† It will, perhaps, be objected, that this same idea has already appeared in the preceding fragment; but such repetitions must occasionally be unavoidable, in a work like this, collected from various detached subjects.

rosary. Another branch might form a falling shower, or drop-ping well, near the grotto; from which the waste-pipes might be led to keep up the water in the park pool, inducing cattle to assemble on its margin; the glitter of this pool might be seen through the stems of trees. But the greatest effect would be obtained from the conduit, or Gothic fountain, near the green-house : this could be thrown up from the well, and the surplus would find its way into the tank beneath. Thus, with actual scarcity, there would be an appearance of great com-mand of water. Perhaps a contrivance might be introduced to filter this water by *ascent*, and make an artificial bubbling fount of the purest and brightest colour. It is not necessary for me to describe the various expedients by which this could be effected, in a place where so much taste and contrivance have already been evinced: all I wish to hint, is, the possibility of making much display of a little water, at the same time losing none. In garden scenery, a fountain is more lively than a pool ; and as the nature of the chalk soil will not admit of those imitations of rivers and lakes, which modern gardening deems essential to landscape; and as, in proportion to the scarcity of anything, it becomes more valuable, it is the duty of the improver to render visible every drop of water that can be obtained : for, besides the pleasure the eye takes in seeing water, we cannot but consider it of the utmost consequence to a garden, where, if the labour of pumping cannot be avoided, it ought to be carried on unseen, lest our choice of the site should be condemned in these words of Isaiah, "And ye shall " be confounded for the gardens that ye have chosen, for ye " shall be as an oak, whose leaf fadeth, and as a garden that " hath no water."

FENCES.

The most important of all things relating to a garden, is that which cannot contribute to its beauty, but without which a garden cannot exist; the fence must be effective and durable, or the irruption of a herd of deer, in one night, may lay waste the cost and labour of many years. Everything at Ashridge is on a great scale of substantial and permanent grandeur, and the fence to the gardens should, doubtless, be the same. The deep walled ha! ha! invented by Brown, was seldom used by him but to give a view through some glade, or to afford security to a terrace-walk; from whence we might see two bulls fighting, without the possibility of danger : this cannot

[Fig. 217. Wire fence called invisible.]

be said of that *wire-bird-cage* expedient, which has, of late years, been introduced to save the expense of a more lasting barrier; and though it may be sufficient to resist sheep, or even cows, for a few years, in the villas near London, yet the mind is not satisfied when a vicious stag approaches it with undaunted eye, and a mien not to be terrified [see figs. 217 and 218]. Add to this, the misery of viewing a landscape through

[Fig. 218. Sunk fence with posts supporting a chain, separating the pleasure-ground from the park.]

a prison-bar, or misty gauze veil ranging above the eye. Besides, iron is a material of which we have had but little experience, except that it too soon decays. For this reason, a line is shewn on the map [*u*, *s*, in fig. 214], which may here-after be adopted; and I must consider the present wire-fence only as a temporary expedient.

I might also add another argument against invisible fences in general (except in short glades), *viz.* that when they divide a park from a garden, they separate two things which the mind knows cannot be united.

In modern gardening it has been deemed a principle to exclude all view of fences; but there are a certain class of flowering plants which require support, and these should be amply provided for in all ornamental gardens. The open trellis-fence, and the hoops on poles, over which creeping and climbing plants are gracefully spread, give a richness to garden scenery that no painting can adequately represent.

The novelty of this attempt to collect a number of gardens, differing from each other, may, perhaps, excite the critic's censure; but I will hope there is no more absurdity in collecting gardens of different styles, dates, characters, and dimensions, in the same inclosure, than in placing the works of a Raphael and a Teniers in the same cabinet, or books sacred and profane in the same library. Perhaps, after all, the pleasure derived from a garden has some relative association with its evanescent nature and produce: we view with more delight a wreath of short-lived roses, than a crown of amaranth, or everlasting flowers. However this may be, it is certain, that the *good* and *wise* of all ages have enjoyed their purest and most innocent pleasures in a garden, from the beginning of time, when the father of mankind was created in a garden, till the fulness of time, when HE, who often delighted in a garden, was at last buried in one.

FRAGMENT XXVIII.

CONTAINING EXTRACTS FROM THE REPORT ON WOBURN ABBEY.

THE improvements I have had the honour to suggest, have nowhere been so fully realized as at Woburn Abbey; I am, therefore, peculiarly obliged to his Grace the Duke of Bedford for permission to avail myself of my original manuscript, by extracting more largely than in any other instance, although such extract can only be considered as a fragment, since the original report consists of ninety pages, elucidated by forty-seven drawings, maps, and diagrams.

TO HIS GRACE THE DUKE OF BEDFORD.

My Lord Duke,

I have the honour to lay before your Grace the following remarks, concerning the further improvement of the grounds about Woburn Abbey. If, in composing this volume, I have had some difficulties, they have arisen less from the nature of the subject, than from my delicacy, as a professional man, making me unwilling to mention, with disapprobation, the works of another. I have, however, endeavoured to do my duty, in conformity to your Grace's instructions, so strongly and so clearly expressed, that I shall repeat the directions by which I have been guided in my consideration of the subject, and which, I hope, will justify my freedom in discussing it.

" Much has been done here, but much remains to be " done, and something, I think, to *undo*. I am not partial to " destroying works recently executed; but sometimes cases " will occur where an alternative is scarcely left. My wish is, " that you should look over everything about the grounds here " attentively, and then freely give me your opinion, as to " what alterations or improvements suggest themselves to your " judgment, leaving the execution of them to my own discre-" tion or leisure."

Such instruction will best plead my excuse for the freedom with which I deliver my sentiments; and if, in many instances, I must condemn what Mr. Holland has done at Woburn, as a landscape gardener, yet, as an architect, the magnificent library, in which this volume aspires to hold a place, will be a lasting monument of his genius and good taste.

The original situation of Woburn Abbey was judiciously chosen in those times when water, the most essential necessary of life, was suffered to take its natural course along the valleys; and before the ingenuity of man had invented hydraulic

3 z

engines to raise it from the valleys to the hills. The great object of the monks was, to take advantage of two small springs, or rivulets, of which the traces are still left in the pools, and shapes of ground, one near the green-house, the other near the dairy. These two streams united a little above the site of the old abbey, contributing greatly to its comfort, by reservoirs and fish-ponds, so requisite to the supply of a numerous ecclesiastical establishment, whose chief food was the fish of fresh water.

It is now too late to inquire why this site was preserved in the present house; or why the residence of a noble family retains the name of Abbey, when every vestige of the original pile has been destroyed. If any mistake is committed, it becomes the duty of the improver to suggest expedients that may retrieve errors, or remedy defects. And since it is impossible to raise the house in reality, or to alter its real situation, we must endeavour to do so in appearance; at least, we should cautiously avoid everything which tends to lessen the magnitude, to depress the importance, or to diminish the character which so obviously belongs to Woburn Abbey, as now altered from a monastic to a ducal residence.

CHARACTER AND SITUATION.

So intimately connected is the character of a place with the situation of the house, that it is hardly possible to separate them in idea; yet it is obvious that, at Woburn, these two circumstances are at variance with each other.

The character of Woburn Abbey (whether we consider its command of surrounding property, its extent of domain, the hereditary honours of the family, the magnificence of the mansion, or the number of its appendages,) is that of *greatness*. To greatness we always annex ideas of *elevation;* and, I believe, in every European language, loftiness of situation, whether literally or figuratively expressed, forms the leading characteristic of greatness; to which we are always supposed to look up, and not to look down. Every epithet applied to it seems to confirm the general opinion, that what is low cannot be truly great; from the exalted sovereign to the kneeling slave, or from the lofty mountain to the humble valley.

But as greatness of character may be distinct from greatness of dimensions, so loftiness of character may exist without loftiness of situation.   The works of art, however great or lofty in themselves, can never be truly so when surrounded by the works of nature, with which they are liable to be compared : thus, the stupendous mass of ruins at Stonehenge is rendered diminutive in appearance by the vast extent of Salisbury Plain.

THE SHAPE OF GROUND.

The surface near the house has been so altered by the various works of art, at different periods, that it is difficult to ascertain precisely what were the natural levels; but it is not improbable that the abbey was originally placed across the valley, or near the conflux of the two small rivulets; leaving a space on one side, if not on both, for the water to take its course towards the west.   As the buildings became enlarged, the valley was lessened, till at length they nearly filled in the whole of the hollow between the two hills [see fig. 219.] *

[Fig. 219.]

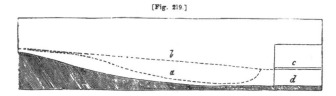

The natural surface is the shaded line.   That natural surface is now altered to the lower dotted line [a]:   the earth had been brought, and filled in, to the upper dotted line [b], making a plane, or, rather, an inclined plane, sloping towards the windows of the south front: [c is the floor of the principal rooms, and d the floor of the basement story.]   If this was done under an idea of giving a natural shape to the ground, the principle was a mistaken one ; for had such been the original shape, we must suppose a hole dug in the ground, in

* Very soon after I had buried the lower story of the house at Welbeck (as described in my volume of " Sketches and Hints, &c.") [see p. 50], Mr. Holland began to do the same thing at Woburn, but never proceeded further than the south front.

which the house had been placed; but the fact is, that, near a large house, the shape of the ground must be made to accord with the building, since no house, however large or small, can be erected without the interference of art, and without disturbing the natural surface of the ground. We must, therefore, study the convenience of the mansion, to which the ground about it must be altered in the way most conducive to its uses and appearance, without fettering the plan by any fancied resemblance of nature. I am quite sure that the old magnificent taste for straight lines, and artificial shapes of ground, adjoining to a palace, was more consonant to true taste and greatness of character, than the sweeping lines and undulating surface of modern gardening.

Such is the convenience derived in the country, from having the principal floor on a level with the ground, that I must highly commend the disposition of the summer apartments at Woburn, where the earth is raised to give a ready communication with the pleasure-ground, without descending a flight of steps. The intention was good, but the mode of execution has proved defective; and had the same idea been continued in the north and west fronts (as once proposed), it would have been fatal to the character of the house, without altering its situation; because it would have reduced it one story in height, a defect for which even the proposed raising of the attics could never have compensated.

VIEWS FROM THE HOUSE.

If the perfection of the art consists in shewing beauties and hiding defects, it must be previously asked, from what point of view any object is to be seen? This may be answered by stating, that the leading features of every place must be considered under the three following heads, or points of view :—

Firstly. As they appear from the windows of the house.

Secondly. As they appear in the approaches to the house; and

Thirdly. As they appear in the walks and drives.

Reversing the order in which these are placed, I shall begin by observing, that, in the last, we are at full liberty to display good features, or avoid bad ones, by altering the course of the drives, &c.

In the approaches we may do the same, yet under certain restrictions, because the roads must lead to the house; but in the prospect from the windows, we have no choice of removing the point of view; it is fixed, and must be stationary: it is, therefore, necessary to study this with peculiar attention, and to ascertain what are the objects most desirable to form this permanent scenery, and how other objects may be introduced, to vary and enliven the same landscape, always seen from the same spot.

THE WATER.

Although the large circular pond at Woburn was originally made by art, yet it has very much the appearance of a Cheshire meer; and I shall, therefore, consider the mode of treating it the same.

There is something so fascinating in the appearance of water, that my predecessor (Brown) thought it carried its own excuse, however unnatural its situation; and, therefore, in many places under his direction, I have found water on the tops of hills, and have been obliged to remove it into lower ground, because the deception was not sufficiently complete to satisfy the mind, as well as the eye. Common observers suppose that water is usually found, and, therefore, is always most natural, in the lowest ground; but a moment's consideration will evince the error of this supposition. Lakes and pools are generally in the highest situations in their respective countries; and without such a provision in nature, the world could not be supplied with rivers: these have their source in the highest mountains; and, after innumerable checks to retard and expand their waters, they gradually descend towards the sea.

If nature be the model for art in the composition of landscape, we must imitate her process as well as her effects: water, by its own power of gravitation, seeks the lowest ground, and runs along the valleys.* If, in its course, the

* Indeed, I have sometimes fancied, that, as action and re-action are alike, and as cause and effect often change their situations, so valleys are

water meets with any obstruction, it spreads itself into a lake, or meer, proportionate to the magnitude of the obstruction; and thus we often see, in the most picturesque counties, a series of pools connected by channels of the rivers which supply them.

THE BASIN.

The large pool, or basin, in front of Woburn Abbey, in its present naked form, is rather an object of splendour than of cheerfulness; yet it is so conspicuous a feature of the place, that it ought not to be given up, without some struggle or endeavour to make it appear more natural.

Could an ample river be obtained through the whole course of the valley, it would, doubtless, be a circumstance worthy of any effort of art to produce it; but the levels of the valleys forbid the attempt: we must, therefore, have recourse to other expedients for retaining the advantage of water, with the least apparent interference of art. The present head, or dam, forms a complete circle of about half a mile in circumference, round which the eye glances in a moment, meeting with nothing to check its progress; and, from the saloon, the ground is seen to fall below the surface of the water. If a tongue of land, or promontory, be formed on the head by the earth to be taken from the south front of the house, two improvements would be carried on at once. This promontory would disguise the dam, and the pool would appear to be the consequence of a stratum of rock, or other hard impenetrable soil, through which the water could not force its way.

The island, the bays, and the channels described on the map and in the drawings, would also contribute to improve the shape of this pool; while the plantations, suggested on the head, would hide the low ground beyond it: for, although water, in nature, is really on high ground, yet, in appearance, it fills the lowest place, because we seldom see ground below

increased in depth by the course of waters perpetually passing along them: thus, if the water only displaces one inch of soil in each year, it will amount to five hundred feet in six thousand years, and this is equal to the deepest valleys. In loose soils, the sides of the hills will gradually wash down, and form open valleys; in hard soils, they will become narrow valleys: but ravines I suppose to be the effect of sudden convulsions from fire or steam, and not made by any gradual abrasion of the surface.

the water; either the descent is so gradual, or the obstruction so bold, as to conceal the different levels, and deceive the sight.

OF BRIDGES AND VIADUCTS.

There are two obvious uses for a bridge; the first is, to pass over; the second, to pass under; the first is always necessary, the other only occasionally so, or where the water under the bridge is navigable; yet, self-evident as this fact may appear, we often observe bridges raised so high, as to make the passage over them difficult, when there is no passage under them required; and this is the case with the present lofty bridge at Woburn. The construction of bridges' has been so often and so ably discussed, that it is dangerous to attempt anything new on the subject; yet I think such a form, adapted to the purposes of passing over, as may unite strength with grace, and use with beauty, is still a desideratum in architecture. The consideration of this subject has led me to insert the following short digression, which will not, I hope, be deemed wholly irrelevant.

Architecture has been classed under two different heads, Grecian and Gothic; the first depends on perpendicular pressure [see fig. 220], the other on lateral pressure [see fig. 221]. By the laws of gravitation, all matter at rest keeps its place by its own weight, or tendency to press downwards; and is only to be removed by superior force acting in a different direction. A perpendicular rock, and a solid wall built upright, will preserve their position so long as the substance, or the materials of which they are composed, retain their power of cohesion; and on this principle all Grecian architecture is founded. Hence have arisen those relative proportions in the different orders, from the heaviest Doric to the most graceful Corinthian, which, after the experience of ages, were deemed fixed beyond the power of improvement; and by these proportions the distances of intercolumniation are regulated according to the strength of the parts supporting and supported.

The arch was rather of Roman than Grecian invention; and, when composed of a semicircle, its construction belongs rather to perpendicular than to lateral pressure; but in every other arch (except the elliptical and the Catanarean arch) there is a great lateral pressure, and this constitutes the basis, or first principle, of Gothic architecture ; to which pinnacles and finials, and other parts, were added for the purpose of strength to the abuttals, although frequently mistaken, and ignorantly copied, as if merely ornamental.

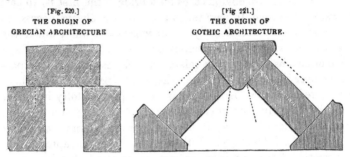

[Fig. 220.]
THE ORIGIN OF
GRECIAN ARCHITECTURE

[Fig 221.]
THE ORIGIN OF
GOTHIC ARCHITECTURE.

The relative pressure of each is shewn by the *dotted* lines.

Hence it is evident that Gothic architecture will admit of larger apertures than Grecian ; and although externally more massive, yet it is internally more capacious. It is requisite that the resistance of the sides, or abutments, should be equal to the lateral pressure of the arch, from the most massive bridge to the lightest roof of an abbey.

Since the discovery of those wonderful excavations in India, of which no date remains, but which have been lately made known to us by the drawings of Hodges, Daniell, and other artists, I have been led to consider, that, besides the Grecian and Gothic styles, there may be a third, distinct from both, the origin of which was very different.

Instead of erecting buildings on the surface of the ground, the people who formed those awful wonders of antiquity began their operations by cutting away the foundation of the rock, to obtain room below, without endangering the superstructure ; and thus, by degrees, the Indian architecture seems to have grown to the most beautiful forms, from the rudest excavations.

The use which I wish to make of this digression, is, to explain the manner of passing the valley by artificial means, where the road does not require a common bridge, but rather a viaduct, to ornament the dam or mound of earth thrown across the valley; and, as this must necessarily cause some obstruction to the water, I think it might be made subservient to the other object in view—that of raising the water to a higher level, instead of digging a deeper channel; thus producing a continued surface of water *in appearance*, though, in fact, the levels *may be very different*.

There being, at present, no architectural form adapted to this purpose, I have ventured to suggest a hint for such a structure as may support the road, and raise the level of the water, rather calling it a viaduct than a bridge [see fig. 222].

[Fig. 222. Viaduct proposed for Woburn Abbey, in order to disguise the dam, or head, which separates two pieces of water on different levels.]

APPROACHES.

As there is no part of the arrangement of the grounds at Woburn where so much alteration seems necessary as in the approaches, I shall take this opportunity of enlarging on the subject, by an inquiry into the cause of the errors so often observable in this most essential part of landscape gardening.

I call the approach the most essential, because it is self-evident, that, if there be a house in a park, there must be a road to it through the park: but the course of the line in which that road should be conducted has been the source of much discussion and difference of opinion. Utility suggests that the road should be the shortest possible: it was for this reason (I suppose) that, in former times, the straight line was adopted, accompanied by rows of trees leading to the front of the house, which was probably the origin of avenues.

4 A

The first grand approach to Woburn was of this kind; but experience having pointed out the monotony of a long avenue, where the house is always seen in the same point of view, Le Nôtre boldly conceived an idea, which was realized at Woburn, at Wanstead, and in the front of some other palaces, *viz.* to obstruct its course by placing a large round basin, or pond, in the middle of the avenue, which not only obliged the road to pass round it, but, by acting as a mirror, shewed the house doubled in its reflection on the surface, and thus increased the importance of its architecture. Such an expedient is beneath the dignity of art, which should display her works naturally, and without puerile ostentation. The straight line in front of a house might be the shortest from the house to the road at one particular spot; but, when it is remembered that approaches are generally necessary from oblique points, it is obvious that they can seldom be brought, with propriety, to one immediately in front.

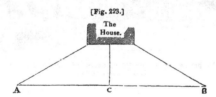

[Fig. 223.]

Those who come from A or B [in fig. 223], will not find C the nearest line to the house : this is sufficient to shew the mistake of some persons, who, in all cases, contend for the old style of approaches by an avenue in front.

When the oblique line was adopted, and a road brought through the park, instead of taking a straight line, it was discovered that, with very little deviation, some interesting parts of the scenery might be shewn in the approach; and, by degrees, its first object, that of being the nearest way to the house, was changed into that of being the most beautiful. Hence have arisen all the absurdities of circuitous approaches, so aptly ridiculed by a modern poet, in describing improvers, who

—————"lead us many a tedious round
To shew th' extent of their employer's ground."

APPROACH FROM LONDON.

Having marked on the ground, and also on the map, the general line, it is less necessary to describe it; but, as there is little difference in the length of the present and proposed lines, it may be proper to assign reasons for the alteration.

The present approach enters through a part of the park, which can be made interesting only by planting all the ground that has been unfortunately cleared of wood; and when the road enters that part of the park, where a few large trees have been left, we perceive that they are too distant from each other, and in an unhealthy state, from the grove having been too hastily thinned. But the most objectionable parts of this approach are the unfavourable circumstances under which the house is first shewn. The first sketch [fig. 224] is a correct

[Fig. 224. Approach to Woburn Abbey, before it was altered.]

portrait of this scene; but, from the difficulty of representing a view down hill, the drawing does not shew the house so low as it appears in reality: it serves, however, to describe the following objections, exclusive of that which gives a bad first impression of the place, from shewing it below the eye.

First. Part only of the south front is visible, which gives an idea of its being a small house.

Secondly. The house is not backed by wood, but opposed to the lawn, which does not form a sufficient contrast to relieve it.

Thirdly. The distant view, though extensive, is not appropriate; it is evidently beyond the boundary of the park, and may be as well seen from many parts of the public road.

Fourthly. The road passes along the side of a sunk fence, and destroys all privacy in the south apartments, which are exposed to every person coming to the house.

Fifthly, and lastly. The immediate and sharp descent near the house, increases the first impression made by its apparently low situation.

The first effect of the house, in the proposed approach, is represented in the second sketch [fig. 225]: after passing

[Fig. 225. Approach to Woburn Abbey, as it has been altered.]

along the great glade, which is terminated by the island in Drakelow-pond, and which, from its length, and the size of the trees, is very magnificent, the road winds among some large oaks, betwixt whose lofty stems the house first appears, partially exciting the attention, till, on our quitting the grove, it is at once displayed to us in all the pomp of greatness, blended with the intricacy of picturesque irregularity. It has no longer the effect of a solitary and inconsiderable edifice, but a palace, of depth proportioned to its front, and accompanied by all the cupolas, and domes, and more elevated parts of those attendant buildings and offices, which it has become the false taste of modern times to hide by plantations. Add to all this, that the whole seems embosomed in a magnificent wood, and, as seen across the valley, it appears elevated and not depressed, while its apparent quantity marks its character as a ducal palace.

THE PLEASURE-GROUND.

By a strange perversion of terms, what is called modern English gardening seldom includes the useful garden, and the name of the ornamental garden has been changed into pleasure-ground. But it is not the name only that has been changed; the character of a garden is now lost in that of the surrounding park; and it is only on the map that they can be distinguished, while an invisible fence makes the separation between the lawn fed by cattle, and the lawn kept by the roller and the scythe. Although these lawns are actually divided by a barrier as impassable as the ancient garden wall, yet they are, apparently, united in the same landscape, and

> ————"wrapt all o'er in everlasting green,
> Make one dull, vapid, smooth, and tranquil scene."
>
> R. P. KNIGHT.

The gardens, or pleasure-grounds, near a house, may be considered as so many different apartments belonging to its state, its comfort, and its pleasure. The magnificence of the house depends on the number, as well as the size of its rooms; and the similitude between the house and the garden may be justly extended to the mode of decoration. A large lawn, like a large room, when unfurnished, displeases more than a small one: if only in part, or meanly furnished, we shall soon leave it with disgust, whether it be a room covered with the finest green baize, or a lawn kept with the most exquisite verdure; we look for carpets in one, and flowers in the other. If, in its unfurnished state, there chance to be a looking-glass without a frame, it can only reflect the bare walls; and in like manner a pool of water, without surrounding plantations, or other features, reflects only the nakedness of the scene.

This similitude might be extended to all the articles of furniture, for use or ornament, required in an apartment, comparing them with the seats, and buildings, and sculpture, appropriate to a garden.

Thus, the pleasure-ground at Woburn requires to be enriched and furnished like its palace, where good taste is everywhere conspicuous.

It is not by the breadth or length of the walk that greatness of character in garden scenery can ever be supported; it is rather by its diversity, and the succession of interesting objects. In this part of a great place, we may venture to extract pleasure from *variety*, from *contrast*, and even from *novelty*, without endangering the character of *greatness*.

THE GARDEN.

In the middle of the last century, almost every mansion in the kingdom had its garden, surrounded by walls, in the front of the house. To improve the landscape from the windows, Brown was obliged to remove those gardens; and not always being able to place them near the house, they were sometimes removed to a distance. This inconvenient part of his system has been most implicitly copied by his followers; although I observe that at Croome, and some other places where he found it practicable, he attached the kitchen-garden to the offices and stables, &c. behind the mansion, surrounding the whole with a shrubbery; and, indeed, such an arrangement is most natural and commodious. The intimate connexion between the kitchen and the garden for its produce, and between the stables and the garden for its manure, is so obvious, that every one must see the propriety of bringing them as nearly together as possible, consistent with the views from the house: yet we find in many large parks, that the fruit and vegetables are brought from the distance of a mile, with all the care and trouble of packing for much longer carriage; while the park is continually cut up by dung carts passing from the stables to the distant gardens. To these considerations may be added, that the kitchen-garden, even without hot-houses, is a different climate; there are many days in winter when a warm, dry, but secluded walk, under the shelter of an east or north wall, would be preferred to the most beautiful but exposed land-scape ; and in the spring, when

> " Reviving nature seems again to breathe,
> As loosen'd from the cold embrace of death,"

on the south border of a walled garden some early flowers and vegetables may cheer the sight, although every plant is elsewhere pinched with the north-east winds, peculiar to our climate, in the months of March and April, when

> " Winter, still ling'ring on the verge of spring,
> Retires reluctant, and, from time to time,
> Looks back, while at his keen and chilling breath
> Fair Flora sickens."
>
> STILLINGFLEET.

The disposition of the gardens at Woburn has now been so far completed, that it would be superfluous to describe their details, because the several objects may be viewed on the spot; but I will briefly enumerate the heads under which they have been classed.

The corridor, or covered-way, by which a sheltered communication is given from the house to the stables, conservatory, flower-houses, tennis-court, riding-house, chinese-dairy game-larder, &c.

The dressed or architectural pleasure-ground, separated

[Fig. 226. The entrance door from the dressed ground to the menagerie at Woburn Abbey.]

from the menagerie by the door [which, on the one side, fig. 226, enters from the pleasure-ground, as] represented in the annexed sketch [fig. 226; and, on the other, opens to a rustic pavillon as in fig. 227].

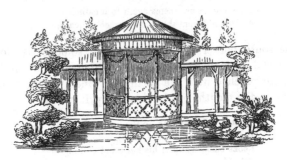

[Fig. 227. Rustic Pavillon in the menagerie at Woburn Abbey, into which the door, represented in fig. 226, opens.]

The forcing-garden has not yet been completed, but, as an object of winter comfort, a sketch [fig. 228] is subjoined.

[Fig. 228. View of the proposed forcing-garden at Woburn Abbey during winter.]

The Chinese buildings were proposed to be decorated by an assemblage of Chinese plants, such as the hydrangea, aucuba, and camellia Japonica.

An American garden was also proposed; and, as an object of curiosity, a botanical arrangement of all the grasses forms an interesting circumstance.

The variety of pheasants and aquatic fowls in the menagerie creates new features in this collection of different scenery.

THE FORCING GARDEN.

If the espalier fruit and common vegetables were provided for elsewhere, one acre of walled garden would be equivalent to eight or ten acres in the common mode of disposing kitchen-gardens. The walls should be placed at such distance from each other, as to admit of a walk near the trees, and of a border for early vegetables between the walk and the part shaded by the next wall; on which a hedge of laurel and lauristinus, or even rhododendrons, might be planted.

The upper part of this garden might be appropriated to every species of forcing: and though it is difficult to ornament the mean, slanting roof of a hot-house, yet, when all other vegetation is destroyed by cold, we may occasionally enjoy the sight of plants protected by art, without disgust at the means by which they are protected.

THE PARK.

So natural is the partiality for extent or greatness of dimensions, that I have constantly been asked, How large is such a park? or, How many miles is it round? And since I visited Woburn, everybody talks to me concerning the length of the park-wall. I can only answer, that I do not estimate places by measurement; and that I never go round the extremity of a place to form an idea of its beauty. With respect to the boundary, whether it be a wall or a pale, my business is to hide it, and not to lead a drive so near as to display it.

In this instance, the fashion of drives has, like all other fashions, passed from one extreme to the other. The ancient drive was in an avenue, through the middle of a park; the modern drive avoids the middle, and skirts round its border; and although two-thirds of the places I have visited, and to which I have suggested improvements, are surrounded by a belt and a drive, yet I must beg leave to repeat my protest against that sort of modern belt by which Mr. Brown's followers have brought disgrace on the genius and good taste of their master.

THE DRIVES.

Before I speak of the drives at Woburn, I shall endeavour

4 B

to trace the progress of fashion in planting;* by which I mean
the various systems adopted at different periods for making
trees artificial ornaments.   The first was, doubtless, that of
planting them in a single row at equal distances [fig. 229, *a*];
and this prevailed in the gardens mentioned by Pliny.

The next step was in doubling these straight rows [fig.
229, *a*], to form shady walks; but fashion, not content with
the simplicity of such an avenue of trees placed opposite
to each other, invented the quincunx [fig. 229, *b*], by
which those straight lines were multiplied in three different
directions.

As the eagerness of adopting this fashion could not always
wait the tedious growth of trees, where old woods existed
they were cut through in straight lines and vistas, and in
form of stars [*c*], and *pates d'oie* [goose feet, *d*], which prevailed at the beginning of the last century.

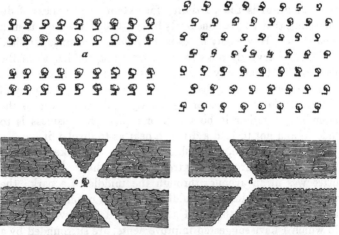

[Fig. 229.   Diagram shewing the avenue, *a*; the quincunx, *b*; the star, *c*; the *pate d'oie*, *d*.]

Fashion, tired of the dull uniformity of straight lines, was
then driven to adopt something new: yet, still acting by
geometrical rules, it changed to regular forms of circles and

[* Several passages in the remaining part of this Fragment, Mr.
Repton had already given in his "Inquiry," &c. (see p. 332); but, as they
are here illustrated by engravings, we have retained them.   J. C. L.]

curves, in which the trees were always placed at equal distances [fig. 230, *e*].   This introduced, also, the serpentine
avenue for a road [fig. 230, *f*], of which there is a specimen in
the approach from Bedford to Woburn Abbey, which is not
unpleasing.

[Fig. 230.]

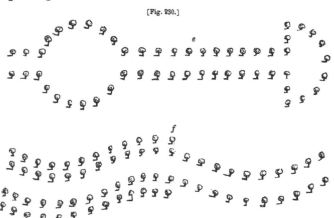

The next bold effort of fashion, was that of departing from
the equi-distant spaces, and trees were planted in patches of
clumps, square [fig. 231, *g*] or round [*h*], alternately shewing
and hiding the view on each side of the road; and where no
view was required, a screen, or double row of trees, entirely
shut out one side [*i*], while, in the other, the view was
occasionally admitted, but still at regular intervals.   This
prevails in the drives at Woburn.

[Fig. 231.]

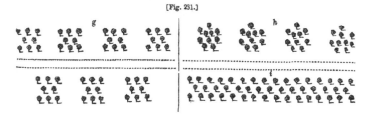

A belt should consist of wood, through which a road may
wind to various points of view [fig. 232: *a, a,* views into the
park; *b,* a view out of the park to the open country]; or

scenery may be shewn under various circumstances of foreground; but a drive should only be among the trees, and
under the shade of their branches; especially where a few large
old trees may help to vary the sameness of a plantation
uniformly consisting of young saplings.

[Fig. 232.]

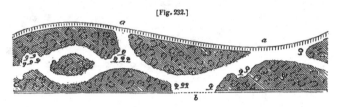

Very different from this is the drive too often adopted,
which is an open drive, so wide that it never goes near the
trees, and which admits such a current of air, that the front
trees are generally the worst in the plantation [fig. 233].
Add to this, that two narrow slips of plantation will neither
grow so well, nor be such effectual harbours for game, as
deeper masses; especially where the game is liable to be disturbed by a drive betwixt them. The belt may be useful as
a screen; but, unless very deep, it should never be used as a
drive; at least till after the trees have acquired their growth,
when a drive may be cut through the wood to advantage [see
fig. 233].

[Fig. 233.]

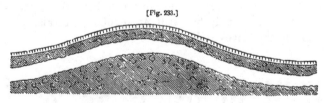

It is not only the line of the modern belt and drive that is
objectionable, but also the manner in which the plantation is
made, by the indiscriminate mixture of every kind of trees,
particularly firs and larches.

All variety is lost, and all contrast destroyed, by different
genera; by the recurrence and monotony of the same mixture
of trees of all the different kinds. And here I must not omit

my full tribute of applause to that part of the drive, at
Woburn, in which evergreens alone prevail: it is a circumstance of grandeur, of variety, of novelty, and, I may add, of
winter comfort, that I never saw adopted in any other place
on so magnificent a scale.

The contrast of passing from a wood of deciduous trees
to a wood of evergreens, must be felt by the most heedless
observer; and the same sort of pleasure, though in a weaker
degree, would be felt in the course of a drive, if the trees of
different kinds were collected in small groups or masses by
themselves, instead of being blended indiscriminately. I do
not mean to make separate groves, or woods, of different trees,
although that has its beauty; but, in the course of the drive, to
let oak prevail in some places, beech in others, birch in a third,
and, in some parts, to encourage such masses of thorns, hazels,
and maple, or other brushwood of low growth, as might best
imitate the thickets of a forest.

It is difficult to lay down rules for any system of planting,
which may ultimately be useful to this purpose. Time, neglect, and accident, will often produce unexpected beauties.
The gardener, or nurseryman, makes his holes at equal distances, and generally in straight rows; he then fills them with
plants, and carefully avoids putting two of the same sort near
each other; nor is it very easy to make him put two trees into
the same hole; he considers them as cabbages or turnips,
which will rob each other's growth, unless placed at equal
distances: yet, in forests, we most admire those double trees,
or thick clusters, whose stems seem to rise from the same root,
and these should be our models in ornamental planting.

FRAGMENT XXIX.

CONCERNING THE LUXURIES OF A GARDEN.

THE fruit or kitchen-garden, as it is generally cultivated,
is little better than a ploughed field, where crops are sown in
drills; for this reason, it has frequently, by Brown, and
always by his followers, been banished to a distance, where it

might no longer be an unsightly object. I have occasionally
found gardens so placed, at two miles from the house, and,
consequently, the choice fruits are removed with as much care
and trouble, in the package and conveyance, as if they came
from Brentford to Covent Garden market.*   What I have to
insert in this Fragment is not the result of any single report,
but is collected from various hints, thrown out at different
places, for the rational improvement of a useful garden, shewing how it may be rendered ornamental; for, though I have
elsewhere asserted that a *ferme orné* is a solecism in language, yet a *jardin orne* may be made one of the most interesting luxuries of a country residence: and this may be
effected in various ways; the most simple, and that which
I have successfully adopted in several villas near London, has
been to surround with a border of shrubs and flowers, three
or four areas of different dimensions, from an eighth to three-
quarters of an acre of garden ground; to raise crops of fruit
and vegetables, perfectly hid from the lawn and walks by the
surrounding screen of flowering shrubs and evergreens;
which, in some cases, will even hide dwarf walls, and pits and
frames for forcing early fruits. By this means, the lawn of
pleasure-ground may be varied in its shapes, and the quantity
of mowing curtailed: and, if we choose to view the interior
of the masses which serve to diversify the landscape, we find
new objects to amuse the eye or gratify the taste, at the time
those fruits are ripe, which are most delicious when gathered
by our own hands.

The interior fence of these patches may be of holly,
roses, gooseberries, or barberries; serving to increase the
mass of screen, and to protect the produce of the garden.
This may be perfectly applicable to places on a small scale;
but, in large establishments, where one garden of several acres
may be preferred, that garden should, if possible, be contiguous to the mansion, and a wood, or screen of shrubs, should
envelop the whole, as it is described on the map of Cobham
Hall [fig. 161, p. 420].

If a kitchen-garden consists of such unsightly crops as we

* I have noticed this error in the extract from the Report of Woburn
Abbey.

see in a common gardener's ground, there will be little inducement to make it one of the visible appendages of a
place; but it may be so arranged as to be highly ornamental;
and, from its sheltering walls, it may always be considered as
a winter garden, when an occasional gleam of sunshine will
invite even the invalid to brave the rigours of the season.

This naturally leads me to observe, how many joys, and
comforts, and luxuries may be preserved, beyond that period
of life when youth and health require no special indulgences.
Having so long dedicated the active part of my professional
career to increasing the enjoyment of rural scenery for others,
my own infirmities have lately taught me how the solace of
garden scenery and garden delights may be extended a little
further, when the power of walking fails, and when it is no
longer possible for decrepid age to reach the ground to
gather fruits, or to pluck, and smell, and admire those humble
flowers which grow near the earth.

The loss of locomotion may be supplied by the Bath chair
with wheels; but, if these are to grind along a gravel-walk,
the shaking and rattling soon become intolerable to an invalid, and, therefore, glades of fine mown turf, or broad verges
of grass, should be provided, as means of avoiding the
gravel; and such grass communications may be so made, as to
increase the interest of the scenery, by varying its features;
for, although a gravel-walk must have its two sides parallel,
or nearly so, yet a grass-walk should never be of any uniform
breadth; it should rather vary in its outline, sometimes
flowing among shrubs, sometimes under trees, as in the
chequered shade of an open grove; and sometimes in one
ample green mall, or terrace, commanding a distant prospect,
a pleasing landscape, or even the curious though confined
combination of rare exotic trees, within the sheltered boundary of the pleasure-ground. All these may be enjoyed by
the cripple, with as much, and perhaps more, satisfaction from
his wheeling-chair, or from a garden-seat, than by those who
can encounter the fields of the farm, or the haunts of the
forest; caring very little for the luxuries of a garden, as
felt under the painful pressure of infirmity. These remarks
are equally applicable to the fruit-garden, the flower-garden,
or the pleasure-ground: they should all be accessible to a

garden-chair on wheels, and all should be provided with
ample grass-walks, to avoid the offensive noise of gravel.

Let us now consider the garden for use, rather than for
beauty, and we shall find that these two objects are by no
means incompatible. The walks of a kitchen-garden are apt
to be uncomfortably exposed to the sun's heat during the
summer and autumn: this may be corrected by training the
fruit-trees of espaliers on hoops over the walks, to make
shady alleys; or covered *berceaux,* from whence the apples,
pears, and plums are seen hanging within our reach; and
grapes so trained will sometimes ripen without artificial
heat. These trellis arcades may be straight or curved, and
the walks may be of gravel or grass, surrounding and enclosing those quarters for garden crops, which, if well
managed, will be scarcely visible from the walks; and a
screen of gooseberries, currants, raspberries, and asparagus
beds, surrounding these, will make a cheerful blind during
great part of the summer months.

If the garden happens to be situated on ground hanging
to the south, it should be formed into terraces one above the
other; and this is particularly applicable to strawberries,
which may then be gathered without stooping: indeed the
same expedient may be used artificially, where the ground is
naturally flat, as represented in the vignette [fig. 234] to this
Fragment. Strawberry-beds may be thus made: confine the
earth at bottom by a brick wall about two feet high, then
slope the mould to the height of three or four feet, and cover
the whole with bricks or tiles, leaving the spaces betwixt for
the roots of the strawberry plants. On the summit, a channel, or trough, is left open in the bricks to receive water,
either from showers or from the watering-pot: the moisture
is conveyed to the roots of the plants without injuring the
fruit, which is, by this means, kept dry and clean, and a little
forwarded by the reflection of the heated bricks: these
should be occasionally secured by mortar, to keep them in
their places. Such a raised bed, when covered with strawberries, either in blossom or in fruit, is one of the most
delightful of garden luxuries; and even when the leaves
begin to decay, it has been observed, in Sir W. Temple's
works, that a most grateful fragrance is produced.

This same sort of raised bed may be used for various purposes, and particularly in a flower-garden, where its form
may be irregular, oval, triangular, octagon, or any other
shape, to raise such diminutive flowers as cannot be seen
without being brought nearer to the eye. The ledges, or
shelves, may be receptacles for ornamental vases, or Maltese
flower-pots; and the lower wall may be covered with jessa-
mine, periwinkle, or such plants as require a little support;
while the upper parts may contain fossils, with rock plants
growing amongst the stones, and falling in festoons over
them. The vignette [fig. 234] only describes a bed of strawberries. Near the end of the bed, a hint is given for training

[Fig. 234. A raised ridge for strawberries.]

gooseberries, currants, &c. to a certain height, to bear their
fruit out of the reach of children, and at a more convenient
height for full-grown persons. In the gardens of Holland,
where such fruits are raised in great perfection, every bush,
as well as every espalier, is trained by hoops, so as to assume
the form of cups, or basins, to admit the sun and air into the
interior, and ripen the fruit. Such attention gives great
neatness to a garden, which ought always to appear trim and
artificial.

The mass of mankind may be so indifferent to the productions of a garden, that they hardly deserve to enjoy its
choicest luxuries. There are many who would not know the
difference betwixt a peach gathered and one that had fallen

from the tree; or betwixt the strawberries plucked from the bed, and those brought from a fruit-shop, perhaps, gathered with unwashed hands the day before.

Of all the places which I have ever seen, where perfect attention was given to the fruit, I was most struck with Wool-laton, in the time of the late Lord Middleton: the grapes were always gathered with a portion of the stalk and leaves; and the bloom of the plums, and other fruit, was preserved from the touch of fingers, by being cut from the trees, and dropped into the baskets in which they were brought to table. The gooseberries and currants, as well as the grapes, were so trained, as to admit branches, loaded with fruit and leaves, to be cut off, and fastened to stands, with iron or wooden hoops, or led in graceful festoons round the dessert, and intermixed with bouquets of sweet-smelling flowers. This may truly be called enjoying the luxury of a garden.

FRAGMENT XXX.

CONCERNING ENTRANCE LODGES AND COTTAGES.

VARIOUS expedients have been suggested, to mark the entrance to a place with importance. A villa, with a few acres, or a park, with an extensive domain, must now be at some distance from the high road, in compliance with the modern custom of placing the house in the centre of the grounds. In such situations, the utility of an entrance-lodge, or cottage, is too evident to require discussing, but its cha-racter may be worth some consideration.

The entrance to a place is generally best marked at any branching off from a public road; and, where the boundary of a park is at some distance from the road, and the entrance a kind of private cross-road, a mere cottage may, perhaps, be sufficient, of any style of architecture, without reference to the style of the house, and a proper gate will distinguish it as an entrance to a place. But, where the gate immediately opens into a park, strongly marked, and bounded by a wall, or park-paling, a lodge seems more appropriate than a cottage: that it should partake of the style and character of the man-sion, seems also to be required by the *laws of unity of design*,

which good taste adopts in every art.   If the architecture of
the house be Grecian, the style of the lodge should be the
same; as in the design for a lodge at Wingerworth House,
page 466, and the annexed sketch [fig. 235], for the entrance

[Fig. 236. Entrance Lodge at Longnor, which is also a school-house.]

to Longnor, where the house is Gothic.   It may be objected,
that the Gothic cottage bears no reference to Woburn Abbey:
but that is not an entrance lodge; it is a cottage near a gate
into a wood, at the distance of some miles from the house.

To mark the entrance to Cobham Hall, the seat of the
Earl of Darnley, built in the reign of Queen Elizabeth, the
style and character of the house proposed to be adopted
in the lodge is not the modern Gothic style, with sharp-
pointed windows, and a flat slate roof just rising over the
battlements, but that which is distinguished by massive
square-headed windows, with pinnacles, mouldings, gables,
escutcheons, and the lofty enriched chimneys of former days,
as shewn [in fig. 236].

To a modern cottage, or lodge, of Grecian architecture,
the gate may either be a light wooden one, between two posts,
or iron folding gates, with brick or stone piers, and it may be
of any fanciful design.   But, if the entrance to a place be
marked by a respectable Gothic lodge, or a correct Gothic
cottage, the gate itself, and even the gate-posts, should also
be of the same correct style of architecture.   I do not mean

the flimsy light deal Gothic gate, frittered with little pointed arches, like a show-box, but the heavy strong oak gate, with massive hinges, and occasionally ornamented with *fleur-de-lys*,

[Fig. 236. Park-keeper's Lodge, Cobham.]

and iron spikes: it should appear to have been constructed at the same period in which the lodge itself is supposed to have been built.

Among the various designs for the entrance into a park, that of an archway is supposed to be copied from those ancient specimens which may still be found near colleges and manor-houses, and in the remains of monastic buildings; but it should be remembered, that such lofty arches are only found when joined on each side by high walls, or attached to buildings surrounding a court-yard. When a lofty archway is seen rising up in the air, being placed at the boundary of a park, and having only a low paling on each side, it is out of cha-racter, and, in fact, bears the appearance of a mere eye-trap, and may be compared to a high gate, or stile, by the side of a gap in the hedge.

The same observation respecting the archway, may also serve for the *gatehouse;* that is, a covered way, with a room over it (which room, in monastic buildings, was called the *scriptorium*). These gatehouses are more appropriate to the court-yards of the mansions, as at Knowle, Penshurst, Hampton-Court, &c., than as entrances to a park. In general, they had large massive close folding-doors, and, sometimes, a small door, or postern, inserted in one of the folds for foot-passengers; and, sometimes, a single door-way separated from the carriage-way, as in the gates of cathedrals, monasteries, colleges, &c. But, with these the modern spruce iron gates will be deemed out of character by all those who have made the antiquities of the country their study, or who consider unity and congruity of design amongst the first principles of good taste.

By J. A. R.

FRAGMENT XXXI.

OF WATER FENCES.

It often happens that a piece of water, whether natural or artificial, forms the boundary of a pleasure-ground near a house, and makes an obvious line of separation betwixt the dressed or mown lawn and the park, or ground fed by cattle. But water is not a sufficient fence on all occasions; for, unless very deep, they will wade through it, and, however deep, deer will swim across. In some cases, as at Woburn and Long-leate, a wall has been built of sufficient depth below the water-surface, to prevent animals from making good their landing: but in the winter this is no fence; and, while the ice remains, they must be either kept out by a temporary line of hurdles, or driven away into a different part of the park; otherwise, in one fatal night, a herd of deer, or a drove of bullocks, will destroy the produce of years in a shrubbery, pleasure-ground, or flower-garden.

The following expedient occurred to me very recently, on considering the view from the windows at Dagenham, in Essex, a seat of Sir Thomas Neave, Bart. The landscape consists of a park, wooded sufficiently, and the distance pre-sents a pleasing offskip; but the most conspicuous feature is a large circular pond, or pool, with naked banks, from which the cattle are excluded by a hurdle, to prevent their poaching the clay shores. The sight of this hurdle is very offensive; but it is rendered doubly so, by being reflected in the water, where it becomes still more conspicuous by its opposition to the sky.   It has long been matter of doubt and difference of opinion, whether it would not be advisable to drain off the water, and make a dry valley, or dell, of this unsightly pool, which, from its situation, reflects nothing but the sky.   Yet, there is something so cheerful in the glitter of water, that we must always give it up with reluctance, however ill-placed, or badly shaped; it was, therefore, decided to preserve this pool, and to make it an ornamental part of the dressed ground near the house.

In the stiff clay soil of Essex, there is great objection to admitting cattle to tread down, or poach, the margin of a piece of water so near and immediately opposite to the win-dows; yet, to exclude them entirely from the pool, would be to rob the water of the most interesting features of which such a pool is capable, *viz.* the reflection of moving objects on its banks, and the glitter of its surface when put in motion. Add to this, if cattle can be kept from browsing the boughs which overhang the water, there is nothing more interesting than the contrast made by dark foliage reflected and opposed to those parts of the margin which reflect only the sky.   An attempt is made, in the annexed plates [figs. 237 and 238], to represent this effect, which also shews the present appearance of the pool, where no objects are reflected except the sky and the line of hurdles [fig. 237].   To realize this landscape, it is

[Fig. 237. Present appearance of the pool at Dagenham, in Essex.]

proposed to fence the opposite bank of the pool by such a line of paling as may sweep round the thickets of thorns, and alders, and brushwood, by which they will be concealed, and then a post and chain should sweep into the pool, just below the surface of the water, admitting cattle to stand on a bank, which should be gravelled, or paved, to prevent their sinking into the soil, and discolouring the water.

[Fig. 238. The pool at Dagenham, as proposed to be altered.]

FRAGMENT XXXII.

CONCERNING IMPROVEMENTS.

I HAVE frequently been asked, whether the improvement of the country, in beauty, has not kept pace with the increase of its wealth; and, perhaps, have feared to deliver my opinion to some who have put the question. I now may speak the truth, without fear of offending, since time has brought about those changes which I long ago expected. The taste of the country has bowed to the shrine which all worship; and the riches of individuals have changed the face of the country.

There are too many who have no idea of improvement, except by increasing the quantity, the quality, or the value of an estate. The beauty of its scenery seldom enters into their thought: and, What will it cost? or, What will it yield? not, How will it look? seems the general object of inquiry in all improvements. Formerly, I can recollect the art being complimented as likely to extend its influence, till all England would become one landscape garden: and it was then the pride of a country gentleman to shew the beauties of his place to the public, as at Audley End, Shardeloes, and many other celebrated parks, through which public roads were purposely made to pass, and the views displayed by means of sunk fences.   Now, on the contrary, as soon as a purchase of land is made, the first thing is to secure and shut up the whole by a lofty close pale, to cut down every tree that will sell, and plough every inch of land that will pay for so doing.   The annexed two sketches [figs. 239 and 240] serve to shew the effect of such improvement; they both represent the same

[Fig. 239.   View from a public road which passes through a forest waste.]

spot; formerly, the venerable trees marked the property of their ancient proprietor; and the adjoining forest, waste, or common, might, perhaps, produce nothing but beauty; now the trees are gone, the pale is set at the very verge of the statute width of road, the common is enclosed, and the proprietor boasts, not that it produces corn for man, or grass for cattle, but that it produces him rent: thus money supersedes every other consideration.

This eager pursuit of gain has, of late, extended from the new proprietor, whose habits have been connected with trade, to the ancient hereditary gentleman, who, condescending to become his own tenant, grazier, and butcher, can have little occasion for the landscape gardener: he gives up beauty for gain, and prospect for the produce of his acres.   This is the only improvement to which the thirst for riches aspires; and, while I witness, too often, the alienation of ancient family estates, from waste and extravagance, I frequently see the same effect produced by cupidity and mistaken notions of sordid improvement, rather than enjoyment of property.   But, to whatever cause it may be attributed, the change of property into new hands, was never before so frequent; and it is a painful circumstance to the professional improver, to see his favourite plans nipped in the bud, which he fondly hoped would ripen to perfection, and extend their benefits to those friends by whom he is consulted.

In passing through a distant county, I had observed a part of the road where the scenery was particularly interesting. It consisted of large spreading trees, intermixed with thorns: on one side, a view into Lord * * * *'s park was admitted, by the pale being sunk; and a ladder-stile, placed near an aged beech, tempted me to explore its beauties. On the opposite side, a bench, and an umbrageous part of an adjoining forest, invited me to pause, and make a sketch of the spot. After a

Fig. 240. View after the forest waste had been enclosed, and the ground subjected to agricultural improvement]

lapse of ten years, I was surprised to see the change which had been made. I no longer knew, or recollected, the same place, till an old labourer explained, that, on the death of the late lord, the estate had been sold to a very rich man, who had *improved it;* for, by cutting down the timber, and getting an act to enclose the common, he had doubled all the rents. The old mossy and ivy-covered pale was replaced by a new and lofty close paling; not to confine the deer, but to exclude mankind, and to protect a miserable narrow belt of firs and Lombardy poplars: the bench was gone, the ladder-stile was changed to a caution against man-traps and spring-guns, and a notice that the foot-path was stopped by order of the commissioners. As I read the board, the old man said,—" It is

very true, and I am forced to walk a mile further round, every
night, after a hard day's work." This is the common consequence of all enclosures: and, we may ask, to whom are
they a benefit?

> " Adding to riches an increased store,
> And making poorer those already poor."

FRAGMENT XXXIII.

EXTRACTED FROM THE REPORT ON SHERRINGHAM BOWER, IN NORFOLK,
A SEAT OF ABBOT UPCHER, ESQ.

SITUATION.

It may appear a bold assertion, to pronounce that Sherringham possesses more natural beauty, and local advantages,
and is more capable of being rendered an appropriate gentlemanlike residence, than any place I have ever seen. I must
here premise, that I do not estimate places by their measure,
or value, each of which may be applied separately to the diamond and the millstone; and, in comparing it with other
places, I must confine myself to those which command views
of the sea, that being always the leading feature of the scenery
of Great Britain, as an island. The most celebrated places of
this description which I have seen, are, Mulgrave Castle in
the north, Tregothnan and Mount Edgecumbe in the west,
and various places in Sussex, and the Isle of Wight in the
south; yet, much of the celebrity of these places may be
derived from the permission liberally given to have them seen
by the public; and, indeed, the boasted beauty of the Isle of
Wight is associated with the moving from one spot to the
other, and the cheerful animation of its visitors and tourists;
for, if we take any one place in that tour, and can suppose it solitary, and divested of this enlivening circumstance, it cannot
be compared with the scenery of Sherringham, where the combination of hill and valley, wood and sea views, continually
remind us of being in that beautiful little island, without the
occasional difficulty of having the water to cross in our return.
Much of the interest in the scenery of the Isle of Wight is indebted to the circumstance of its being visited only in summer,

when the gay decorations of the gardens, whether belonging to a palace or a cottage, present an assemblage of elegance and comfort, in which Sherringham is at present woefully deficient; but which it is the object of these pages to provide. I recollect, when I first visited the Isle of Wight, a continued series of fine weather, amidst the profusion of roses, and other fragrant shrubs, operated on my senses like a charm, till, on opening a door in one of the most delightful retreats, the sight of cloaks and umbrellas made me exclaim, " Can it ever rain in Paradise?" In considering Sherringham as a permanent residence, and not as a mere summer villa, we must recollect how it may appear in winter.

THE SEA.

The view of the sea at Sherringham is not like that of the Bay of Naples, or even the southern coasts of England: the sea is to the north; and, however delightful the summer visitors of the Norfolk coast may deem such a view, those who have experienced the cold north winds of winter, and even those who contemplate their effect on some of the oaks opposed to their violence, will be fully aware, that a view of the sea, from the house, ought not to be the first consideration; and will pity the bad taste of any one who should recommend a site for a mansion looking towards the sea—

> " Qui . . . . semper amabilem,
> Sperat nescius auræ
> Fallacis."
>
> [" Who . . . . always amiable,
> Hopes thee, of flattering gales
> Unmindful."]   MILTON's *Trans.*

THE WOODS.

The effect of these woods I consider as accidental, rather than designed, since it is obvious, that the only rule observed, was, to plant such land as would not bear corn, and, consequently, all the hills have been clothed boldly; and, fortunately for their beauty, the value of the timber was not so much considered as that of the land; since it may be observed, that the trees in the valleys have grown much taller

than those on the hills, and the walks and drives have been
made through these woods, yet their comparative interest is
hitherto unknown, since few can distinguish betwixt what
they do see, and what they might see. Some trifling changes,
in the course of these walks, would prove how easy it is, by
the proper line of a path, to make it not only beautiful
in itself, but a beautiful display of other beauties; or,
as De Lille expresses it,

> " Les sentiers, de nos pas guides ingenieux,
>   Doivent embellir en nous montrant ces lieux."

[The paths, those ingenious guides of our steps,
In shewing, ought to embellish the different points to which they lead.]

For this reason we may assert, that the treasures of Sherringham are yet in the mine; and, from the present site of
the house, almost useless.

PLANTATIONS.

Some have asserted, that it is more pleasant to make
improvement by the *axe* than by the *spade;* but I consider it
a fortunate circumstance that some further planting is necessary, since I may venture to affirm that, after a few years, the
proprietor will derive more real satisfaction from the trees
planted by himself, than from those which have long existed.
All planters delight in woods reared by themselves, as parents
are most fond of their own progeny.

In making new plantations, some useful hints may be
taken from that great variety which at present exists, either
from accident, by being planted at different times, and of different sorts of trees, or, perhaps, from the influence the sea
may have had in destroying some and checking others; for,
although in some places the sea has proved an enemy, yet
it points out what trees are best adapted to the situation and
exposure.

THE SITE FOR THE HOUSE.

This is an object most important in landscape gardening,
yet there is none so often mistaken, or misunderstood, because
mankind are apt to judge by the eye, rather than by the
understanding, and oftener select objects for their beauty,
rather than for their use or intrinsic worth. The experience

of the inconveniences to which most beautiful situations are liable, has induced me to view the subject in all its bearings, and well to weigh against each other all the advantages and disadvantages which ought to influence our choice: these I have generally classed in the following order:—the *aspect*, the *levels*, *objects of convenience*, and, lastly, the *views from the house*.

Firstly. The *aspect*. There can be no doubt that a southern aspect is the most desirable for rooms which are to be occupied throughout the year, because the sun, in winter, is always acceptable, and, in summer, it is so much more elevated, that it is rarely objectionable, and easily shaded. This is not the case with the eastern or western aspect, where the rays, being more oblique, are not to be shaded but by obliterating the prospect; and, as the prevailing winds, with rain, generally come from the south-west, a little turn towards the south-east is to be preferred. This I propose at Sherringham, and, for two other reasons, it makes the view towards the opposite woods more central; and it gives more room for the offices and appendages proposed towards the west. A northern aspect is seldom advisable, except in mansions on a very great scale, or in Cornwall, or on the southern coast, where it is generally preferred to the sea exposure; it will, therefore, I trust, be acknowledged, that the site is *perfect* as far as it relates to *aspect*.*

Secondly. The *levels*. This is an object of much more importance than is generally supposed. We frequently see houses placed, for the sake of the prospect, so high, that they are annoyed by every wind; and others, for the sake of shelter, so low, that they are flooded by every heavy fall of rain, or by the sudden melting of the snow. The site here proposed is on a sufficient eminence to enjoy prospect, and yet to be sheltered from the sea-winds: the ground, by *nature*, falls gently from it in every direction, except towards the north; and, in that direction, it will easily be made to do so by *art*: this is necessary to prevent any damps from the hill, and to provide a sufficient drainage for the house and offices,

---

* The reasons for a south-east aspect were before given, in Fragment No. 22, although here repeated to preserve this Fragment entire.

all of which will require very little cost, or labour.   Thus, I
trust, I may pronounce that the site is perfect with respect to
its *levels*.

Thirdly.   *Objects of convenience,* of which the first is the
supply of water.   This is an object of great importance, yet I
have frequently seen large houses placed where no water can
be had, but by aqueducts, or distant land-carriage ; and, as it is
not only for the constant use of the family, that water is essential, but as a security in case of fire, some great reservoir, or
tank, ought always to be provided near the house.   From the
situation of the two ponds near the site, we have reason to
expect water will not fail.

Fourthly.   *Sufficient space* to contain all the numerous
appendages of comfort and convenience, as offices and office-
courts, stables, and yards for wood, coals, linen, &c., all which
should be near; and others, at no great distance, such as
kitchen-garden, melon-ground, poultry-yard, timber-yard, ice-
house, &c.   These, if possible, should be on the contrary side
of the house to the flower-garden, conservatory, and pheasantry, which are naturally connected with pleasure-ground.

Fifthly.   *Relative objects,* or such as, though not immediately belonging, must be considered as relating to the place,
and, therefore, must be properly connected with it, *viz.* the
post-towns, the church and village, and the sea ; to all which
there must be roads, and these may be made highly ornamental, useful, and convenient, or the contrary.   It fortunately
happens, that the three roads from Aylsham, Holt, and
Cromer, all meet at the summit of the hill, from whence the
public roads descend steeply towards the two towns of Upper
and Lower Sherringham; and, at this spot, I propose to enter
the premises, and proceed to the house along a line of approach, the most easy, natural, varied, and beautiful ; and, as
it is nearer than the public road, it places the new site exactly
at the same measured distance with the old one from the three
post-towns, while the apparent distance will be shortened by
a mile, because we are apt to consider ourselves arrived at a
place as soon as we have passed the gate of the grounds, or
woods, or parks, belonging to it.

Sixthly.   *View from the house.*   Although, with many,
the views from a house form the first consideration, yet I am

not so infatuated with landscape as to prefer it to any of the objects already enumerated. Perhaps, a natural habit of cheerfulness operates too powerfully on my mind; but I have ever considered the view of trees and lawns *only*, as creating a certain degree of gloom, which, I am convinced, is oftener felt than acknowledged by the possessors of places admired for their solitary grandeur.   We are apt to lament the desertion of such family mansions for the residence of London, in winter, and watering places in summer; but we should consider the difference betwixt the country gentleman's seat, when only separated from his neighbours and dependants by court-yards or garden-walls, and the modern fashion of placing the house in the middle of a park, at a distance from all mankind,

> " Where only grass and foliage we obtain,
> To mark the flat, insipid, waving plain;
> Which, wrapt all o'er in everlasting green,
> Make one dull, vapid, smooth, and tranquil scene."
>
> KNIGHT's Landscape.

To this might be added, that,

> " Now not one moving object must appear
> Except the owner's bullocks, sheep, or deer,
> As if his landscape were all made to eat;
> And yet he shudders at a crop of wheat."

For, in the present taste for park scenery, a corn-field is not admissible, because every fence must be removed, except those which are most offensive, such as separate woods and lawns.   In the principal view to the south, this modern taste may be indulged to the greatest excess, by " lawning a hundred good acres of wheat;" but I should not advise the extending the verdant surface too far, as I consider the mixture of corn-lands with woods, at a distance, more cheerful than grass, because, at certain seasons, at seed-time and at harvest, it may be enlivened by men as well as beasts.   I hope I may be here allowed to indulge my favourite propensity for humanizing, as well as animating, beautiful scenery, by a hint respecting the future occupation of Sherringham.   It has already been observed, that the places most celebrated for their beauty are most known to the public; but " many a gem

of purest ray serene" is locked up in the casket, lest man
should breathe upon it; let me hope to unlock this treasure
occasionally to strangers. I do not mean to destroy the comfort
and privacy of Sherringham, by admitting, near the house,
all the tourists and felicity-hunters of Cromer and the coast,
but at such a distance as the temple. One day in the week
might be granted them to share in the beauties of this spot.
This occasional glitter of distant moving objects, with the
sight of carriages coming to the house, would furnish lively
features to contrast with the quiet, yet appropriate view from
the house towards the south.

THE VILLAGE.

Notwithstanding the modern fashion of placing a house in
the centre of a park, at a distance from the haunts of men, or
even the habitation of its own dependants and labourers, yet
there are numerous objects belonging to a village with which
the mansion must be connected, such as the church, the inn,
the shop, the carpenter, the blacksmith, and other tradesmen,
to which may here be added, the farming premises and the
steward's house.

The vicinity of a village is very differently marked in different parks. In some, I see lame and blind beggars moving
sorrowfully towards the hall-house, where I know, and they
fear, no relief will be given; in others, I see women and children, with cheerful faces, bearing their jugs, and milk, and
provisions, at stated periods: and I know, before I enter the
house, which are the happiest families. In some places, I
hear complaints, that the neighbours are all idle thieves and
poachers; in others, that all the inhabitants of the neighbouring villages would rise at night to serve their liberal
patron: and I have been often led to consider the source of
this difference. Formerly, the poor labourers on an estate
looked for assistance, in age or sickness, to the hand that paid
for their work when they could work; now, they are turned
over to the parish-officer, and prisons are erected, under the
name of workhouses, for those who are past all work. A common farmer, who works as hard as his labourers, and with them,
is considered as one of themselves; but, when a very opulent
gentleman farmer told me that, by rising at four o'clock every

day, and watching his men all day, he could get more work done, I thought he paid dearly for it; and whether the poor slave is urged on by the lash of the negro-driver, or the dread of confinement in a workhouse, he must *feel* that man is not equal, though he may be taught to *read* that he is so.

I consider the proximity of the town of Sherringham as a mine of wealth, a source of infinite interest, more valuable than the interest upon interest of the usurer. The workhouse, instead of an object of terror to the poor, and of disgust to the rich, may be made to look more like an hospital, or an asylum, and less like a prison: the street may be improved, and a cheerful village-green, with benches and a maypole, may be laid open to this house of industry: this will remind us that happiness may be extended to all ranks of mankind. The labourers' cottages, belonging to the estate, may be marked by neatness, and decorated by those ornamental shrubs and creepers, which make the whole Isle of Wight a garden so enchanting to strangers.

Instead of forbidding all access to the poor, in some places, I have observed it is customary, one day in the month, or oftener, if necessary, particularly after any storm of wind, to admit into the woods, but under the eye of the keeper, all persons belonging to the parish, to pick up dead wood for firing; and, in these places, no wood is stolen, and no trees are lopped and disfigured. With respect to the game, which is everywhere, and particularly in Norfolk, the perpetual source of suspicion and temptation, I foresee, that at Sherringham it will be one source of conferring happiness; for there is a great difference betwixt shooting and coursing: one is a selfish, the other a social, enjoyment. The villagers will occasionally partake in the sport, like those where the games of cricket or prison-bars are celebrated; thus promoting a mutual endearment betwixt the landlord, the tenant, and the labourer, which is kept up with little expense, securing the reciprocity of assistance of each to the other, by a happy medium betwixt licentious equality and oppressive tyranny.

Although the local advantages of Sherringham may not be deemed of general interest, yet their consideration has had great influence with me, in the preference I have given it over every other in which I have been consulted.

4 E

Firstly. It is situated within half a mile of the sea, without being exposed to it.

Secondly. It is within a morning ride of the capital of the county, and within an hour's drive of the several post-towns and market-towns of Holt, Aylsham, North Walsham, and Cromer; at each of which are balls and book-clubs, besides the monthly meetings of magistrates, and annual fairs and festivals, where society and amusement may occasionally be had, to vary the monotony of rural life.

Thirdly. The soil is delightful for habitation, being neither so light and sandy as to be barren, nor so rich and wet as to make the roads impassable.

Fourthly. It is within five or six miles of the sea-port of Blakeney, to the west, and of Cromer, to the east, where those who do not object to the mixed company of a watering-place, may partake of its variety.

Fifthly. There is no manufactory near; this, for the comfort of habitation, is of more importance than is generally supposed; manufacturers are a different class of mankind to husbandmen, fishermen, or even miners: not to speak of the difference in their religious and moral characters: the latter from being constantly occupied in employments which require bodily exertion, and their relaxations being shared with their families and friends, become cheerful and contented; but the former lead a sedentary life, always working at home, and seeking relaxation at their clubs, the birth-place and cradle of equality, discontent, and dissatisfaction.

Sixthly. There are some who consider that no place can be perfect without water, while others do not consider the sea as water belonging to a place, but as its boundary: in answer to both these fastidious hypercritics, it may be observed, that here two lively brooks flow through the estate; and, that, in a distant recess in the woods, a small pool exists, which might be increased to any size.

I cannot help considering Sherringham as deriving a degree of advantage from what some will think the reverse, because they not only look to increasing the value, but the quantity, of an estate. This is bounded by property to the west, that is not to be purchased; and, therefore, like the boundary of the sea to the north, it so far fixes a boundary to

our wishes. Men are apt to indulge the vain hope of making
all they see their own, like children who cry for the moon, or,
like dogs who bark at it.

WALKS, DRIVES, AND STATIONS.

The natural shape of the surface is so infinitely varied,
that it is impossible, from a map, to form any idea of the
scenery; and drawings can but feebly represent a few of those
landscapes which change at every step. The peculiar character of Sherringham is beauty, without any of that sublimity which is derived from horror, as on the brinks of
rocky precipices in mountainous regions: yet there is a sublimity attached to the sea, which is here softened into the
character of beauty, and forms the leading feature of Sherringham. For this reason, I do not advise that degree of
softness and artificial smoothing, described by Mr. R. P.
Knight, in his attack on Brown's followers,—

> " To improve, adorn, and polish they profess;
> But shave the goddess whom they came to dress."

There are some few trees which have felt the terrific effects
of winter winds; these cannot be removed without endangering others, to which they have long been the advanced guard;
and, so far from wishing them trimmed, or otherwise reduced
to a softer character, I consider them like a dry rugged channel of a winter's cataract, leaving, in summer, sublime memo-
rials of the power of Nature's mighty agents: they form a
contrast to the generally prevailing forms of beauty; and, for
this reason, I have recommended, near the same spot where the
ridge of the hill is to be cut down a little, to ease the descent
of the road, that its banks should be left steep and abrupt, and
not smoothed and turfed over; since a road is an artificial object, and may be avowed, in such cases, as a work of art. As
this chasm, dividing the land from the sea views, will be one
of the most striking stations at Sherringham, I have called it
the Scalp, from a noted scar of this kind in Ireland. The annexed view [figs. 241 and 242] is supposed to be taken from
this spot.

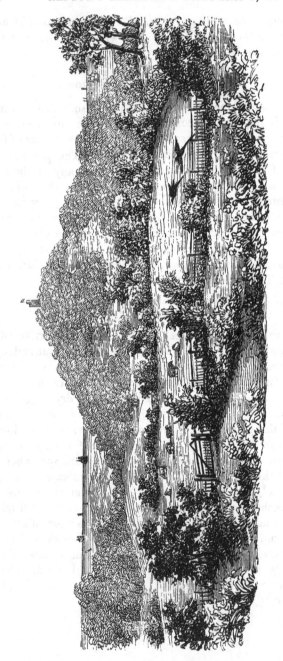

[Fig. 241. View of the grounds at Sherringham Bower before Mr. Repton's improvements were commenced.]

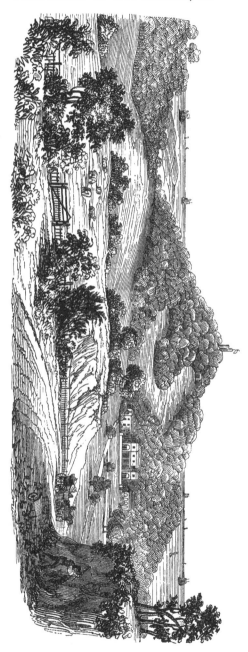

[Fig. 242. View of Sherringham Bower after Mr. Repton's improvements were completed.]

INTERIOR ARRANGEMENT.

It is remarkable, that, of the many thousand plans which have been made, in different ages and countries, no one has ever been deemed so far perfect as to become the model for any future design.   In digesting the arrangement for the house at Sherringham [figs. 243, 244, and 245], we have proposed a plan [fig. 243], in many respects differing from other houses, for which we plead the following reasons :—

Firstly. The wishes of the proprietor.

Secondly. The adaptation of the house to the situation, character, and circumstances of the spot.

Thirdly. The style in which it is supposed to be inhabited.

Most modern houses, as well as those of former times, are too large to be occupied with economy.   It is not a large house, but a large room, that is most comfortable to live in; yet, many such large rooms tend to increase the expenses, if constantly lived in, and the miseries of life, if only used occasionally.   Let us, then, consider what are the rooms required for a house, on the scale here proposed.

These may be thus enumerated,—

1st. One large living-room, to contain books, instruments, tables, and everything requisite to modern comfort and costume [see the section fig. 246].

2nd. An ample eating-room, to be used in the morning for breakfast, and not to have a fire lighted five minutes only before dinner.

3rd. An entrance, with such vestibule and passage as may impress a certain degree of importance, without useless waste of space or expense, the ancient hall not being necessary.

4th. A room on the ground floor, which I call a parlour; it may serve various purposes, besides that of the proprietor's own study ; it should have a bed, in case of age or infirmity ; or, it may be occupied by an eldest son ; but its chief use is to give the proper number of bed-rooms on the floor above, which would otherwise be defective, in a house in the country.

5th. The gentleman's own room, connected with the offices, to give access to persons on business, without admitting them into the body of the house.   This may be gun-room, justice-room, &c.

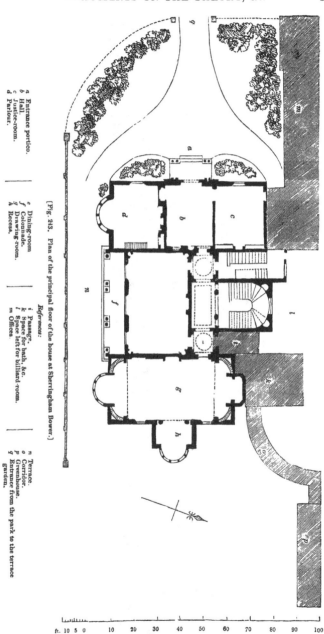

[Fig. 243. Plan of the principal floor of the house at Sheringham Bower.]

References:

a Entrance portico.
b Hall.
c Justice-room.
d Parlour.

e Dining-room
f Colonnade.
g Drawing-room.
h Recess.

i Passage.
k Space for bath, &c.
l Space left for billiard-room.
m Offices.

n Terrace.
o Corridor.
p Greenhouse.
q Entrance from the park to the terrace garden.

ft. 10 5 0　　10　20　30　40　50　60　70　80　90　100

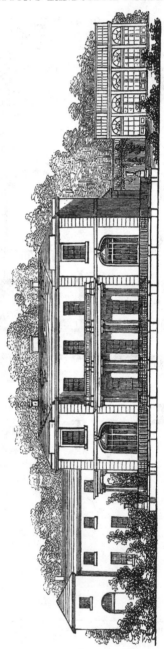

[Fig. 244.   General elevation of the south front of Sherringham Bower, as proposed to be altered.]

6th. The lady's own room, or boudoir, up-stairs, and connected with the wardrobe and bed-room, on the same floor, having a *degagement,* or private stairs, although the approach for strangers is by the principal staircase.

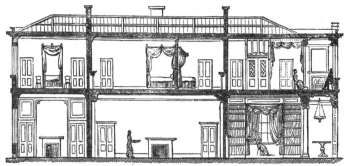

[Fig. 245. Section through the house, shewing the parlour, the dining-room, and the drawing-room.]

7th. The bed-rooms to have dressing-rooms, in which sofa-beds may occasionally be used.

8th. The rooms over the offices, to be used as nurseries, should have large folding-doors to admit air, and may be connected, on the same level, with the hill at the back of the house, for exercise to the children.

[Fig. 246. Section from north to south, through the living-room; shewing two bed-rooms and wardrobe over.]

The eating-room is of that proportion, about two by three, which is now considered the standard of perfection : indeed,

in many modern houses, every room partakes of the same shape
and dimensions; such a room requires tables and sofas to fill
up its area, and create that sort of intricacy which is so ad-
mirably conspicuous in the old houses of the date of Queen
Elizabeth, where large bow windows and deep recesses give a
degree of comfort worth copying in a modern room.   With
this idea, the windows of Sherringham [see fig. 244] are proposed to take a new character, as applied to Grecian architecture, which, in fact, has no more to do with a modern sash
than with a large Gothic window.

In the centre of this room, and opposite to the fire-place,
is a deep recess, which will be one of the most interesting and
striking novelties, admitting a small company to live in the
room, or out of the room, at pleasure, and commanding a delightful view of the flower-garden, with just so much of the
sea as will be sufficient to announce its proximity, without
exposing the room to its baneful effects.   The view to the
east, from this window in the recess, will be so peculiar, that
it may, perhaps, be advisable to exclude all views from the
windows on the sides, only leaving the upper part for transparent blinds, or stained glass.

FRAGMENT XXXIV.

EXTRACTED FROM THE REPORT OF ENDSLEIGH,

A COTTAGE ON THE BANKS OF THE TAMAR, IN DEVONSHIRE, BY PERMISSION OF
HIS GRACE THE DUKE OF BEDFORD.

SITUATION AND CHARACTER.

THOSE who have sailed on this beautiful river, near Plymouth and Saltash, will figure to their minds one of those
calm sequestered retreats, reflected on the smooth surface of a
broad expanse of waters, very different from the river scenery
of Endsleigh: to explain this difference it will be necessary to
describe the Tamar.

There are hardly two things in nature more contrasted,
than a river, near its source, in a mountainous country, and

the same river when its becomes navigable, and spreads itself into an estuary, like the Tamar at Plymouth. Nothing can be more delightful to those who have braved the storms of the ocean, than to sail between the romantic banks of the Tamar, whose echoing rocks often repeat the music, which, from pleasure boats, enlivens its peaceful surface; and a cottage, on the banks of the Tamar, will naturally suggest such tranquil scenery. Very different is that of Endsleigh. Here, solitude, embosomed in all the sublimity of umbrageous majesty, looks down on the infant river, struggling through its rocky channel, and hurrying onwards with all the impetuosity of ungoverned youth, till it becomes useful to mankind. This idea often occurred to me while contemplating the river on the spot. The Tamar, like all mountain streams, however it may amuse the eye with its frolic motion, by not being navigable, or passable, becomes a barrier, and seems to serve no other purpose than that of dividing the two counties of Devon and Cornwall; but, even in this apparently useless state, it is busy in collecting the "little streams which run among the mountains;" and, on tracing its progress, we find that it soon becomes more and more useful to man, till, at length, it is acknowledged as the great source of the harbour of Plymouth, to which England owes much of its glory and its commerce.

In speaking of the course of the Tamar, I should wish to make a distinction betwixt the *channel* and the *bed* of the river, if I may be allowed so to use these two words. By the *channel*, I mean the whole flat surface over which a river spreads its waters during the floods of winter, extending to the foot of the hills which form the valley. By the *bed*, I mean the narrow channel to which the water is confined during that "*belle saison*," when all Nature presents her beauties to advantage; when all rivers sleep in their beds, and even the most turbulent are restrained within their narrowed limits. Let us now consider the process of Nature in forming this bed. *Light* never moves but in straight lines. *Water* always takes some degree of curvature. The rays of light may be broken by reflection, or refraction, but can never be bent. Water, on the contrary, may easily be bent, but cannot *be broken without changing* its fluid character to froth. The course of a river is never straight, and seldom along the middle, or lowest part of the flat, but it shoots across, from side to side, increasing its utility by thus retarding its progress : this observation applies to all rivers, though I was first led to examine the subject by the tortuous course of the river Manyfold, in Derbyshire.

[Fig. 247.]

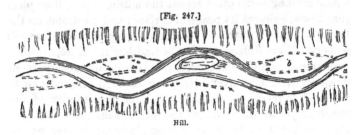

In this sketch [fig. 247], which I have supposed to represent the course of any river, two dotted lines [*a a*] shew a second track, which the water seems to mark out during floods, and which leaves, occasionally, swamps or pools of water [*b*], in summer, after the river has subsided. This sort of channel may be observed at Endsleigh, in the shape of the ground on the left bank of the river.

Sometimes a river forsakes its bed entirely, and takes one of these new channels: and I have frequently had occasion to assist, or retard, this operation of nature, by an interference of art; but, in the present case, I shall only revert to the difference which I have endeavoured to establish betwixt the *channel* and the *bed* of the river Tamar, so far as it relates to the best means of crossing it *without* the sort of bridge aptly described by Cowper,—

> " That, with its wearisome but needful length,
> Bestrides the wintry flood."

If a substantial bridge were necessary or expedient, no person would be so competent to construct it, as my scientific and experienced friend, Mr. Rennie, who has been consulted on the subject; but I shall beg leave to make some observations, first, on the uses of a bridge ; and, next, consider its effect on the scenery. The wood opposite being now annexed to Endsleigh, access is required to it ; but, if there were no other expedient, I should doubt the accommodation being equivalent to the difficulty and expense of such a bridge as might be passable at all seasons, requiring, at each end, an embankment, and arches, on dry ground, above the level of the highest winter floods. This would be necessary, if it were a public road; but, on the contrary, it would be a private bridge, seldom used, and might be dispensed with during floods; therefore, such a bridge is not absolutely necessary. With respect to its effect on the scenery, it would present an object totally at variance with that calm sequestered retreat which forms the striking characteristic of Endsleigh : since a great bridge announces a great road, and a great road destroys all solitude, both real and imaginary, there is, also, another objection to a conspicuous bridge in the situation proposed.

The part of the Tamar forming the chief view from the house [fig. 248], is so nearly in a straight line, that it would more resemble an artificial canal than a natural river, if the extremity, now forming a graceful curvature, were to be terminated by a bridge, and especially one so large as to rob the river of all its importance. In addition to these objections, it may be added, that, if a crossing can be effected more immediately near, and opposite to, the house, it would, doubtless, be a better situation for a bridge, in point of convenience ; and, as an object of beauty, it might assume a picturesque character more in harmony with that of the place [see fig. 249].

OF THE PICTURESQUE.

This word has, of late, excited considerable interest and controversy; but the word, like many others in common use, is more easy to be understood than defined; if it means all subjects capable of being represented in a picture, it will include the pig-sties of Moreland, as well as the filthy hostels of Teniers and Ostade : but the absurdity of representing all that is visible, without selecting what is most beautiful, cannot be better exemplified than by the following fact. One of our most eminent landscape painters was desired to make a portrait of a gentleman's seat : he saw the place during a land flood, and, when the whole valley was covered by vapour, he made a beautiful picture of a fog, after the manner of Vernet; and thus he painted an atmospheric effect, when he should have painted a landscape. In like manner, a beautiful woman, represented during a fainting fit, may display great ingenuity

[Fig. 248. General view from the south and east fronts of the cottage at Endsleigh, before it was altered.]

[Fig. 249.   General view from the south and east fronts of the cottage at Endsleigh, as proposed to be improved.

in the artist; but, surely, this is sickly picturesqueness.   The
subjects represented by Salvator Rosa, and our English Mor-
timer, are deemed picturesque; but, are they fit objects to
copy for the residence of man, in a polished and civilized
state?   Certainly not.   This, naturally, leads to the inquiry,
how far we may avail ourselves of picturesque circumstances
in real landscape.   These circumstances may be classed under
three heads:—steepness of ground—abrupt rocks—and water
in rapid motion; for we may consider wood, and lawn, and
smooth water, as common to all landscapes, whether in Cornwall or in Lincolnshire.

In the drives through Leigh Wood, some advantage has
been taken of the steepness; but it should be shewn as an
object of beauty, from the precipitous side of the road, and
not as an object of terror, by making the roads too steep.

There are many places in which romantic rocks are now
totally hid by brushwood; these, doubtless, require to be
brought into view.   But, of all picturesque objects, there is
none so interesting as water in rapid motion; and it is the
duty of art to avail itself of every opportunity to force it into
notice.   In a mountainous country there hardly exists a dell,
or dingle, in which some stream, that might be drawn forth
to form a conspicuous part in the picturesque landscape, does
not steal its way, unseen, amongst the long grass or foliage
of brushwood, and is, therefore, entirely lost.

In all mountainous countries, it is common to place troughs
to receive the water which flows from the neighbouring hills.
These, by the road side, as drinking places for cattle, form interesting circumstances in the landscape, peculiar to romantic
scenery; while the interest is considerably heightened, by reflecting, that the supply is not the scanty produce of human
labour and mechanism, but flows from that source whence the
most mighty rivers derive their existence: perhaps there is
hardly a more striking example of the inexhaustible bounty of
Providence than may be drawn from the never-ceasing overflow of " *L'abreuvoir des Montagnes.*"   [Watering place of
the mountains.]

THE WEIR.

Instead of a magnificent and costly bridge at a distance
from the house, I shall propose crossing the water immedi-

ately opposite, by means of a weir.   The surface of this weir
may be levelled, and paved with flat stones, to the width of
fifteen or twenty feet, at about eighteen inches or two feet
below the common summer height of the river, making a safe
ford for carriages ; and, by inserting large blocks of stone, a
bridge of timber, or cast-iron, may be thrown over for horses
and foot passengers, above the common summer's flood, but so
strongly secured as to bear the winter torrents to flow over the
whole.

A stream from the river may be brought, sufficient to turn
the under-shot wheel of a corn-mill, or it may be worked by
an over-shot wheel: to supply which, a channel, or feeder, may
be brought through the wood, from so high a level up the
river as to produce the sort of water-fall which I have hinted
in the sketch [fig. 249], where such a stream is supposed to
pass over the face of certain rocks now hid by brushwood.   It
is hardly necessary to remark, how much the view from the
house would be enlivened by the smoke of a cottage on the
opposite side of the water; and, if this cottage were to be a
mill, the occasional traffic, and busy motion of persons crossing
the Tamar, would add to the picturesque effect of a landscape,
which, at present, wants a little more animation.

Perhaps it might be possible to give much additional interest to the Tamar, at Endsleigh, if this weir could be converted into a salmon-leap, of any height, that would not
require much embankment to preserve the meadow from being
overflowed by common floods.

THE COTTAGE.

Having considered the situation and natural character of
the country, I must now consider the artificial character by
which Endsleigh is made habitable : for, without the aid of
art, the most romantic or picturesque scenery in nature is a
desert, and only fitted to the habitation of wild beasts.   The
first question that obviously occurred was, what style of house
will best accord with this landscape ?   Here the good taste of
the noble proprietor of Endsleigh was directed by what he
saw.   An irregular farm-house, little better than a cottage,
backed by a hill and beautiful group of trees, presented an

4 G

object so picturesque, that it was impossible to wish it removed and re-placed by any other style of building that
architecture has hitherto invented, *viz.* a castle, or an abbey,
or a palace, not one of which could have been so convenient
and so applicable to the scenery as this cottage, or, rather,
group of rural buildings. With respect to the manner in which

[Fig. 250.   View of Eudsleigh Cottage, on the banks of the Tamar]

the design has been executed, I shall only say, it is such as
will do credit to the *name of Wyatt*, when time shall have har-
monized the raw tints of new materials.   The design and outline are so truly picturesque, that I must regret my inability
to do them justice [see fig. 250].

PLEASURE GROUND.

One of the subjects to which I was instructed to direct
particular attention, was the fence and line of demarcation
betwixt the lawn to be fed, and that to be mown and dressed
as pleasure-ground.   The general fall of the ground from the
house to the valley, is an inclined plane, with the exception
of a small rise in the centre, which had been artificially converted into a kind of bastion; this left a considerable space to
be covered with flowers and shrubs; but, when I began to
mark the situations of clumps and patches, I placed persons at
different stations, and found that, in every part of the surface
of this lawn, beyond the distance of twenty-five feet from the
house, any shrub of six feet high would hide, not only the
meadow below, but also that line of river which, by an unin-

terrupted continuity of glitter, constitutes the leading feature of the place. This is very different from the stagnant *sheets of water* (as they are called) which require masses of planting to hide the mechanism of their artificial deception. One obvious advantage of removing the fence so much nearer the house, will be that of introducing the appearance of cattle to animate the landscape, and, by their perspective effects, to shew the distance of lawn betwixt the house and the Tamar; and, perhaps, a certain portion of the opposite bank might be thrown into pasture, with the same view, when access can be had to it by means of the weir proposed.

COMFORTS AND APPENDAGES.

If houses were built only to be looked at, or looked from, the best landscape painter might be the best landscape gardener; but, to render a place, in all seasons, comfortable, requires other considerations besides those of picturesque effect.

It is during the winter, and in the shooting season, that a residence at Endsleigh will be most desirable; but the climate and south-western aspect of a mountainous district will expose it to the rains, winds, and fogs, which are the natural concomitants of all lofty and picturesque stations. In spring, it has become a fashion to desert the country, and in summer, every field is a garden; but, in autumn and winter, we experience the truth of a maxim which I have endeavoured to inculcate, and must again beg leave to repeat, that *a garden is a work of* ART, *making proper use of the materials of* NATURE. A well cultivated fruit-garden requires shelter to secure its produce in autumn; and this same shelter may be extended to the comfort of its inhabitants, during that season when a walk along a south wall, while the sun shines, though " ten times repeated," will please, more than the richest landscape in the most romantic country, when stripped of foliage and exposed to driving winds, and covered with its wintry garment of snow. For this reason, a garden becomes the chief appendage of comfort, and should never be at a distance; and, though it may be offensive, when enclosed in the usual way, with lofty scarlet walls,—yet, if the walls were to be disposed in terraces, and rendered ornamental by piers, or arches, for each tree recessed, the garden at Endsleigh might be made no unsightly feature; but, from the relative situation of the cottage, the proposed conservatory and the plantations, it would be very little seen. The same intervening objects which tend to hide the walls from the view in the valley, will also tend to intercept the current of air, during the sweeping gusts of wind and fog from the south western mountains, at the same time that, from their declivity, the sun's rays will act with uninterrupted force.

We read of the hanging gardens of Babylon, and I have heard described (by an eye-witness) something similar in the gardens still existing near Damascus.* Of all the comforts belonging to a garden, there is none more delightful than the covered way, or rustic corridor of Woburn: such a line of communication naturally suggests itself here, from the cottage to the conservatory, and from thence to the forcing-houses, terrace-garden, &c.

The long sketch [fig. 249], is supposed to be taken from the windows of the dining-room; the terraces, grass, and gravel, seem to justify the boldest interference of art in the accompaniments of this garden scene. The style of conservatory, the alcove in the children's garden, and the fountain and artificial trimness of the parterre, must all be considered with reference to the noble occupiers, rather than to the humble character of a cottage. Since contrast and variety are not less sources of pleasure than uniformity, the trim character of this garden of art, will act like the frame to a natural landscape. At the end of the gravel terrace is a quarry, which might be converted into a grotto-like receptacle for specimens of the fossils and ores abounding in the neighbouring mountains.

* From these hints, I will confess, that, in two other situations, I have recommended a similar disposal of garden in terraces : but, with this difference,—at Beaudesert (Marquis of Anglesea's), the shape of the ground requires the walls to be *straight;* at Sherringham Bower (Mr. Upcher's), the walls were proposed *convex,* and the ground behind the cottage at Endsleigh requires the walls to be *concave:* thus, the same expedient may be varied, to suit various situations, but all contributing to the comfort of habitation.

CONCLUSION.

So interesting and so picturesque a subject, makes me regret the inadequate efforts of my pencil in representing, as well as the difficulty of my progress in viewing it. I will, however, indulge the hope that the preceding pages may not only be useful in improving the scenery of Endsleigh, but in furnishing employment and amusement to its noble possessors for many years to come; and having, in a manner, provided against the rigours of winter, I will not be unmindful of that winter of life which must alike assail the cottage and the palace. With this in view, I will venture to advise, that all the walks be made sufficiently wide to admit a carriage; and having, myself, lost the power of gathering a flower, or picking up a fossil from the ground, I have found great comfort in banks raised to the height of three or four feet on a face of ornamental pebbles, to bring nearer to the eye those lesser rock-plants, or delicate blossoms, which are too minute to be seen from the ground. At this enchanting retreat, the most pleasing attention has been paid to the comforts of infancy and youth, of which the children's cottage is one of the most perfect examples. Let the same attention be extended to solace the infirmities of age.

It is with peculiar satisfaction that I have been called upon to exercise my utmost skill on this subject, since everything that can contribute to the enjoyment of its scenery, I know must also contribute to the improvement of the neighbouring country in its agriculture, its mineralogy, its civilization, and the general happiness of all who dwell within the influence of this *cottage on the banks of the Tamar.*

FRAGMENT XXXV.

CONCERNING HOUSES OF INDUSTRY.

VERY soon after the Sherringham report had been written, in which some hints respecting the treatment of the poor were introduced, my attention was again called to the subject by an

application from the parish of [Crayford], in Kent, to give a
design for a new workhouse, when the following report was
addressed to my son, at that time the officiating minister
there :—

To the REV. EDWARD REPTON,* *at* [*Crayford,*] *in Kent.*

MY DEAR EDWARD,

Your letter, communicating the wishes of your parishioners, that I should give my opinion respecting the plan and
situation for a new workhouse, or house of industry, with all
proper attention to the comforts of the poor, has excited my
heartfelt satisfaction; and, as this may be amongst the last
efforts of my professional duties, I feel the subject peculiarly
interesting to me.

The present wretched building is so unhealthily placed in
the low and wet marshes, that the first consideration seems to
be, the choice of a wholesome spot, on a dry soil.   This may
be found on the edge of the heath about to be enclosed, and
near the side of the high road from London to Dover, where a
large gravel-pit presents a bold terrace full facing the south,
and so formed by the excavations already made, that very
little more digging would be necessary to make a secure fence
to enclose the premises.

My idea of the design, or plan, and certain regulations
for the future comforts of its inhabitants, will, I hope, be understood by a reference to the annexed sketch [fig. 251].   The
building, as before mentioned, is supposed to front the south,
and to have an ample platform, or terrace, betwixt it and a
steep bank of the gravel-pit.   The centre consists of one long
and lofty room, for the paupers to live in, and to take their
meals;   this is flanked by two buildings, which contain the
governor's and matron's dwelling, kitchen, store-rooms, and
other useful apartments.   At the back of the premises, towards
the north, is a square yard, on each side of which is a work-
shop, with bed-rooms for the paupers over them.

[* At this time, 1816, curate of Crayford, in Kent, and now, 1839, prebendary of Westminster.   In 1808, Mr. Edward Repton published "The
Works of Creation," a series of discourses for Boyle's lectures, in one
volume 8vo.—J. C. L.]

The difference betwixt the cold darksome gloom of the north quadrangle, and the warm cheerful appearance of the site towards the south, may easily be imagined ; and suggests the idea of taking great advantages of the contrast.

Let the back-yard be considered as a sort of punishment for misbehaviour and refractory conduct, where, shut up between four buildings, nothing can be seen to enliven the prospect : while, on the contrary, from the south terrace, cheered by the sun, the view of the country will be delightful; since the immediate foreground consists of a garden, and the perpetually varying and moving scene which is presented by the great road to Canterbury, and the coast.

In addition to the usual employments of the paupers in the work-rooms, it were to be wished that more wholesome and useful labour might be taught to the children, than spinning, and other manufactures. This might be considered as the reward of good conduct : the children, supplied with

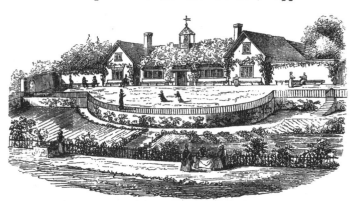

[Fig. 251   Design for a workhouse proposed for the parish of Crayford, in Kent.]

spades, and hoes, and tools, proportionate to their strength, should be taught and exercised in the cultivation of the garden, and, perhaps, drilled to become the future defenders of their country.

The sketch [fig. 251] will, in some degree, explain the effect of this scene, as viewed from the high road. We may suppose the warm benches, along the front of the building, occupied by the aged and infirm, who may there enjoy their few remaining days of sunshine, without being totally shut out and lost to the world. On the warm tiles of the central building, some vines may be trained; and the produce of these, and every part of the garden, such as fruits and flowers, may be exposed to sale on the public road, and the profits of these commodities might be the reward of extraordinary industry, or good behaviour.

In this plan there is no pretension to ornament in any respect that may incur unnecessary expense, except, perhaps, in the small cupola for the bell; and this appendage, trifling as it is, gives to the whole that characteristic feature which distinguishes it as a public building.

To you, my dear Edward, it is unnecessary to remark one circumstance, which you may, perhaps, find an opportunity of inculcating to your parishioners; that, in providing for the future comforts of the poor, they may possibly be anticipating the future happiness of themselves, or their descendants; since we too often see the hard-hearted opulent oppressor, in the vicissitudes of life, reduced to look for support to those public institutions to which he has reluctantly contributed.*

It may, perhaps, be objected to the design, that something more ornamental might have been proposed, perhaps adopting the Gothic style; but the answer is obvious,—the first consideration in a poor-house is economy.

The prevailing taste for fragments of Gothic architecture is apt to display itself in the doors and windows of a dairy, for which there can be no plausible reason assigned; but, if the Gothic character be introduced in any small building, there is none more appropriate than the schools, either public or private, which, of late, have been erected, sometimes by the parishes, and sometimes by individuals, as ornamental appendages to their parks; under the latter circumstances, a more costly design may be recommended. Having made several for this purpose, one of them is annexed (see vignette p. 563), which has been proposed for a school, endowed and patronized by Mrs. Burton, at Longnor, near Shrewsbury.

* This plan was, at first, highly approved by the leading persons in the parish, till it was discovered that the situation proposed was so desirable, that the site, occupied in private houses, would produce more profit, and, therefore, the poor, for the present, continue in their former unwholesome abode; but, as a late orator observed of negro slaves, compared with eels flayed alive—" they are used to it."

FRAGMENT XXXVI.

HARESTREET.

OF QUANTITY AND APPROPRIATION.

ALTHOUGH, during a long and active life, my efforts have contributed to the happiness of some hundred individuals, and the employment of some thousands; I trust, that not a single instance can be adduced in which useless expenditure was advised, for unreasonable gratification of vanity; but wealth is never so well employed, as in improvements that display the genius of art, and call into active employment the labourer and artificer. To demonstrate the little consequence of quantity or value, when speaking of the beauty of scenery, many places have been mentioned, which may, perhaps, appear too inconsiderable in a work that treats of dukedoms and royal domains: but I wish to evince, that, in many cases, great effect may be produced by a very contracted quantity of land, and, not unfrequently, that almost everything depends on the foreground. Thus, in the villas on the margin of the lake of Geneva (like that of Gibbon), nothing more is necessary than a terrace, or a few shrubs and flowers, to form a frame to the picture: thus, also, it frequently happens, that, by the enclosure of a common, or the grant of a small piece of land from a forest, the most essential benefit may be derived, although the quantity of land acquired be very trifling; and I have often observed, that the cupidity natural on such occasions, generally leads to the obtaining more land than can be rendered useful; since it is either too small to be fed, or too large to be kept under the scythe and roller.

In my former volume, I used the word *appropriation*, to describe that sort of command over the landscape, visible from the windows, which denotes it to be private property belonging to the place.

A view into a square, or into the parks, may be cheerful and beautiful, but it wants appropriation; it wants that charm which only belongs to ownership, the *exclusive right* of enjoyment, with the power of refusing that others should share our pleasure: and, however painful the reflection, this pro-

4 H

pensity is part of human nature. I have too frequently
witnessed a greater satisfaction in turning a public road, in
stopping a foot-path, or in hiding a view by a pale and a
screen, than in the most beautiful improvements to the
scenery; and sometimes have contended in vain against the
firs and poplars, which, on the verge of a forest, presented
more agreeable objects to the proprietor than the scenery of
the forest itself; one acknowledged that he would rather look
at a young sapling of his own, than the most venerable oaks
belonging to the Crown.

The propensity for appropriation and exclusive enjoyment
is so prevalent, that, in my various intercourse with proprietors
of land, I have rarely met with those who agreed with me in
preferring the sight of mankind to that of herds of cattle, or
the moving objects in a public road, to the dull monotony of
lawns and woods. Of these few, I cannot resist mentioning
one venerable nobleman, who enjoyed health, cheerfulness,
and benevolent feelings, more than eighty years, retaining to
the last his predilection for the scenery of a garden, rather
than that of a park; and who used, at his villa, on Ham Common, to enjoy the sight of the public passengers from his
garden-seat, surrounded by roses. To this rare instance of
benevolence in the noble Viscount Torrington, may be added
that of his friend and cotemporary, the late Duke of Portland, who gave leave to all persons to pass through the park
at Bulstrode, and even encouraged the neighbouring inhabitants to play at cricket on the lawn. How different is this
from the too common orders given at the gates and lodges of
new places, recently purchased by strangers, and only visible
to themselves and their own inmates! For the honour of the
country, let the parks and pleasure-grounds of England be
ever open, to cheer the hearts, and delight the eyes, of all who
have taste to enjoy the beauties of nature. It was, formerly,
one of the pleasures of life to make tours of picturesque
inquiry; and to visit the improvements in different parts of the
kingdom: this is now changed to the residence at a watering-
place, where the dissipation of a town life is cultivated in a
continual round of idle, heartless society; without that home
which formerly endeared the life of a family in the country.
And, after all, the most romantic spot, the most picturesque

situations, and the most delightful assemblage of nature's choicest materials, will not long engage our interest, without some appropriation; something we can call our own; and if not our own property, at least, it may be endeared to us by calling it *our own home.*

I will conclude these Fragments with the most interesting subject I have ever known; it is the view [figs. 252 and 253],

[Fig. 252. View from Mr. Repton's cottage, at Harestreet, before it was improved.]

from the humble cottage to which, for more than thirty years, I have anxiously retreated from the pomp of palaces, the elegances of fashion, or the allurements of dissipation : it stood originally within five yards of a broad part of the high road : this area was often covered with droves of cattle, of pigs, or geese [see fig. 252]. I obtained leave to remove the paling twenty yards further from the windows; and, by this *appropriation* of twenty-five yards of garden, I have obtained a frame to my landscape; the frame is composed of flowering shrubs and evergreens; beyond which are seen, the cheerful village, the high road, and that constant moving scene, which I would not exchange for any of the lonely parks that I have improved for others [see fig. 253]. Some of their proprietors, on viewing the scene I have described, have questioned my taste; but my answer has always been, that, in improving places for others, I must consult their inclinations; at Hare-

street, I follow my own.   Others prefer still life—I delight in
movement; they prefer lawns fed by their own cattle—I love
to see mankind; they derive pleasure from seeing the sheep
and oxen fatten, and calculate on the produce of their beef
and mutton: perhaps, they might not object to the butcher's
shop, which I have taken some pains to hide, giving the preference to a basket of roses. This specimen may serve to shew
how much may be effected by the foreground; how a very

[Fig. 253. View from Mr. Repton's cottage, at Harestreet, as improved by him.]

small object, aptly placed near the eye, may hide an offensive
object ten times as large; whilst a hedge of roses and sweet-
briars may hide the dirt of a road, without concealing the
moving objects which animate the landscape. Such is the
lesson of quantity and appropriation. And, as I have now
approached near the end of my labours, and am still permitted,
though with difficulty, to collect my thoughts on a subject
most interesting to my feelings, I will add a lesson of far
greater moment.   When I first appeared before the public, in
1794, in a work which has long been out of print ["Hints, &c."
pp. 25, 26], the Introduction began with these words:—

    " My opinions on the general principles of landscape gar-
" dening have been diffused in separate MS. volumes, as op-

" portunities occurred in the course of my practice : and I
" have often indulged the hope of collecting and arranging
" these scattered opinions, at some future period of my life,
" when I should retire from the more active employment of
" my profession; but that which is long delayed, is not, there-
" fore, better executed; and the task deferred to declining
" years, is frequently deferred for ever; or, at best, performed
" with languor and indifference."

Twenty years have now passed away, and it is possible
that life may be extended twenty years longer, but, from my
feelings, more probable that it will not reach as many weeks;
and, therefore, I may now, perhaps, be writing the last
Fragment of my labours. I have lived to see many of my
plans beautifully realized, but many more cruelly marred:
sometimes by false economy; sometimes by injudicious extravagance. I have also lived to reach that period when the
improvement of houses and gardens is more delightful to me
than that of parks or forests, landscapes or distant prospects.

I can now expect to produce little that is new; I have,
therefore, endeavoured to collect and arrange the observations
of my past life: this has formed the amusement of the last
two winters, betwixt intervals of spasm, from a disease in-
curable, during which time I have called up (by my pencil)
the places and scenes of which I was most proud, and mar-
shalled them before me; happy in many pleasing remem-
brances, which revive the sunshine of my days, though sometimes clouded by the recollection of friends removed, of scenes
destroyed, and of promised happiness changed to sadness.

The most valuable lesson now left me to communicate is
this: I am convinced that the delight I have always taken in
landscapes and gardens, without any reference to their quantity or appropriation, or without caring whether they were
forests or rosaries, or whether they were palaces, villas, or
cottages, while I had leave to admire their beauties, and even
to direct their improvement, has been the chief source of that
large portion of happiness which I have enjoyed through life,
and of that resignation to inevitable evils, with which I now
look forward to the end of my pains and labours.

While I was actually writing this last page of my work, I
received a letter from one of the ablest statesmen now left to

the country, in which are these words:—" The best comment
" upon what has been the leading pursuit and employment of
" your life, is to be found in the relief and solace which, at
" this time, you derive from it. ' *Quid purè tranquillet ?'*
" [What can bestow pure tranquillity ?] has long been a phi-
" losophical question ; religion answers it. But I have always
" thought that the sort of taste which you have eminently
" contributed to form and diffuse, has a peculiar tendency to
" soothe, refine, and improve the mind ; and, consequently, to
" promote most essentially the true and rational enjoyment of
" life."

Feeling the full force of this just remark from one of the
most pious and benevolent of men, I will finish with the remark of one, who possessed more wit than real worth, who,
after enumerating various experiments to obtain happiness,
concludes with these words,

" *Allons mes amis, il faut cultiver nos jardins.*"

[Come along, my friends, and let us cultivate our gardens.]

THE END.

GENERAL INDEX.

4 K

WHITEHEAD AND Co., PRINTERS,
76, FLEET STREET, LONDON.

Printed in the United States
By Bookmasters